Published 2020 by River Publishers
River Publishers
Alsbjergvej 10, 9260 Gistrup, Denmark
www.riverpublishers.com

Distributed exclusively by Routledge
4 Park Square, Milton Park, Abingdon, Oxon OX14 4RN
605 Third Avenue, New York, NY 10017, USA

Library of Congress Cataloging-in-Publication Data

Langlois, Pierre.
World ESCO outlook / Pierre Langlois and Shirley J. Hansen.
p. cm.
Includes index.
ISBN 0-88173-675-9 (alk. paper) -- ISBN 978-8-7702-2301-0 (electronic) -- ISBN 978-1-4665-5814-4 (Taylor & Francis distribution : alk. paper) 1. Energy industries. 2. Energy industries--Case studies. I. Hansen, Shirley J., 1928- II. Title.

HD9502.A2L36 2012
333.79--dc23

2012015016

World ESCO Outlook / Pierre Langlois / Shirley J. Hansen
First published by Fairmont Press in 2012.

Routledge is an imprint of the Taylor & Francis Group, an informa business

ISBN 0-88173-675-9 (The Fairmont Press, Inc.)
ISBN 978-1-4665-5814-4 (print)
ISBN 978-8-7702-2301-0 (online)
ISBN 978-1-0031-5170-8 (ebook master)

Table of Contents

Introduction

The energy service industry is not static. Internal and external stressors emerge. Energy efficiency (EE) is now promoted as a fundamental solution to climate change and many different policies and instruments are being developed to increase its use in all markets.

Customer demands force different responses from the industry. Energy service companies (ESCOs) all live with their own *force majeure,* surrounded by a rapidly changing environment. All of which prompts new opportunities as well as major changes in the industry. The 2011 World Bank report on ESCO development in China and the 2011 National Association of Energy Service Companies' report are indicative of the constant need ESCOs, energy end-users and financial institutions have for current industry information. We are exceedingly pleased to add this comprehensive guide to your library.

These changing forces remind us that ESCOs are much like the hermit crab. Since the hermit crab does not have a hard external shell, it must occupy someone else's shell in order to survive. The hermit crab searches for a new home much as ESCOs are constantly looking for new customer shells.

If one were to speculate as to the criteria a hermit crab might use to select its new shell, high on the list would be *fit*. The shell needs to be big enough to allow growth, but not too big to carry. Similarly, the industry must keep an eye on "fit" problems brought about by new policies, legal frameworks and changes in financing structures. Another criteria would be: no sand. Imagine how hugely uncomfortable irritating sand could be over the years. If the ESCO industry does not look carefully enough, it can experience irritations/risks over the years as well.

If the new environment under which the ESCOs operate is not adequate, the companies' capability to get a good fit to meet the market's evolving demand will permeate every corner of the operation. Further, if our industry does not stay on top of the changing environment, it could have a negative impact on the use of the energy performance contracting (EPC) concept and its ability to effectively serve the global initiative it supports. In recent years, varied international experience and more sophisticated risk management have changed how EPC is perceived and used worldwide.

World ESCO Outlook offers a unique and abbreviated review of the ESCO industry, the EPC concept and how it is used in diverse cultures. It builds on the research and work performed for *ESCOs Around the World: Lessons Learned in 49 Countries*, published in 2009. We are incredibly fortunate to have the most knowledgeable players around the world share how they perceive the industry's evolution in their respective countries. Their experience and insights offer a context to understand what EPC really is in practice. With the support of these in-country reports, *Outlook* lays out for all to see, the real ways the concept is implemented as well as the difficulties it encounters in various cultures. For those who wish to have a command of this evolving industry, *Outlook* is structured to give the reader an undisputed reference.

We wish to acknowledge the tremendous value these amazing contributing authors bring to *World ESCO Outlook*. They are listed with brief biographies and contact information in Appendix A.

We also wish to acknowledge several others who provided exceptional assistance and support along the way. The book in hand never fully reveals the "behind the scenes" work that makes it all possible. In particular, we'd like to cite the fine work of Jacky Tremblay, Ron Allan Go-Aco and the editing team of Econoler, and Jim Hansen of Hansen Associates. Our thanks to each of you for your diligent efforts in trying to make us look good. We would also like to thank the folks at The Fairmont Press who continue to believe in our efforts to promote the EPC market through our publication efforts.

Of necessity, ESCOs must be very sensitive to the world around them and flexible enough to meet changing conditions. In short, the EPC industry is constantly changing and adapting. Keeping track of these changes is a huge challenge. It is our hope that *Outlook* will provide a valuable reference guide as well as offer key benchmarks to help gauge the industry as it changes and grows.

Shirley J. Hansen
Pierre Langlois

Section I

Energy Efficiency Makes Money

EE is such a good idea, it should sell itself. EE conserves our finite energy resources and makes money while reducing pollution. ESCOs should not have to sell EE. Instead, they should offer solutions to all the barriers met in implementing a project. We struggle to get EE recognized for its value by the market and worry that it is essentially ignored by too many policy makers. It is a problem that plagues the energy community around the world.

One of our biggest challenges is to change the focus. For some reason, the emphasis and the resources are being dedicated to the "high maintenance" alternative. Under the rubric of "renewables," it seems only the most costly are getting invited to the ball. Governments promote "green" and frequently use it synonymously with renewable. Even when EE is recognized as the best solution, it is still perceived as being a more complicated approach. Governments, even in these tight economic times, tend to look for ways to incentivize the expensive alternatives. Many utilities and power plants are required by law to include a certain percentage of renewables in their energy mix. Corporations, which brag about their sustainability efforts, typically banner the renewables they rely upon.

Somehow when we weren't looking, EE lost the green war. The great benefits of EE have been lost on the politicians, the press and the general public. *The greenest electron is the one not generated.* EE is the most economical answer to our pollution concerns. EE is typically less expensive than renewables. EE can actually be a revenue source. To repeat: EE makes money while reducing pollution. We can even use that revenue to buy down the cost of more expensive renewables. Why is EE such a hard sell?

An *Outlook* on ESCOs cannot overlook the vital core of our business. If we cannot make the business case for EE, we will have a difficult time making EPC the recognized valued approach that it is. Throughout this section and the book, ways to put the focus back on EE and to make the sale are identified.

Chapter 1

The Global Perspective

If we are to wrap our arms around the full perspective of today's energy performance contracting (EPC), it helps to get a sense of where we have been and exactly what steps have brought us to this point. As we look over our shoulder to trace the path we have taken, the process will likely give us a glimpse of the future and, if we are lucky, it is apt to reveal those ingredients that will lead to success for the ESCO of tomorrow.

LOOKING BACK

An amazing number of people think EPC was a concept created in North America about 30 years ago. Others profess to "know" the whole idea was a product of US utilities in the early 1980s. They are all wrong.

So Long Ago in France

The basic concept originated in France over 100 years ago. Compagnie Générale de Chauffe (CGC) is credited with a guaranteed savings idea while working with district heating firms. Scallop Thermal, a division of Royal Dutch Shell, took CGC's idea and modified it to fit the demand side of the meter. Then, Scallop took it to the UK and the US. The concept was well received. In an era of climbing energy prices, the idea that a firm could identify energy conservation opportunities for a client and install the measures for a share of the cost savings found a market niche, but all did not go smoothly.

Emerging Problems

Opposition came from some surprising sources—governments, lawyers, engineers and procurement people.

It quickly became evident in the UK that a law existed which made energy performance contracting impossible. Mr. Anees Iqbal and his colleagues talked to a young man in accounting, John Majors (later UK Prime Minister) into revising the law to permit EPC. Anees was very persuasive.

In the United States, the Department of Energy, particularly the Atlanta Support Office (ASO), was dead set against the EPC concept and waged near war to prevent its acceptance. ASO launched a broad investigation into one of its principal advocates. By law, such an investigation requires documented evidence of "waste, fraud or abuse" on the part of the person being investigated. No evidence was ever documented. The DOE Inspector General's office, however, still warned government officials not to talk to or be seen with the lead advocate. State energy offices, former employers and potential contractors were all approached in an attempt to discourage the EPC concept and to discredit the lead advocate, Dr. Shirley J. Hansen.

There was a need in the market place and ESCOs were tenacious. Gradually, attitudes changed and today the US government is a major customer of what it calls energy savings performance contracting.

In Canada, at about the same time as EPC was emerging in the US, Jean Gaulin, who was a high-level official in an investment branch of the utility in the province of Quebec, taught the idea of selling savings and getting paid back in part from the savings generated. A few months later, having joined forces with the leading EE consulting firm in the country, the first Public Private Partnership related to EPC and the first ESCO in the country was born. Many were skeptical, so it took a few years to demonstrate that the idea worked and could bring incredible value to the market.

Historically, ESCO development research has revealed that governments can have a major impact on the acceptance of EPC. Some governments have been instrumental in enabling EPC, while others have vehemently spoken against it and even held back its acceptance. Others have simply done nothing to help develop it. Governments that stand silent on the issue tend to create uncertainty and have made potential ESCOs reluctant to enter into an unknown field.

Since the US became one of the pioneers in this EE financing approach, it offers many examples of obstacles a growing industry can face.

Many not familiar with the concept, especially those who would be forced to change the way they work, or think, presented a strong opposition. **Lawyers** probably headed the list. Since they were not familiar with the contract language, they did not want to appear uninformed to their clients; so they took the easy way out and declared it illegal. Such tactics kept clients (Anne Arundel County Public Schools—and therefore the taxpayers) from benefiting from the huge sources of revenues available through reduced energy costs. A few lawyers, who took the time to understand the process, like Jeff Genzer, became strong advocates and were instrumental in furthering ESCO industry development.

In some instances, the legal problems were a legitimate concern. In Tennessee, the procurement laws provided for acquiring equipment and purchasing consultation, but did not speak about the combined equipment/consultation/services that EPC offered. The law had to be changed. With the support of the attorney general's office, new language was passed that allowed for the combined services EPC offered.

One of the leading ESCOs at the time, Honeywell, hit a major bump in the road when the firm's general counsel declared that EPC was illegal in the US. Honeywell already had EPC projects in most states. The response of the Honeywell rank and file was to draft legislation that would make it "legal" and work diligently to have such laws adopted. The "Honeywell Law," which too often was not particularly well drafted, did become the precursor to legislation in many US states, and facilitated the implementation of EPC projects.

While lawyers remained a significant obstacle, they gained considerable support against the concept from **procurement** people. People charged with purchasing in an organization were very comfortable with the black and white of direct purchase, bid/spec procedures and the least cost approach. They were most reluctant to leave their comfort zone. Numbers on paper were easy to defend. Selecting a firm based on qualifications or the perceived merits of a project was a grey area.

Procurement people wanted to have their engineers conduct the audits, list the equipment that the engineer said was needed to reduce energy consumption and then put it out to bid. They did, however, like the idea of having the ESCO guarantee the results specified in the contract. Unfortunately, the bid/spec approach did not fit EPC. No ESCO in its right mind would guarantee an unknown engineer's work. To gain some risk protection, the ESCO would carry out its own audit. The potential client would, therefore, end up paying for two audits.

If the procurement office held sway in the organization, the equipment list approach was used. This approach cost the organization in many ways. Objective analysis shows that the organization:

- Had only the opinion of a single engineer and lost the richness of ideas that could be obtained from multiple proposals from firms that specialized in EE.
- Subjected itself to proposers who played with numbers. It was amazing how great the benefits could be made to look with just a little finessing.
- Lost much of the post-contract savings. If the bidders know what they are doing, they are very careful to provide the client with absolutely nothing more than what is asked for in the specifications. A really good bidder will look for the cheapest way to meet the minimal requirements of a bid and nothing more. Bid/spec by definition seeks *barely acceptable work*.
- Would get just the equipment listed in the Request for Proposals (RFP); not a comprehensive energy savings program.

The process did not identify an ESCO with the qualification to provide the client the most cost-effective approach to achieving comprehensive EE. Instead, the ESCO incurred greater risk by being forced to play with somebody else's scorecard. Greater risk meant the ESCO had to hedge. As a consequence, the client got less project and less savings for the same level of investment. Ultimately, end-users learned to use procedures that satisfied the EPC process and the procurement office.

The preferred approach has been a decision matrix to determine which ESCO best met their selection criteria. The matrix also afforded the procurement people a paper trail to defend the selection process. Further, this approach typically used an evaluation committee to review the proposals, so the selection responsibility was shared.

In the 1970s and early 1980s, **engineers** were not comfortable with this new means of financing EE measures. In fact, the majority vociferously opposed it. EPC would force engineers to survey a facility, submit a report, and *be held accountable for the results of their recommendations*. In a time when the whole energy auditing effort was still relatively new, engineers were not comfortable with this new approach. Engineers were also discouraged by the fact that EPC interjected a layer between themselves and the client. The engineer or his/her A&E firm found themselves working for an ESCO, not the client. As a consequence, they lost control of the projects. Consulting engineers did not like this aspect. The head of the energy committee at that time for the American Consulting Engineers Council (ACEC), Don Carter, helped turn things around. A very persuasive, competent engineer, Don was able to show his colleagues how this new approach just changed the engineer's role. He even got the ACEC to include some positive comments about the use of EPC in its brochure.

A more difficult bump in the road was posed by **financiers**. They remain a problem in many countries today. Financiers are in the business of renting money. Their job is to be sure they get the money back plus the "rent." To protect themselves from the risk the process poses, they typically ask for some collateral. This collateral generally takes the form of "owning" valued equipment or relying on the client's balance sheet. This approach is referred to as asset-based financing. When asked to finance EE, they were asked to retreat from this secure approach and place their trust on predicted energy savings. The collateral became nebulous cash flows, which were based on engineering calculations and proposed risk mitigation/management strategies. This was an approach that bankers were not trained to understand and were decidedly uncomfortable with. It does not take too much imagination to picture the expression on the banker's face as he said, "You want us to do what?!"

ESCOs got their knuckles bloody knocking on bankers doors trying to find financing for EPC. Furthermore, interest rates are based on perceived risks. The perception of risk based on some engineer's idea of future energy savings was very high; so the cost of money was also high.

After the price of oil dropped in the late 1980s and most EPC agreements in North America changed from "shared savings" to "guaranteed savings," the financing picture changed dramatically. The ESCO was no longer the principal borrower, the customer secured the financing with the secondary backing of the ESCO. With the surety of the ESCOs, which often belonged to big companies, or security insurance offered for smaller ESCOs, financiers began to see EE as an attractive investment. In some instances, insurance companies went so far as to line up the financing.

Eventually, the investors would be found at national ESCO conferences where they, in effect, would ask ESCOs to "Please take our money." The investors, such as Energy Capital and Dana Commercial Credit, would even host receptions to get new ESCO customers.

Without a doubt, the biggest bump in the road turned out to be **the price of energy**. When oil prices fell in the mid-1980s, it appeared on the surface to be a good thing. When you have, however, entered into several contracts based on certain payback calculations as did active ESCOs, who never saw the price fall coming, it can become a nightmare. The prevailing way to conduct EPC across the world in the early 1980s was through "shared savings." The basic shared-savings concept was that the ESCO would put its equipment in a facility and would get paid back through a share of the cost savings. This concept worked very well as long as the price of energy stayed the same or went up, but when it dropped, the payback periods became much longer. Sometimes the payback period was longer than the contract. ESCOs went out of business. Owners were left with the obligation of paying the subcontractors. The whole idea of shared savings got a very bad name. The industry nearly died.

THE ADVENT OF GUARANTEED SAVINGS

Out of the ashes, there emerged a new concept: guaranteeing the amount of energy saved. When ESCOs stepped back and looked at the old shared-savings model, they realized they had been betting on the future price of energy—a very dangerous game and one over which they had no control. The new model guaranteed the amount of energy to be saved and secondarily guaranteed that the value of that energy would pay the customer's debt service obligation, provided the price of energy did not go below a certain floor price.

In this new model, the customer would secure the financing. The ESCO would no longer have to carry the credit risk as well as the performance risk. Dropping the huge risk tied to guessing the future price of energy plus new strategies for sharing other risks made for a more doable package. During this time of transition, customers frequently asked for dual proposals: one with shared savings and one with guaranteed savings. Backing out the numbers through careful analysis revealed that shared savings financing costs were much greater than the same project with guaranteed savings. In the new model, the customer would get more services and savings for the same level of investment.

In its efforts to establish roots and grow, the industry has experienced similar problems in other countries. Each country and its culture are unique, but the industry around the world has experienced some very similar problems. The commonality of the problems faced has led us to some very valuable lessons learned. The research for *ESCOs Around the World: Lessons Learned in 49 Countries* revealed seven dominant lessons. They are shared below. The research for *World ESCO Outlook* has reinforced some of these lessons and brought new concerns to light. These findings are shared in Chapter 10.

UP TO NOW: LESSONS LEARNED AROUND THE WORLD

Among all the unique qualities of each nation and its ESCO industry runs a surprising amount of commonality. Working on *ESCOs Around the World*, an incredible wealth of good information was compiled from people in the ESCO business with diverse situations. While conditions were, and are, unique, there is much in common than can benefit us all. The contributors to *ESCOs Around the World* were very generous in sharing the obstacles they had encountered, some solutions they had tried and the opportunities their respective country's ESCO industry had identified.

The Global Marketplace

Before addressing individual lessons learned, it seems pertinent to address overseas opportunities. In the earlier global review of ESCOS and in this one, our colleagues offer some exciting choices. If you are wondering if these opportunities are for you, consider a comment once made by Mr. Don Smith, a pioneer in US performance contracting. When Smith was asked why he was looking at foreign markets for Energy Masters, he replied rather bluntly, "Because that's where the market is, stupid!" Indeed, in today's world, opportunities are everywhere for EE and EPC.

A word of caution, however. There is not sufficient information in the country reports offered in Chapter 9 to take you beyond US shores. An incredible amount of additional homework is needed. As

Jim Hansen writing in, *Performance Contracting: Expanding Horizons,* stated, "An in-depth investigation of selected countries with limited risks can allow an ESCO to bring together the ingredients that can lead to success. An ESCO, which does not have the internal resources to make a careful business evaluation in a selected country, should secure outside consultation or stay home."

In the process of assessing how things are done and how different markets evolve, one can get a new understanding of the broad potential of EPC. From reading all the information from the contributing authors, you will discover what is currently happening in many countries can present new opportunities or provide valuable lessons that can be applied closer to home.

The reports also reinforce the idea that many societies are a patchwork of religious, linguistic and cultural diversity. A critical component in developing a successful business is becoming informed regarding local conditions and becoming part of the local scene.

Assessing the development of EPC in any environment reveals prevalent concerns:

- **A working legal system, cultural constraints**. A prevailing problem is the need for a functioning legal system. This is especially evident in developing countries where contracts are rarely enforceable. Work typically proceeds with the close relationship of the parties and a handshake. This is not as foreign as it seems. Currently, EPC has been brought to a standstill in one southern state in the US, as a client apparently feels no compunction to comply with the contract terms to which it originally agreed, and unbelievably, the courts are presently expected to okay a unilateral change in contract language. A handshake would probably have been more binding.

  One can never be too cautious about understanding the legal limitation of a contract when working in such a complex and comprehensive approach as EPC.

  In China the greatest concern, which is closely related to contract enforcement, is collecting payments from customers. In Shandong Province, an ESCO cleverly hired individuals, who were close to specific customers, to collect those payments. That person's only job has been to collect the payments on a regular basis. His pay is based on making that collection. It works.

- **Government leadership**. In the ESCO community, there has been a prevalent concern about the lack of government leadership. Many in the EPC industry would like their respective governments to define what an ESCO is and does. They would also like the government to describe the parameters of operation. Our colleagues have repeatedly made it clear that the absence of government action or the wrong action has caused huge obstacles in some countries.

  Closely related to the reliance on government leadership for industry growth is a growing need, according to in-country experts, for accreditation and certification. In Austria, the government has developed the "Thermoprofit" quality labeling procedure. In the US, some ESCO accreditation has been done by the National Association of Energy Service Companies of its members, and the US government periodically prequalifies ESCOs as eligible to perform work in federal agency buildings under an abbreviated selection process. In Canada, the federal government works through a predefined list of approved ESCOs.

  Since an ESCO is often viewed as being only as good as its internal resources or its last project (which throws certification into question), there is a growing trend to certify the expertise of specific ESCO experts. The certification programs offered by the Association of Energy Engineers are filling a great need in this regard. In particular, the Certified Energy Managers program and its joint program with the Efficiency Valuation Organization (EVO) to certify Measurement and Verification Professionals are gaining traction around the world.

- **Measurement and verification (M&V)**. Our earlier research revealed that French and Tunisian stakeholders had a strong need to assess EE investments. China and Malaysia have stressed the need for M&V to "document," and sell success. Initial work by the US Department of Energy led to the creation of EVO, which has become the recognized international organization on M&V guidance. EVO's International Performance Measurement and Verification Protocol (IPMVP) has become the basis for internationally accepted M&V practices in most developed countries and is gaining incredible momentum in emerging ones.

- **Financing**. Financing ESCO work has proven to be a perennial problem. Several countries continue to suffer from a lack of interest by commercial banks in activities related to EE and even more specifically to EPC. In developed economies, new ideas are developing and banks are beginning to be interested in the potential market EPC represents for them. In emerging economies, multilateral banks, such as the European Bank of Reconstruction and Development and the World Bank have been very active supporting EE and the use of EPC as an effective implementation mechanism. However, not enough has been done to help commercial banks see the need for, and the most effective procedures for, loaning money to ESCOs. The International Energy Efficiency Financing Protocol (IEEFP), developed and promoted by EVO, has taken some initial steps in this regard.

- There is a growing use of **guaranteed savings** as a financial model. Outside of North America, however, **shared savings** is still frequently used as an EPC model. The popularity of shared savings is based on its clear advantage when introducing a new concept to an area without asking the customers to incur any debt obligations. In addition, there is appeal for customers who are having difficulty establishing creditworthiness in a transitional or developing economy. Finally, some large companies promote shared savings as they have a market advantage over small ESCOs, which cannot establish a significant market presence without becoming too highly leveraged. Still, the shared savings approach has shown its limits in emerging economies, where most local ESCOs have to use their own equity to finance projects, seriously limiting their growth in increasingly demanding markets.

- A persistent concern is the lack of customer awareness of the opportunities ESCOs offer. Despite efforts in many countries over the past two decades to inform the public, evidence suggests that the potential client base still does not really understand the process. A number of factors feed the prevalence of ignorance including a growing range of problems, such as excessive claims of success by ESCOs, tariff correction "rip-offs," inadequate knowledge of, and lack of investment in, Investment Grade Audits (IGAs), weak M&V and project failures. In such instances, bad news travels much faster than good news, and problems are readily believed.

- **Importance of O&M.** Operations and maintenance (O&M) practices greatly affect whether or not a piece of equipment will operate near design—and whether it will deliver predicted savings. A study conducted for the US Department of Energy revealed that up to 80 percent of the savings in an effective energy management program could be attributed to the energy-efficient practices of the O&M personnel. This data reveals the huge risks ESCOs take if they ignore O&M people and the training they should have. Conversely, it also reveals an inexpensive way to manage risk.

As one ESCO observed, "EE pays for green." This succinctly describes a very good reason for ESCOs in all corners of the globe to become more involved in the effort to help customers move toward sustainability. EE can foot the bill and EPC can address the implementation barriers that are present in the market.

LOOKING FORWARD

As we examine the current conditions and activities of EPC around the world, we find some interesting surprises. The EPC world is indeed changing. Since the concept is so attractive and addresses many of the same issues policy makers and institutional program designers have been trying to resolve for a long time, EPC has a great future. As the goal to develop a more energy-efficient world becomes more and more critical, EPC will see many new opportunities. As the reports from the contributors reveal, EPC is already going further than many expected under different shapes and structures. A prevalent belief in the industry is that this is just the beginning.

Chapter 2

ESCOs and EPC Models

In mature markets, ESCOs have become sophisticated energy efficiency (EE) project developers responsible for an unusually wide spectrum of tasks. Typically, they can identify, design and help secure project financing; install and supervise the maintenance of equipment; measure and verify project energy savings; and assume the risk that the project will reduce the customer's energy and operating costs to a level, sufficient to repay the investment.

ESCOs distinguish working capital from project financing. Working capital is used for general corporate purposes and front-end development of projects prior to construction, while project financing is used to pay for the ESCO's cost to implement the specific project. These costs include design, development, equipment acquisition, subcontractor services and the ESCO markup. Project financing is further divided into construction and long-term financing. The former is used to pay for project implementation, which is later converted to long-term financing when project installation is complete and commissioned.

The energy performance contracting (EPC) model benefits energy end-users by reducing technology risks they would otherwise incur if they did it themselves. The ESCO provides engineering-based savings projections and financial pro-forma estimates (the majority of which they are willing to guarantee). Another benefit is that many EE projects include infrastructure-related improvements, which the customer needed to make anyway. This enables end-users to eliminate or reduce their internal capital budget, which had been allotted for these investments.

EPC is understood differently from one country to another. This limits comparative analysis. The problem is further compounded by the fact that experts within the same country frequently use different terminology and may have different perspectives of the process. Nevertheless, we can present the most basic models that are used around the world, for they typically serve as the bases for all models now used throughout the countries where EPC is present.

BASIC ESCO MODELS

ESCOs are generally classified into the following four categories based on their composition and ownership:

- Independent ESCOs—ESCOs not owned by an electric or gas utility, an equipment manufacturer or an energy supply company. Many "independent" ESCOs concentrate on a few geographic markets and/or target specific market segments.

- Building equipment manufacturers—ESCOs owned by building equipment or controls manufacturers. Many of these ESCOs have parent companies with an extensive network of branch offices, which provide a national (and international) footprint. Frequently, the ESCOs can channel through the parent company's sales forces with its specialized staff, which already offers packages of EE, renewables and distributed generation "solutions" to specific market segments.

- Utility companies—ESCOs owned by regulated or non-regulated electric or gas utilities. Many utility-owned ESCOs currently concentrate on regional markets or focus on the service territories of their parent utilities.

- Energy or engineering companies—ESCOs owned by international oil companies, non-regulated energy suppliers or large engineering firms.

The nature of an ESCO's composition and ownership does not necessarily typify the ESCO offering. ESCOs affiliated with an equipment manufacturing firm may supply equipment from competing companies. Utility and energy supply firms are not limited by territory or supply source. For example, US-based Duke Energy's former ESCO was active in South Africa. Energy engineering firms, however, do tend to focus on the technology related to a project, and frequently need more emphasis on the financial aspects.

On a limited scale, ESCOs can be differentiated on the basis of their marketing structure or approach, which focuses on:

- technologies (boilers, controls, lighting, etc.);

- vertical markets (schools, hospitals or steel plants, etc.); and/or
- utility /energy suppliers (electricity, heating/cooling or compressed air, etc.).

However, ESCOs rarely limit themselves to a narrow market in any country as they try to make the most of their presence to tap into a market. ESCOs are more apt to be regionally focused than market-focused.

BASIC EPC CONCEPTS

The two dominant EPC models in the world are shared savings and guaranteed savings. A third approach, which is gaining in popularity, is chauffage.

Shared Savings

In shared savings, the ESCO organizes the financing of the total upfront capital cost of the project and is totally responsible for repaying the lender as illustrated in Figure 2-1. The client pays the ESCO a percentage (or a fixed amount) of its achieved cost savings from

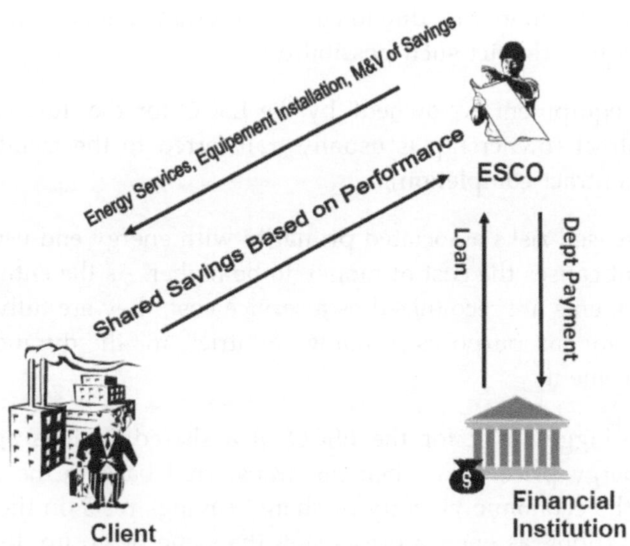

Figure 2-1: Shared Savings Financial Model

the project. The intent is for this payment to be large enough for the ESCO to repay the project investment to its lenders, cover M&V costs and fund any other associated costs. The energy end-user assumes no direct contractual obligation to repay the lender, only the ESCO has this obligation.

For a long time, this model was the one most closely associated with EPC. This structure, however, creates a lot more risk for the ESCO than do other models because the ESCO not only assumes the project performance risk but also assumes the credit risk and the energy price risk.. The shared savings approach frequently requires an equity investment. This investment, in combination with the higher risks assumed by the ESCO, means there is a higher capital cost with this model than with the guaranteed savings structure.

The primary typical characteristics of shared savings can be summarized as follows:

- the energy end-user and the ESCO share a predetermined percentage of the energy cost savings;
- ESCOs carry both the performance risk and the credit risk;
- financing for energy end-users does not negatively impact their credit capacity and can be off-balance sheet (this is becoming more and more rare due to current international accounting practices that restrict such possibility);
- the equipment is "owned" by the ESCO for the duration of the contract (ownership is usually transferred to the facility owner at contract completion);
- increased risks associated primarily with energy end-user repayment causes the cost of money to be higher. As the entire project payments are recognized as a service cost, they are fully deductible for tax purposes in many countries, for the duration of the agreement.

The biggest risk for the ESCO in a shared savings agreement is the energy price risk, when not transferred back to the end-user. Indeed, the economic viability of shared savings rests on the price of energy. As long as energy prices stay the same or go up, the project will typically pay for itself.

Guaranteed Savings

In the guaranteed savings model, the client finances the project and makes periodic debt service payments to a financial institution. The ESCO may not bear any direct contractual obligation to repay the lender. However, secondary risks and surety concerns rest with the ESCO and may become a direct contractual obligation. Because of that specificity, it is increasingly common in the US, for example, to have securitization requirements placed on the ESCO.

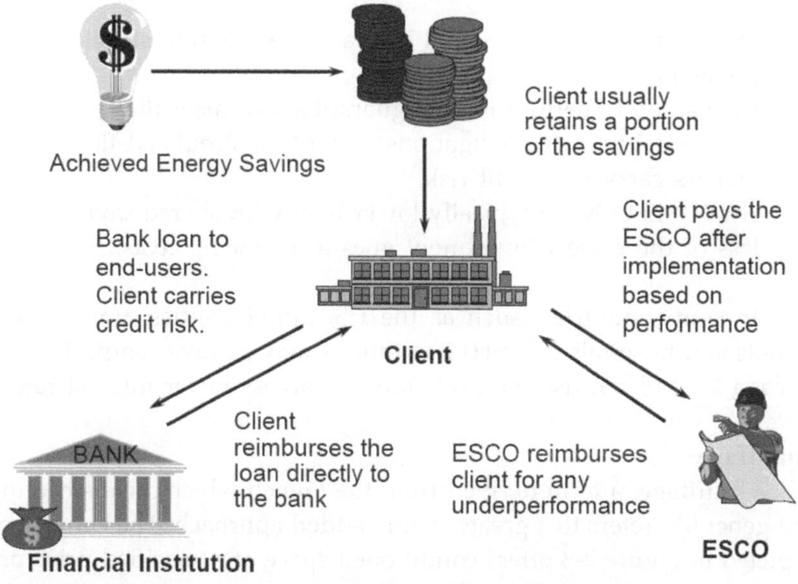

Figure 2-2: Guaranteed Savings Financial Model

The ESCO guarantees energy savings performance to the energy end-user and secondarily guarantees that the value of the energy saved will be sufficient to meet the customer's debt service obligation, provided the price of energy does not go below a stipulated floor price. In developed markets, the attraction of EPC has been where an ESCO might provide a performance guarantee. In such performance guarantee, the realized project savings are expected to cover all the related project costs, including debt service obligations of the lender, M&V fees to the ESCO and any other incremental costs (maintenance, etc.) incurred by the project during the contract period. On the other hand, it is possible that some services offered, such as maintenance, may be kept outside of the guarantee. If the achieved savings fall short of the ESCO guaranteed savings amount, the ESCO will reimburse the energy end-user for such a shortfall. If the realized savings exceed the guaranteed amount, the ESCO may share a portion of the excess, depending on how the contract has been structured. The ESCO's portion of the excess savings is usually predicated on the risk taken and the extent of ongoing services provided to the customer.

The primary characteristics of a typical guaranteed savings agreement can be summarized as follows:

- the amount of energy saved is guaranteed, within stipulated parameters;
- the value of energy saved is guaranteed to meet the customer's project debt service obligations down to a stipulated floor price;
- owners carry the credit risk;
- risks to ESCOs are typically lower than with shared savings; and
- less of the project investment goes to financing costs;

In some countries, such as the US, municipalities, universities, schools and hospitals (MUSH) institutions may be tax exempt. In such cases, a MUSH can use its legal status to access lower interest rates.

Chauffage

Chauffage, a term derived from the French which means heating and generally refers to a greater value-added approach. The concept as depicted in Figure 2-3 offers conditioned space at a specified price per square foot (or square meter) through a supply and demand contract(s) offered by the ESCO. The ESCO manages all energy supply and demand efficiencies.

Following the patterns, which have evolved in the telecommunications industry, ESCOs are showing an increasing tendency to bundle or unbundle their services. Ultimately, increasing numbers of ESCOs are apt to be selling conditioned floor space, which will provide ESCOs and energy end-users with a more effective and efficient means of guaranteeing the return on investment. Such an advent would bring us full circle back to the first guaranteed energy efficiency agreement developed in Europe: "chauffage"—an integrated supply/use efficiency solution.

In practice, ESCOs sometimes focus only on supply efficiencies

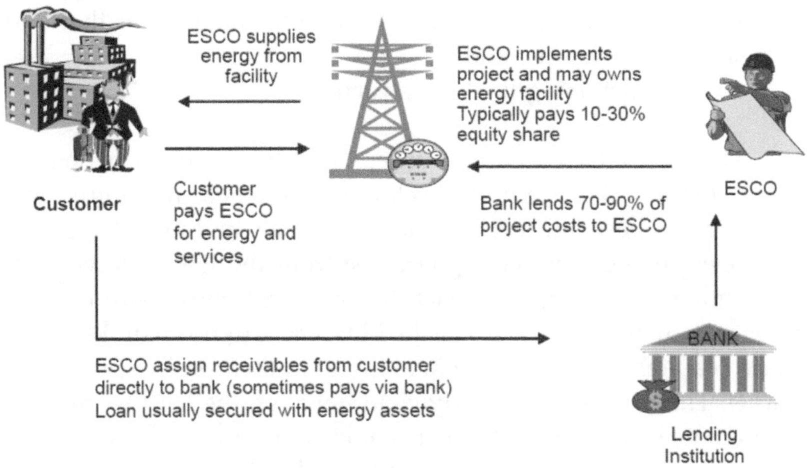

Figure 2-3: Chauffage Financial Model

and refer to the contract as chauffage. In other cases, it may include some demand-side energy efficiency measures, then some type of ownership, a part or totality, of HVAC systems may be owned by the ESCO. In all cases, the contract typically provides for some means of making adjustments for energy prices on an annual basis.

Chauffage is more apt to be used in countries where there is a large heating and/or cooling load, as it is more focused on internal supply-side EE. Furthermore, it is often associated with transfer of obligations related to the operations and maintenance of a facility where energy savings are but one component. The approach is often used in Central and Eastern Europe with municipal district heating plants.

EPC MODELS—CONTRASTED

While over 90 percent of the EPC agreements in North America are currently structured for guaranteed savings, chauffage is an increasingly popular model in Europe. Shared savings is still the favored approach in countries where ESCO markets are emerging.

The guaranteed savings model is mainly adopted in countries where there is:

- a high degree of familiarity with project financing;

- sufficient technical expertise within a banking sector that understands EE projects; and
- a well-functioning banking structure.

The guaranteed savings concept is difficult to use when introducing the ESCO concept in developing markets as it requires energy end-users to be creditworthy and to assume investment repayment risks, even if they receive a guarantee from the ESCO. Nevertheless, the guaranteed savings approach favors long-term growth of ESCOs, because it enables newly established ESCOs with no credit history and limited capital resources (unable to invest in their projects) to enter the market. This entrance is usually conditional on the ESCOs willingness to guarantee the savings to energy end-users. In North America, the guaranteed savings model evolved from the shared savings model in the 1990s. As end-users became more comfortable with energy saving technologies, they were willing to incur more risks and welcomed the opportunity to significantly reduce interest costs in exchange for the acceptance of more risk. In Canada and the US, the primary instigator in the development of guaranteed savings was the drop in the price of energy and the consequent extension of payback periods. ESCOs learned they could not bet on the future price of energy. In developed countries, the public sector usually opts for this structure in order to optimize the amount of infrastructure investments made in its facilities from a performance-based contract.

In developing markets, the shared savings concept is a good introductory model as energy end-users assume no financial risk. This addresses the issue faced by energy end-users in transitional economies with respect to satisfying creditworthiness criteria put forth by banks. Another reason lies in the fact that a new concept, such as EPC, is easier to establish in a country if the customer does not have to incur debt, or can avoid the political/legal procedures in this regard.

The shared savings concept relies heavily on ESCOs' borrowing capacities, and may create serious difficulty for small and even large ESCOs, which lack access to financial resources. By incurring debt on even a limited number of projects, an ESCO may find itself too highly leveraged to obtain financing for the implementation of additional projects. This may constitute a key barrier to industry growth. Indeed, it basically constrains ESCOs to continually raise substantial amounts of equity, so their balance sheets are similar to those featured by banks and leasing companies rather than those characterizing service companies. Consequently, the shared savings concept limits the long-term market growth and competitiveness of small ESCOs. The high financing costs resulting from the shared savings approach helps explain the prevalence of short payback measures only, often referred to as cream skimming. The result is an increased number of "lost opportunities" for significantly reducing energy consumption and cost in energy end-user facilities.

The difficulties with shared savings have prompted the use of the Special Purpose Vehicle (SPV) [or Special Purpose Entity (SPE)] model. This approach creates a SPV for each project. The SPV is owned jointly by the lender and the ESCO. The SPV funds the project and collects/distributes the revenues. This model keeps some of the costs off the ESCO's books and keeps it from being so highly leveraged.

WHICH CONCEPT IS BEST?

Critical issues to address when selecting financing approaches include:

- On whose balance sheet are the project assets?
- Who is really at risk in terms of project performance?
- Is the financing project-specific?
- Why should one care about the questions?

Off-balance sheet financing can preserve a customer's access to capital and facilitates project approval by an organization (this appealing approach is less and less possible due to recent changes in international accounting rules). As far as the risk component is concerned in pay-from-savings, shared savings and chauffage contracts, the ESCO typically takes the savings risks head-on. Alternatively, in guaranteed savings the customer guarantees to repay the debt obligation while the ESCO guarantees the customer that the savings required to make the payment will be met within specified parameters. As a consequence, a new question arises: what is the difference between the ESCO taking the risk and the customer taking the risk and being indemnified? Careful analysis of the relative risks and the associated costs is needed.

The decision to establish project-specific financing is a risk diversification issue. Depending on the structure of the deal, project-specific financing may have the repayment predicated on the receipt of a specific stream of revenue. For example, if a project is financed by a general revenue bond, it does not matter to the investor whether the project performs. The bonds still have to be repaid, regardless. Many energy performance contracts are financed using project-specific financing while payment is predicated on project performance. In the event that these projects fail to perform, the ESCO may not be repaid. If financing is not project-specific, the cost of capital for a specific energy end-user could be lower due to risk diversification.

FUTURE MODEL APPROACHES

ESCOs have largely ignored or done little work in many markets. Faced with credit issues, ESCOs usually shy away from large residential and commercial projects that are highly leveraged and owned by limited partnerships.

ESCOs are usually successful in arranging long-term project financing, which can be a non-recourse arrangement to them by assigning contract rights to lenders. However, large residential and commercial project owners frequently have all of their collateral pledged to their lender and the ESCO is often unable to get a first security interest in the equipment scheduled to be installed. Such a financing structure may be difficult to develop since the ESCO has no clear ability to place claim on the equipment.

Since many public entities cannot obligate future administrative boards, the ESCO also has difficulty securing funds for public sector projects unless they include a "non-appropriations" clause in the contract. This clause allows the public board to meet its fiduciary responsibilities, but also offers some protection to the ESCO. It recognizes that the board cannot obligate future boards, but also imposes customer constraints as it allows the ESCO to remove the installed equipment while it prohibits the board from replacing that equipment. (This approach is becoming less and less possible in the public facilities in many countries where ownership of any installed equipment in a facility has to remain in the hands of the public.) The key to securing ESCO protection in these circumstances is to install something that is essential to the facility's operation.

Leases and other non-recourse financing vehicles used by ESCOs typically have a very small component of equity. Lenders can be comfortable with such large leverage for three reasons. First, they screen the people to whom they lend and lend only to people they believe will have no trouble paying them back. Second, the terms and conditions of financiers' contracts strongly favor them in the event of the energy end-user's default. Finally, financiers often diversify credits by pooling or syndicating them.

Consequently, it is easy to understand why highly leveraged large residential and commercial properties are not generally appropriate for these types of financing vehicles. One alternative would be to use financing vehicles that have a much higher blend of equity (30 to 50 percent) and which have terms and conditions that provide the lender with 50 to 70 percent reasonable security even in the event of a partial default by the energy end-user. These approaches have been used by several ESCOs with varying levels of success. As more ESCOs target the numerous project opportunities available in large residential and commercial sectors, these vehicles are apt to become more common. However, it is unlikely that this approach will be more broadly applied in the future as (i) it is more costly and complex; (ii) investment property will become more financeable as it is moved into real estate investment trusts; and (iii) owners will become more willing to invest in their own funds as they recognize the effect of increasing net operating income on overall asset value.

In the large residential sector, one method to facilitate financing is to use the utility bill for collection, with the long-term obligation migrating with ownership. Several years ago, PacifiCorp in the US, pioneered using the utility bill to collect payments for ESCO services in its FinAnswer program. Under this program, the utility implemented and financed facility improvements. Subsequently, it billed the occupant over many years for the work, including interest. In the event that the property was sold, or changed hands, the obligation either had to be paid off before the sale or the obligation moved to the new owner. Obviously, the ultimate recourse was that the utility would cease supplying power. There were no significant defaults by the energy end-users under that program. Hence, use of the utility bill as a collection mechanism has been seen as improving the credit quality of energy end-users. This approach can be attractive to a large number of competitive energy providers as they realize the benefits of putting these items on the energy bill. Due to legal constraints in many countries, utilities are not permitted to bill anything other than their actual delivered energy. If a public sector utility is able to use such an approach, there is a distinct need in such cases to put a solid wall between the non-regulated ESCO and its regulated parent.

As transaction costs may make projects in this sector uneconomical, ESCOs usually elect not to target small residential, commercial and industrial facilities. The transaction costs associated with residential and small commercial energy end-users can be reduced if credit pre-qualification is simplified and if credit and performance risk can be treated on a portfolio basis. Credit pre-qualification might consist of reviewing data already collected in electronic format by a third party. Portfolio risks can be taken on a portfolio basis when very few customers actually require that their particular project be evaluated for savings. In such cases, these types of transactions might look like simple lease or rental agreements with long-term savings guarantees. For instance, an ESCO might provide an appliance and guarantee related savings. ESCOs must weigh the transaction costs/administrative burden against the potential profitability of a project. Most ESCOs look for a potential customer to have a high annual utility bill, which can vary from USD100,000 to USD1 million for a comprehensive demand efficiency project to be economically viable. Savings may be based on a typical use installation that the ESCO maintains for future inspection and proof to consumer watchdogs. Alternatively, an ESCO may agree to measure and verify the savings on the actual equipment of the energy end-user with the provision that if the savings are confirmed the customer will pay for the assessment. Alternatively, if the savings are not confirmed, the ESCO will not only pay for the assessment but also make a cash settlement representing the benefit which the ESCO had guaranteed to provide. Such arrangements can reduce transaction costs involving a smaller energy end-user to the point where these programs become economically viable.

The ESCO industry is showing an increased proclivity to learn from other firms, both within and outside the energy service community. This movement paves the way to modifying existing ESCO financial models and customer approaches. The reports from other countries around the world presented in Chapter 9 offer a glimpse of how rapidly the industry in changing and new models are emerging.

Chapter 3

Financing

A strategic and profitable marriage of expertise and opportunity manifested itself when Compagnie Générale de Chauffe (CGC) determined the company could make money by correcting supply inefficiencies. From that time forward, energy efficiencies became a source of financing. Ironically, financing has also been a major barrier to energy performance contracting (EPC) as the industry develops around the world.

One of the most significant of these barriers is a lack of commercially viable financing, which is not caused by a lack of available funds per se, but rather an inability to access existing funding capacity at local banks and financing institutions (LFIs) on commercially-attractive terms. This lack of access is caused by a "disconnect" between the traditional lending practices of LFIs and the financing needs of energy efficiency projects. LFIs typically apply their traditional "asset-based" corporate lending approach for energy efficiency projects that is limited to their lending a maximum of 70%-80% of the value of assets financed (or collateral provided). Unfortunately, there is often little or no collateral value in the EE equipment once installed in a facility; rather, the value is the cash flow generated from the equipment after installation. To date, most LFIs (due to lack of knowledge) have not recognized nor appear to believe that meaningful cash flow can be generated from EE projects, or that such cash flow can be relied upon to repay the related loans. Consequently, LFIs generally assign no value to the cash flow generated, thus requiring Hosts to encumber their credit capacity to finance energy efficiency.

The International Energy Efficiency Financing Protocol (IEEFP) has reported, LFIs do not seem to recognize this increased credit capacity, which is caused, in large part, by the fact that they:

- do not appreciate the broader business opportunities and economic benefits of EPCs;
- are not familiar with the unique complexities of EPCs;

- do not have the internal capacity to properly evaluate risks and benefits of EPCs; and
- are unwilling to invest in building internal capacity to properly evaluate EPCs due to the relatively small dollar size of the investment.

Financing issues related to EPC cannot be considered in isolation. They impact every aspect of a project, and go beyond the typical budgetary considerations. A good example is the way EPC contracts were constructed in Slovakia. The manner in which the value of the product is stated in the contract can make a huge difference. Given the typical approach EPC offers, the initial predictions of the "cost of investment" can be soft. If, however, there was a contract dispute in Slovakia, a soft number could take years to adjudicate while a firm fixed price could be settled in months, even weeks.

Since EPC is a front-loaded approach, the absence of a fixed price can leave the ESCO very vulnerable until the price has been set through an investment grade energy audit, and the audit recommendations accepted by the owner. In such cases, an agreed upon "place holder" value may be assigned until a firm figure has been set. This kind of protection, however, can only happen if the ESCO is aware of the problem. This awareness typically comes from experience and the benefit of insights from an in-country partner.

No matter which way we turn in EPC, financing confronts us as a critical issue. Most customers are driven to seek performance contracts because: a) the financing is arranged or provided by an ESCOs; b) the process cushions their risks through the financial structure of a project; and c) it enables them to secure the most favorable financing terms, which affects the quality of a project for both the customer and the ESCO. Different EPC models meet different project needs. Clearly, financing is a major element in performance contracting and ESCO operations.

The shared savings model is based on a predetermined percentage split of the energy cost savings. Such a split between the ESCO and the customer worked well when the price of energy remained stable or increased. In the late 1980s, however, the price of energy dropped dramatically and the paybacks from the energy conservation measures often became longer than the contract. ESCOs awoke to the fact that the shared-savings model required them to bet on the future price of energy. As ESCOs are primarily technical risk management business, they realized that trying to guess the future price of energy was unacceptable. In any case, the value to develop and bring the needed financing directly to a client remains quite attractive worldwide, mainly in countries where the banking sector is not very sophisticated.

The energy price dilemma associated with shared savings prompted a search for an alternative. A new model to marry expertise and opportunity emerged in North America: guaranteed savings. ESCOs determined they should be guaranteeing the amount of energy saved; not the amount of energy cost savings achieved. To make a project work financially, they secondarily agreed that the amount of energy saved would have sufficient value to pay for the cost of the project. As the ESCO expertise grew and the industry became more sophisticated in risk management, the guarantee was conditioned on the monetary value of the energy saved. A floor price was agreed upon and the customer contractually accepted the risk if the cost of energy went below the floor price. This risk was acceptable to the customer for dropping energy costs freed money to cover project costs. The developed concept simplified the issue for bankers; however, it limited the use of this approach to clients willing to accept the risks of financing directly.

Before we turn to financing specific EE projects, a couple of thoughts on broader financing strategies for the ESCO should be addressed:

- Financing is so basic to EPC that every ESCO needs someone qualified to oversee the company's financing strategies and to analyze the financial component of every project during preliminary assessment.

- The qualities of the financial officer need to be multi-dimensional with an understanding of total business operations, individual project financing issues and factors that influence financing, including finance models and project risks.

- Proving to financiers that the risks have been identified and effectively managed is crucial to project financing. Whether real or perceived, bankers peg the interest rates on what they deem to be the level of risks associated with the project. Presenting a project in the most favorable way, therefore, not only gets the funding, but gets the most favorable terms.

Bottom line: ESCOs, which get the best terms can deliver better projects for the same level of investment, have more satisfied customers, become more competitive in the marketplace—and make more money!

When we follow the money trail, it becomes evident that getting a project financed should be a shared effort between the ESCO and the customer. Part of the ESCO's responsibility is to be sure the customer understands that this cooperation is important to the bank and will often result in a larger project for the amount invested. But perspectives vary. It is the ESCO's responsibility to put together a bankable project. The ESCO typically arranges the financing. Its reputation and history often adds surety, which offers financiers added confidence that they will get their investment back in a timely manner. Since the customer usually incurs the debt, owners need to know exactly what financing options are available and the implications of each option.

CREATING BANKABLE PROJECTS

ESCOs, which have been in this business for a few years, remember knocking on the financial doors until their knuckles were bloody. Today, at least in developed economies, financiers knock on ESCOs' doors ... if, and it's a big **IF**, ESCOs can put together bankable projects. In North America, and in some European Union countries, ESCO financing has evolved to the point that, should an ESCO not be able to get a project financed, it is time to rethink the project.

What is a "bankable" project? Simply put, it is a clearly documented economically viable project. Building a bankable project starts with sorting out the pieces that make a project economically viable. The first step is to examine the key components, make sure each aspect and any associated risks are properly assessed, and develop a plan which effectively shows how those risks will be managed. Every measure, each component, typically carries a risk factor, and the management/mitigation of each risk factor is apt to carry a price tag. An effective ESCO knows how to assess the components and how to package them into a project that can be financed.

THE CUSTOMER

Pre-qualifying customers is an art. The critical factors for the ESCO are: a) developing the selection criteria; b) asking the right questions; and c) learning to walk away when a "lucrative" project doesn't match those criteria. Ironically, one of the major drivers of EPC is the owner's need for financing; so it seems like a dichotomy that a primary pre-qualification for a customer that needs financing is to be creditworthy. But a customer can be cash poor and creditworthy at the same time. In fact, a potential customer, who is creditworthy and cash poor is an especially promising candidate for EPC. A school system, for example, is typically creditworthy and legally backed by the state, but its revenue stream is often sparse.

Most ESCOs have an understanding with a financial house (or houses) as to what constitutes acceptable credit standing for a customer. Some even have prescribed forms for the ESCO's salespeople to fill out; so all the pertinent information is acquired and presented in a routine fashion. The credit check at this stage is like most others. Financiers want the information that can reasonably assure them that the loan will be paid back. The range of information, which a financial house will need regarding a potential customer, typically includes:

- the type of transaction proposed; e.g., equipment title provisions, purchase options, payment terms, and the performance contracting financing model to be used;
- the organization's tax status (if appropriate to financing conditions);
- longevity of the potential customer's organization; ownership;
- its business prospects;
- evidence that the customer can keep the savings, which will provide the all-important revenue stream from which the payments will be made as well as an incentive for the owner to participate;
- financial condition with three years of complete and current financial statements, i.e., bond rating, audited financial statement; and
- preliminary project calculations.

The critical financial information needs to be adequately documented. No matter how charming, persuasive and attractive a potential customer may be, the financials must be in print—and signed. In their zeal to make a sale, ESCO salespeople are sometimes tempted to take the customer's word for its credit standing. But the financier will not. ESCOs can become blinded by the "savings opportunity" and spend a lot of money developing a project based on false financial assurances only to eventually learn that the owner cannot meet the necessary financial criteria. This is an expensive lesson for ESCOs. Unbelievably, some need to learn it more than once.

In addition to the customer's creditworthiness, financiers are more inclined to loan money when larger ESCOs are involved. Their size and track record often offer the surety needed to provide lower interest rates. Smaller firms, however, need not be discouraged by this apparent market advantage. In some countries, the small firm can either get performance bonds or insurance to cover the savings guarantees. Smaller firms have also found that they can just propose to guarantee a smaller part of the global savings to be generated by the project. Furthermore, even with these added costs and higher interest rates, the small firm can still compete with the margins charged, for example, by an ESCO affiliated with a manufacturer.

Once the ESCO is satisfied with the customer's creditworthiness, consideration can be given to other criteria which will be used to weigh the customer's partnership quality, including the administrative commitment to the project, the attitudes and abilities of the operations and maintenance people, etc. These "people factors" and other critical concerns are generally folded into a scoping audit that assesses project potential. The scoping audit is little more than a walk-through audit with a very educated eye. The purpose is to be sure that further pre-qualification and marketing efforts are warranted. Start-up ESCOs, or "WISHCOs"—a name too often earned—are inclined to overlook the risks associated with the people factor. Since it is vital to project success, it cannot be over-emphasized.

Once the other pre-qualification criteria have been met and the potential customer has accepted the concept, then a full feasibility study is needed. Before the ESCO incurs the expense of a premium quality energy analysis, an agreement to cover the costs of the audit if the project does not go forward is increasingly used to protect the ESCO's investment. The content of this planning agreement is discussed in the following chapter pertaining to contracts.

Energy Audit Quality

A standard energy audit with its "snap shot" approach, which does not sufficiently assess long term variables, as well as mostly unmeasured data collection of equipment is not good enough for performance contracting. These audits typically assume that present conditions will prevail for the life of a project. When an ESCO bets money on predicted *future* savings, these assumptions must be tested through careful risk assessment procedures. Only an investment grade energy audit (IGA), which adds specific risk appraisals to the standard calculations, will meet performance contracting needs. In recent years, energy engineers have learned to look at a facility, assess its mechanical/electrical/thermal conditions, and determine the ability of the remaining equipment subsystems to accept the recommended measures. An IGA goes beyond these engineering skills and requires the art of assessing people, including the level of commitment of the management to the project, the extent to which the occupants are informed and supportive and the O&M staff's abilities, manpower depth and attitude.

A key aspect of a quality IGA is a carefully detailed base year with the average energy consumed over several years *and the operating conditions*, which caused that consumption. The ESCO that consistently delivers a quality IGA, which accurately predicts potential savings, builds a track record that financiers find very heart warming. A good IGA is at the heart of a bankable project. When the total project plan is wrapped around a quality IGA and delivered by an ESCO, which can back its predictions with a solid history of successful projects, financiers smile.

For the reader, who must perform, oversee or assess the quality of an audit, it is strongly recommended that he or she read *Investment Grade Energy Audits: Making Smart Energy Decisions*, which is available from The Fairmont Press, in order to understand in greater detail all the issues and subtleties of performing good IGAs.

Equipment Selection and Installation

Predictive consistency, the hallmark of quality audits, rests on knowing what works. And what does not! To support a guarantee, ESCOs must have considerable control over the equipment specifications and the selection of the installation subcontractors. Generally, this control manifests itself in order of preference from the ESCO's point of view in (1) working as a general contractor or construction manager, which supplies all the equipment and installation; (2) having primary responsibility for developing the specifications in cooperation with the owner and making the final equipment selection; and (3) preparing specifications in cooperation with the owner and identifying acceptable bidders for the owner's final selection. For the owner, these options offer progressively more control and increasingly transparent costing. The more control an owner exerts, however, the more risk the ESCO assumes, the lower the project economic viability becomes and the project bankability drops accordingly.

A financier's due diligence carefully assesses the ESCO's ability to make good on its guarantee and to control the variables that threaten the savings and the guarantee. As always, money follows risk. It's worth repeating: interest rates are directly related to the project risks as perceived by the financier. For both parties, the predicted benefits must outweigh the expected risks or the project is not bankable. It follows that the control exercised by the owner directly affects the project benefits—inversely. The owner control level translates directly into ESCO risks, project vulnerability and viability—and higher interest rates. Money that goes to pay the interest is not available to buy services and equipment, which produce the savings. Owner control, therefore, has a price tag that has a negative impact on the project.

PROJECT MANAGEMENT

One of the great appeals of EPC is the extent to which the ESCO's fee, and profit, rides on the project's success. A truly successful ESCO knows the project is only beginning once the construction/installation/commissioning is done. There are three key components to managing a project, which are closely related to its success. They are: a) key ingredients for a planned effective partnership; b) a strong maintenance component; and c) a project manager.

It is impossible to overstate the key role a good project manager plays in achieving energy savings and in fostering a strong sense of project partnership. A project manager and the project he manages cannot function effectively without a carefully planned communications component. At some point in any communications strategy, there must be a quantification of project progress and its success. This invariably rests on the financial data. It is critical that all parties talk the talk and walk the walk on energy efficiency financing.

MONEY MATTERS

EPC is primarily a financial transaction. Return on investment is usually the motivating force for the end-user, the ESCO and the financing source. Conversely, the money that will be lost if management does not act should be a major factor in setting financial priorities. A key consideration is how much the organization will spend if it does not reduce energy consumption.

Understanding the financial implications of all energy actions, or lack thereof, are key. Bosses have fired energy managers and managers have become furious at ESCOs—all because the organization was not achieving predicted "savings," when, in fact, energy savings had been eaten up by increases in the rate schedule. The "front office" may not understand how changing rate schedules can destroy predicted savings. The concept of cost avoidance is essential to weighing and communicating performance contracting benefits, especially when energy prices are volatile. The guidelines for weighing cost-effectiveness, calculating cost of delay and computing/graphing cost avoidance are vital to understanding and communicating project needs and accomplishments.

Cost-Effectiveness

Because every business, every organization, uses a significant amount of costly energy, a wide array of EE measures are available to them. However, these opportunities must be weighed to determine which measures offer the greatest financial benefit. This requires not only an evaluation of the cost-effectiveness of the measures, individually and in combination, but a broader financial analysis as well.

Time was when cost-effectiveness ruled nearly every EE decision. Payback was the critical parameter in auditing. Today, improving the work environment and determining which measures will bring the greatest value to existing assets have become important considerations.

Two tin cans and a string are probably still more cost-effective in short distances than a cell phone, but few would opt for such an alternative. We have "tin can and string" opportunities to save energy, but they seldom add much value to the customers' physical assets and are not apt to have the useful life owners want.

Other operation and facility considerations also need to be considered. A new roof is not apt to be the most cost-effective measure. However, if the old roof is leaking, a new roof with increased insulation may be the most critical need. Or, replacing an old boiler that is not only inefficient, but unpredictable and demands a lot of maintenance, may take precedence over a more "cost-effective" controls option.

Incorporating all costs and savings associated with a purchase for the life of the equipment is increasingly being used as a means of judging cost-effectiveness. This approach, Life-Cycle Costing (LCC), may appear to administrators in government to be the antithesis of the required low-bid/first-cost procurement procedures. If *specifications* call for LCC as a means of determining cost-effectiveness, then LCC can be compatible with low-bid procedures.

Cost-effectiveness cannot be assigned to measures without consideration of the conditions within which the measures must operate, including the capabilities of the operations and maintenance staff, the condition of existing energy-related equipment, and the costs associated with mitigating identified risks. As we become more sophisticated, we recognize that any payback calculation must look at all investments required to deliver the savings. Initially, in calculating paybacks, the investment considered only acquisition and installation. Today, risks that represent impediments to realizing the savings are identified and the mitigation/management costs associated with those risks are a key component of the payback calculation formula.

Cost-effectiveness is one measure of economic feasibility. It is an essential ingredient in performance contracting. It answers the question: "How soon can we get our money back from this investment?" There are various ways to calculate the time necessary to recoup the cost of the original investment. These range from simple payback and adjusted payback to the more complicated LCC.

Cost of Delay

Some things can be put off without a loss of revenue. EE work cannot. Every tick of the clock, every day that passes, represents dollars an organization may have wasted by consuming needless energy. Every hour of delay forces an owner to give money to the utility that, through energy management, could have been used to educate students, train sales representatives, offer patients additional services, launch a media campaign, meet constituent needs, make a bigger profit, etc.

Those working in energy management have become accustomed to weighing options by calculating cost-effectiveness as discussed above. The rapidity with which energy savings recover initial investments should be a major factor in weighing energy retrofit *vis-a-vis* other investments. The cost of delay is almost the mirror image of the simplified cash flow formula. The same factors that contribute to energy saving cost/benefit analysis affect cost of delay calculations, but in a negative sense.

Even when using internal resources, calculating the Cost of Delay related to various financing options makes good sense. Limited resources for other organizational needs may, in the long run, prove to be more costly. If those needs are paramount and the budget is tight, then getting the work done through performance contracting becomes an even more viable option.

An organization is frequently reluctant to use performance contracting, because it prefers to do the work itself and "save the service costs." Reality seldom meets expectations. Since O&M work is so very cost-effective, organizations often aspire to doing this work themselves so they can reap the savings before awarding a performance contract. Unfortunately, the O&M work has a way of getting postponed and delays, often in years, are apt to far exceed the potential "do-it-yourself" savings.

When an organization is considering the use of in-house labor and funds to save money, it should carefully weigh the Cost of Delay first. And the delayed period should be based on how long it took to get similar work in place in the past—from conception to acceptance of the installation—and not with regard to what someone hopes might be the case this time. In addition to the cash flow impact, it pays to also recognize the benefits to the organization of improving its capital stock. Not only is the capital stock value enhanced, but the company's overall fiscal condition benefits, and typically O&M costs are reduced.

Cost Avoidance

Rising prices can wipe out all the dollar gains that have been made by reducing energy consumption. The key is to identify what an organization *would have been paying* if it had not cut back! In order to communicate energy management benefits to others as costs rise, it is important to be able to talk about what would have been the costs had energy consumption not been reduced.

The joy of counting the dollars that would have gone to the utility makes cost avoidance very real and very gratifying. In order to calculate cost avoidance, a base year must be established. The base year consumption multiplied by the current price per unit will reveal "what it would have cost." Cost avoidance is what it would have cost minus current costs. Most top management or board members seldom have the time or inclination to wade through a pile of numbers; it will pay to graph the data when cost avoidance for more than one year is involved.

ESCOs need to employ these concepts as part of their marketing technique. Cost of Delay is particularly effective. It is also imperative that ESCOs take the time to keep customers informed of their cost avoidance and cumulative benefits over the life of the project. It is recommended that such information accompany any invoices for services to the customer. Cost avoidance numbers, preferably graphs, should be a regular feature in the monthly billing process. Without it, customers quickly forget what the utility bill used to be and why they are paying the ESCO. This becomes particularly critical when there has been a change in administration or top management.

It is important to never lose sight of the total facilities/processes as part of the organization's investment portfolio. New equipment, modifications and energy services should all be designed to improve that portfolio and enhance the work.

THINK MONEY while developing a project. An ESCO should never focus solely on the technical feasibility and viability of a project. It has to be concerned from the first moment of the development with the financing side of the proposal. The ESCO has to know in advance what the financing organization is looking for and what are its conditions for approval. Therefore, keeping the financing party informed about the development of the project from the initial stages is fundamental to project success.

When the various assessments are complete and the pieces assembled, it is time to think about the official presentation to a financier. To do that, it helps to think how the banker will view a project.

It is not as mysterious as some believe. Taking a proposal for funding to a banker or financier is just a variation on a theme. It is another sales job.

A few rules of the road might help.

- Know yourself, your company and your project before you approach an investor.
  - Be aware of your strengths and weaknesses.
  - Build on your strengths.
  - Acknowledge your weaknesses and how you intend to minimize/compensate for them.
  - Keep both aspects in perspective.
  - Know why someone should invest in your project and in you.

- Be aware that every strategy has financial implications and risk parameters. During proposal preparation, the task is to determine:
  - Is the strategy financially sound?
  - Have the risks been adequately addressed?
  - How will investors react to what you propose?

- Remember, bankers rent money. That is the business they are in. You are the customer. The days of going hat-in-hand for a loan are past. *Bankers need a quality project to invest in every bit as much as you need the funds.*

- We need to go beyond finding the project financing to finding the best deal on the money. Terms vary and quality projects should get the very best terms.

- But the generalizations stop here. We need to avoid over-generalizing about investors. Your challenge is to find financiers who want to invest in what your company can deliver. If they understand your business, it will temper their perception of risk and you will get better terms.

While writing a proposal, it is well to remember the difference between the bankers' focus on "return on investment" and the risks ESCOs regularly work with. Returns are relatively easy to quantify, but risks typically are not. It helps to show investors the risk management strategy you have and how you intend to execute that strategy. The approach is to acknowledge that the risks exist, while taking complex issues and simplifying them for investor consumption. In short, show them you know what you are doing, but do not belabor it.

THE OWNER'S PERSPECTIVE

The first step for an owner in achieving the most effective financing is to get an ESCO that can deliver a bankable project. The ESCO's track record and its bank relationships can tell the owner a lot about that.

Roughly 95 percent of the performance contracts in the United States are currently structured for guaranteed savings with the owner typically accepting the debt through third-party financing (TPF). TPF is especially attractive if the owner qualifies for tax exempt financing. Since the debt will be on the customer's books, owners have some important choices to make regarding that financing.

- Leases are often found to be an attractive way to finance performance contracts and are available predominantly in two forms: operating leases and capital leases.

  If a debt ceiling or greater indebtedness is a problem, an operating lease, which is "off balance sheet" and can have less of an impact on the owner's future capacity to borrow, can be attractive. But the qualifications for an operating lease are pretty narrow; as a result, a certified public accountant or some in-country person, who understands local financing needs to be consulted prior to the agreement.

  The majority of energy equipment leases are capital leases. Conditions vary by country. In the US, for example, if a lease meets any of the following criteria, it is considered a capital lease:
  - the lease term meets or exceeds 75 percent of the equipment's economic life;
  - the purchase option is less than fair market value;
  - ownership of the equipment is transferred to the customer (lessee) by the end of the lease term; or
  - the present value of the lease payments is equal to 90 percent or more of the fair market value of the equipment.

Conversely, if in the US a leasing arrangement meets any of the above criteria, it cannot be an operating lease.

Leases work very effectively with guaranteed savings programs. Articles in the popular press too often imply that leasing is available only with shared savings. Not so. Shared savings is only one type of performance contract. Any performance contract can be structured to use lease financing. The owners' choices, therefore, include whether or not to use leasing and, if used, what type of lease should be employed.

Other financing options may be presented to an owner. Unless the owner or someone on the staff is very comfortable with such concerns, retaining the services of a financial consultant is advisable ... and may be the cheapest option.

THE BUY-IN, THE BUY-DOWN

As a final cornerstone to this financing business, owners should not overlook the value of taking an equity position in the project. It is a way to get non-energy-related projects incorporated, and/or reduce ESCO and financier risks. A little owner equity can be a powerful leveraging force and make a bigger project possible. The bottom line for owners seeking to finance energy efficiency is: ask your banker. Find out what the men and women with the money need. Then, use their guidance to develop a project. The financier's due diligence, in the end, is the ESCO's and the owner's best guarantee that the project is doable.

A PRIMER OF FINANCING

There are a number of financing mechanisms available. They differ based primarily on the type of owner, the purposes for which the financing can be used and in the legal steps required to effect them. Some are only available to government agencies. Some are more applicable in certain countries and may be modified to fit local conditions and laws.

Everyone is familiar with conventional loans available from commercial banks. The other major types of financing vehicles are discussed below. Some do not lend themselves as easily to performance contracting and are cited here mostly to serve as a basis of comparison.

General Obligation Bond

Definition: GOs are bonds secured by a pledge of a government agency's full faith, credit and taxing power.

General obligation bonds are payable from ad valorem property taxes, and typically require voter authorization. Local laws stipulate the conditions to be met.

Special Assessment and Mello Roos Bonds

Definition: Bonds issued to fund projects conferring a benefit on a defined group of properties. The bonds are payable from assessments imposed upon the properties (in the former case) or from special taxes levied upon the properties (in the latter case) which receive the benefit.

Special assessment financing is generally used for infrastructure projects—e.g., roads and sewers—while Mello-Roos bonds fund facilities and services, such as libraries and library services. These types of bonds, when available, have been, and continue to be, very controversial in the eyes of the general public.

Revenue Bond

Definition: Bonds secured by a specified source of revenue or revenue stream.

Revenue bonds have numerous uses. Bonds for water, hospitals, airports (among others) are all examples of revenue bonds, where the revenue from a specific source; e.g., airport, water enterprise, hospital is pledged to repay the bonds. Sometimes a third party is established to collect revenues and to administer the promised repayment for a fee.

Since revenue bonds are dependent on a revenue stream, they are analogous in this regard to EPC and may be adapted to EPC financing.

Lease-Based Financing

Definition: Financing in which the fundamental legal structure is a lease. These include Certificates of Participation, Lease Revenue Bonds, and privately placed municipal leases.

Lease-based financing differs from debt financing primarily from a legal perspective. The obligation to make debt payments is *unconditional*. Lease payments, on the other hand, are *conditional*: they need only be made if the lessee has full use and possession of the asset being leased. Restrictions on issuing debt vary by state and by country and may impose significant conditions. These typically do not apply to leases. All lease-based financings share an underlying structure, as described below.

Lease financing is the most common type of local government financing in most states. Generally, the lessee is an owner; e.g., a municipality, with a project to fund. From a legal perspective, the lessee undertakes the project; i.e., buys the equipment or makes capital improvements, on behalf of the lessor. The lessor leases the project to the owner, which makes regular lease payments. When the term of the lease is over, the owner purchases the project for a nominal sum, often a token dollar, from the lessor.

Investors fund the lease made by the lessor, in exchange for which they receive the lease payments made by the lessee. This may be done through certificates of participation, lease revenue bonds or the lease document itself. The money investors pay for these instruments goes to a lease administrator (for simple municipal leases) or an underwriter, who deposits it (less the underwriter's fee) with the trustee bank. The lease administrator or underwriter, in turn, makes the funds nominally available to the lessor but, in fact, makes them available to the lessee for projects.

As mentioned above, the lessee is not required to make lease payments until and unless it has full use and possession of the project. When the project has been completed and there is something in place to lease, the owner begins to make scheduled lease payments. The lessee deposits its lease payments with the trustee bank, which makes the required interest and principal payments to the investors.

Special provisions are offered should the leased equipment or building be damaged. Under such conditions, the lessee may stop payments, until the project is repaired or replaced. The legal documents require that the repair be made as quickly as possible, so that investors wait as short a time as possible for repayment to commence. Because of this abatement risk, lease-based financing carries a higher interest rate than other types of financing.

Lease-based Financing Vehicles

Municipal Leases

This term is often applied to leases even when a municipality is not involved. This may be nomenclature for tax-exempt entity and is frequently used as a legal definition. It is usually a simple lease, which is funded by one investor—typically a bank or credit company. The bank funds the lease and the lessee makes the lease payments to the bank or credit company.

Master Leases

This "umbrella" lease is a variant of the municipal lease with general terms and conditions. As the lessee makes individual purchases or begins individual projects, leases or lease schedules are funded and appended to the master lease agreement. If a performance contract is being done in phases; i.e. 10 buildings at a time on a 60-building campus, a Master Lease may be used in order to make use of the funds from each successive phase. The Master Lease also serves well when the timing or amount of funds to be needed are not yet known.

Certificates of Participation (COPs)

This mechanism allows investors to purchase certificates, which offer evidence of their participation, and enables them to participate in the stream of lease payments being made by the lessee to the lessor. Certificates of participation have much higher costs of issuance than municipal leases, but carry lower interest rates. They are well suited for larger, longer term projects. In performance contracting, the COP approach may be used as pool financing, which can fund several projects. For the investors, this approach spreads the risk over several projects; thus, diminishing the risks associated with just one project.

Lease Revenue Bonds

Lease revenue bonds are similar to a COPs, except that instead of a corporation serving as lessor, one government agency acts as lessor while the jurisdiction needing funding serves as lessee. In these cases, a lessor government issues the bonds, enters into a lease with the lessee jurisdiction and the lease revenues are pledged as repayment of the bonds.

LEASE AMOUNT

The lease amount begins with the project cost, but it does not end there. In general, the following are added to that cost to arrive at the final lease size.

Capitalized Interest

It is the amount of interest that becomes due during the acquisition or construction period and is sometimes referred to as "interim construction financing" in performance contracting deals. Because the lessee cannot be compelled to make lease payments until it has full use and possession of the project, investors are concerned about being paid during the acquisition or construction period. Investors, therefore, require that the interest amount be "capitalized," or borrowed, through the lease, and set aside to be used to make interest payments during that period. The longer the construction/acquisition period, the more capitalized interest is needed. For example, the interim construction interest on a $4 million performance contract project may be $200,000, so the amount financed to include the capitalized interest would be $4.2 million.

Occasionally, the lessee uses internal funds during construction to avoid this interest cost.

Reserve Fund

It is an additional amount (usually one year's interest and principal payments) added to the lease amount and deposited with the trustee bank. This fund is used to make interest and principal payments to investors if the lessee is late or fails to make its lease payment. A reserve fund is often required for COPs and Lease Revenue Bonds, but is usually not considered necessary with municipal leases.

Cost of Issuance

Costs of attorneys, financial advisors, consultants, and incidentals, are usually funded though the lease.

When all is said and done, a $1,500,000 lease may make only $1,280,000 available in project costs with capitalized interest, reserve fund and costs of issuance all taking their toll.

It should also be noted that at least one ESCO, which brags about not sharing the savings, takes its entire fee from projects in an approach similar in process to using the Costs of Issuance. Owners should be aware that any ESCO making such claims is getting its fee up front from the financier. The ESCO enjoys reduced risks, avoids the depreciated value of the money and the owner pays the financing for this fee for the life of the project. While promoted as great value to the owner, the advantage all belongs to the ESCO.

The above primer gives a flavor of the options an owner can consider. The goal is to raise awareness as to the range of financing considerations. It is not sufficient information for an owner to decide all financing issues related to a given project. Unless the owner has personnel on staff comfortable with all aspects of financing, consultation with a certified public accountant (CPA) or the owner's banker is an excellent precaution. The due diligence of the project financier will benefit an owner, but it will not guarantee that the selected financing scheme is the mechanism that would best serve the owner's needs.

FINANCING IN EMERGING MARKETS

Over the past 10 years, hundreds of millions of dollars have been invested in international finance for EE projects and the development of ESCOs. Despite the work of numerous international agencies, including Multilateral Development Banks (MDBs) such as the European Bank of Reconstruction and Development (EBRD) and the World Bank (including its private sector branch the International Finance Corporation—as well as different bilateral organizations and development organizations such as the UN—only some progress has been made in removing project financing as the single largest barrier to the widespread implementation of EE and the ESCO industry around the world. This in no way is intended to be a criticism of the intent of any of the international agencies or MDBs, but rather to point out the difficulty in creating viable financing structures for EE in general and EPC in particular. The investment made by some MDBs has focused on developing guarantee mechanisms for local banks and financial institutions within targeted developing countries. Unfortunately, training of commercial bank personnel has not been as thorough.

International agencies have made substantial investments in developing countries intended to create and develop ESCO industries. A lot of the funding was used to pay independent consultants to create and conduct technical training workshops for local engineering, construction and other types of small consulting and contracting companies within the target countries to become ESCOs. The training programs focused primarily on teaching them how to develop and implement savings-based EE projects and included such things as how to perform energy analyses, quantify savings opportunities, create energy baselines and produce M&V plans. These programs also tried to teach the commercial and financial skills needed to apply the performance contracting model and become an ESCO.

MARKET-BASED INCENTIVES

Too often, projects are viewed by private end-users as a low-priority "infrastructure" investments versus an investment in the growth and development of their core business. Unless EE projects can meet very aggressive hurdle rates providing attractive simple paybacks, they have been unable to compete for the limited capital of private businesses. Discerning the difference between redirecting funds now paying for wasted energy in contrast to new budget allocations for increased production needs to be made.

The complexities associated with EPC projects, such as M&V, create a somewhat apathetic attitude by end-users and financial sources. Financial incentives are, therefore, highly desirable if we accelerate the widespread implementation of sustainable energy efficiency. It is important to note that no matter what incentives are developed, they are only to be provided to supplement and not to actually finance or fund EE projects. The actual financing needs to be underwritten by the local financial community in order to create an infrastructure that will be sustained over a long time.

Chapter 4

Making Contracts Work

Energy performance contracting (EPC) is by essence a contractual relation between a project beneficiary and an ESCO. EPC rests on a series of legal documents. Critical to this is the degree to which the contracts are enforceable. Unfortunately, in some countries, contracts are not worth the paper they are printed on. The "backbone" of a contract is the degree to which the provisions can be enforced. The efficiency and strength of the justice systems are what give contracts value.

Legal procedures, therefore, can dictate content. For example, in one eastern European country court cases have been resolved in weeks/months if the monetary value of the project was clearly set forth. If no specific value for the project is clearly stated, contract resolution can often take years. Since EPC initially rests on ESCO qualifications, not on a stipulated financial amount, ESCOs in any country can remain vulnerable for the already spent project development costs for years. This high risk exposure may persist until the customer has accepted the investment grade energy audit recommendations and a dollar figure has been assigned. In such cases, an agreed upon "place holder" value might be assigned until a firm figure has been established. This kind of protection, however, can only happen if the ESCO is aware of the problem. And this awareness typically comes from in-country partners.

Contract conditions can vary significantly as a result of negotiations. Therefore, owners and managers should never lose sight of the fact that energy service agreements (ESAs) are negotiable. Some ESCOs hand out contracts as if they are cast in bronze and the word processor has yet to be invented. Any firm that comes in with a "take-it-or-leave-it" contract and *attitude* is not a firm that will work **with** its customer to achieve the best results. An owner should take this as an indicator that it is past time to look for another ESCO! This chapter will conclude with some thoughts regarding effective negotiations and the unique role they play in EPC.

The caveats cited above should not lessen the perceived importance of contracts themselves. In this chapter, we present a reliable framework for contracts in the EPC process with the full understanding that every aspect needs to be reviewed and, as appropriate, modified by an in-country attorney to be sure the contract language fully complies with national, regional and local laws.

Contrary to fears engendered by some, increased EE can be achieved without any negative impact on the work environment. In fact, the contract conditions can, and should, stipulate the provisions that will create an enhanced productive environment for workers and occupants.

Attorneys, who are not comfortable with the performance contracting concept, can be a major impediment to achieving an agreement. For them, a piece of the puzzle may be missing. If an organization's attorney does not normally provide counsel in contract law, it is prudent to seek additional or outside counsel. Even though performance contracting has become a more common approach to EE financing, this type of contract may still be without precedent in the attorney's experience. So, even if the lawyer has contract law experience, it will expedite the process if he or she is provided with references of attorneys of record, who have successfully implemented energy performance contracts. The state energy office, the organization's consultant or the ESCO can usually supply such information. It will also facilitate procedures if the attorney is brought into the process early. An effective way to do this is to provide him or her with sample planning agreements and ESAs to review.

CONTRACTS: LAYING THE GROUNDWORK

Establishing the criteria, preparing the RFP and evaluating the proposals should lay much of the groundwork for a contract. Neither the solicitation document nor the proposal should be considered as all-inclusive, *or binding*. Lawyers in the public sector have been known to insist that no new conditions can be considered that have not been previously introduced in the RFP or the proposal. If ever faced with such bureaucratic "mumbo-jumbo," it may pay to bring attorneys with some experience in EPC to dissuade anyone from believing such false decrees.

Items not in the solicitation or proposal can be placed on the table for discussion during negotiations. Modifications in what an organization asks for or the firm proposes to do are commonplace. Should the proposal or parts of it become part of the contract, by included language or by reference, a statement should be made in the contract indicating the order in which related contract documents prevail.

The solicitation can be used to state that the customer reserves the right to make the proposal part of the contract. All organizational conditions upon which the proposal was based, however, should remain consistent or the ESCO should not be expected to comply with the stated provision.

It is strongly recommended that in preparing a request for qualifications (RFQs) or proposals (RFPs), copies of an EPC contract recently executed with a similar organization by an ESCO under consideration should be requested and then carefully reviewed to get a sense of what the firm really expects. At the time of negotiation, contracts from the two or three ESCOs under final consideration should also be reviewed carefully by the prospective customer's legal counsel.

There are typically three contracts in a performance contracting "package." They are the planning agreement, energy services agreement (ESA) and financial agreements. Most aspects of financial agreements are not unique to performance contracting and, in fact, are typically common to any financing arrangement. With that in mind, financial contracts are not treated here.

PLANNING AGREEMENT

Brutal experience has taught ESCOs that they cannot afford to give away their energy audits. At one time, the audits were thought to be an effective sales tool. They were certainly effective in showing the owner the existing savings potential. All too often, however, the ESCO did not receive a contract to implement the proposed measures.

The loss is greater than the labor and overhead, as auditors who can perform a quality investment grade audits (IGA) are still in short supply. Since a reliable audit is fundamental to a guarantee, an ESCO cannot afford to waste this talent. ESCOs are in the business of selling projects, not audits. An IGA, which does not evolve into a project, denies the ESCO its auditor's time—time which could have been used by the ESCO to open up a real project opportunity.

Owners have also been on a learning curve and increasingly recognize that an IGA is a premium grade audit and is a critical component of a quality project. Rather than the usual "snapshot" approach, which assumes that existing conditions will remain the same, an IGA will give the owner a better understanding of the way certain energy efficiency (EE) measures will behave over time in his or her facility. An owner, who can get an IGA for the price of a traditional energy audit, or even for a slightly greater cost, is smart to do so.

To protect the ESCO's interests and to preserve the auditor's time for IGAs that will lead to projects, the Planning Agreement has been introduced into the performance contracting process. The Planning Agreement, also referred to as a project development agreement, is a short contract of three or four pages, which addresses:

- The objectives which have been agreed to by both parties;
- The conditions which the IGA must satisfy for the owner;
- A statement that if the objectives and IGA conditions are met and the project is not forthcoming, the owner will pay the ESCO a specified amount for the audit;
- A statement that if the IGA does not meet the agreed upon objectives and conditions, the owner pays nothing; and
- The cost of the audit will be rolled into project costs if the project moves forward.

The objectives, which are listed to protect the owner's interest, usually specify the working environment that is necessary and state that the recommended measures must not have a deleterious impact on that environment.

The audit conditions in the Planning Agreement generally stipulate the expected range of savings as well as any audit procedures and parameters that are key to the facility manager and owner.

Since the IGA is a superior audit and serves as a critical bridge to contract implementation, the IGA cost, as specified in a Planning Agreement, normally carries a premium over the cost of a traditional audit. That premium can be as high as 50 percent above the typical energy audit cost.

The Planning Agreement typically does not require the ESCO to engage in the project. Obviously, the ESCO has pre-qualified the customer and believes a good opportunity exists; or it would not tie up its engineer's time, nor incur the costs. Even with the best pre-qualifications, however, nasty surprises do emerge and ESCOs should not be obligated to proceed with a project that is not economically viable. Under such conditions (if all other provisions are met), the owner usually pays for the audit, but at a reduced fee, as provided in the Planning Agreement.

ENERGY SERVICES AGREEMENT

Fundamental to any energy performance contract is the recognition that the project must perform over time. It is savings performance in the out years that deliver the greatest benefit to the owner and ESCO. A quality contract, therefore, will establish a cooperative framework that functions well over time. This means that a properly prepared and executed contract will facilitate a project's progress with a minimum of misunderstanding between the ESCO and the customer. If the language is clear and well understood by both parties, and if the terms are fair to both sides, the foundation exists for a project that will benefit both the ESCO and the customer. A poor contract invites controversy and bad feelings, often leading to project failure.

Topics generally addressed in an ESA are:

- financial terms and conditions;
- existing inventory and conditions;
- equipment/building modifications and services;
- user and ESCO responsibilities;
- risk parameters, such as default/remedies; and
- the construction contract provisions.

These items may all be covered in one comprehensive document or, separate schedules pertaining to work in specific buildings, or clusters of buildings. In large projects, schedules are typically used so that measures can be added to the contract as work progresses. The one document approach is generally reserved for small projects of limited scope.

Generally, an ESA is divided into two parts. The first part of an ESA typically states that the two parties agree that the firm will supply services to the customer and, in broad terms, outlines the services as well as the payment for these services. The second part is usually an agreement to agree on broad conditions.

Typically, an ESA is the basic contract and serves as a master agreement for multi-phase projects. In a large project, neither the customer nor the ESCO wants to wait until all the measures are implemented before benefiting from the savings. This is why the multi-phase approach has enjoyed increasing popularity. For a multi-phase effort to work, blocks of work are agreed to and attached to the ESA as "schedules" (may also be referred to as attachments, annexes or addenda). These schedules spell out exactly what is to be done in specific buildings or processes in the agreed upon scope of work.

The schedules become a part of the contract and describe measure-specific conditions, such as how savings will be calculated. With a large campus or installation, these schedules are then repeated for each phase of the project. For example, 40 buildings may be involved in the total project but the plan is to do it in phases of five buildings each. A set of schedules will then be developed for each set of five buildings. Each schedule is signed by both parties for each phase as work moves forward.

Third-party financing is typically used in a guaranteed savings model in North America, with the ESCO carrying only the performance risk and not the credit risk. Another project financing model is emerging where the lender and ESCO create a Single Purpose Entity (SPE), which carries the credit and helps keep some credit liability off the ESCO's books. In all cases, a parallel agreement with the ESCO can then be entered into to install and maintain the equipment, provide other services and guarantee that savings will cover required payments.

The term, performance contracting, emerged as the accepted label since the energy service company must perform to a certain standard (level of savings) as a condition of payment. These performance considerations are integral to the contract components and are implied throughout most contract provisions.

TYPICAL ENERGY SERVICES AGREEMENT COMPONENTS

- Recitals (traditional, but not essential)

- Equipment considerations
  — ownership
  — useful life
  — installation
  — access
  — service and maintenance
  — standards of service
  — malfunctions and emergencies
  — upgrading or altering equipment
  — actions by end user
  — damage to or destruction of equipment

- Base year conditions/calculations, baseline adjustment provisions and a re-open clause

- Measurement and verification procedures
  [In multi-phase projects, the 2nd, 3rd, and 4th bullets are addressed in the schedules and general terms of reference are used in the master contract.]

- Other rights related to ownership
- Commencement date and term renewal provisions
- Compensation and billing procedures
- Late payment provisions
- Energy usage records and data
- Purchase options; buyout conditions
- Insurance
- Taxes, licensing costs
- Provisions for early termination
  — by organization
  — by firm
  — events and remedies
  — non-appropriations language (for government entities)

- Material changes
- Conditions beyond the control of either party (*force majeure*)
- Default
  - — by organization
  - — by ESCO
- Events and remedies
  - — by organization
  - — by ESCO
- Indemnification
  - — for both parties
- Arbitration
- Representations and warranties
- Compliance with laws and standards of practice
- Assignment
- Additional contract management terms
  - — applicable law
  - — complete agreement
  - — waivers
  - — severability
  - — further documents, schedules
- Schedules (by designated group of buildings or project phases)
  - — description of premises; inventory of equipment
  - — energy conservation and/or efficiency measures to be performed
  - — base year conditions and calculations, variables and baseline adjustment provisions
  - — savings calculations; formulas
  - — measurement and savings verification procedures
  - — projected compensations and guarantees
  - — comfort standards
  - — contractor O&M responsibilities
  - — O&M responsibilities of the owner
  - — termination, default value, buyout option
  - — existing service agreements

- — calculation of other savings; e.g., existing service/maintenance contracts
- — training provisions
- — construction schedule
- — approved vendors/equipment

A contract offered by an ESCO is designed to ensure that its interests are protected. It is not necessarily designed to protect the interests of the owner. As in all contract negotiations, it is up to the customer to make sure its interests are protected. Prior to negotiating a contract, both parties need to consider the implications of the various key components and the latitudes within which an item can be negotiated. In other words, decide what is not negotiable, what conditions can serve as "trading stock" and in what priority.

The best contracts are tailor-made for a specific client and a particular project. No one "model contract" will suffice. The key contract considerations offered below provide a foundation for developing the contract.

KEY CONTRACT CONSIDERATIONS

Any good performance contract will provide a clear understanding of each key contract element. The following discussion offers a context in which these key matters can be considered with an attorney.

Equipment Ownership

The financing scheme used and the point at which the organization takes equipment ownership can affect the organization's net financial benefit and may affect depreciation benefits.

In an energy saving project, the useful life of the proposed equipment is a key factor in post-contract benefits.

ESCOs and/or their financiers usually insist on a first security interest in the installed equipment or collateral of equivalent value.

In the case of buy-out provisions, termination and default values, procedures for establishing capitalized equipment worth may be set forth in the original contract. Terms, such as fair market value, need to be carefully defined. The buy-out provisions will typically be greater than the value of the equipment, as the ESCO's fees for services, compensation for risks incurred and potential savings benefits need to be factored in.

Malfunctions

Provisions for immediate, and back-up, service in the case of malfunctions need to be spelled out. This is especially important if the contractor is not a local firm. Local distributors for the selected equipment frequently serve this function with further back-up provided by the ESCO. Maximum downtime needs to be considered. The allowable emergency response time will vary with the equipment installed and with how essential it is to the operation.

ESCOs need to establish an understanding with the distributor or designated emergency service provider as to the timing and the extent to which emergency services will be provided before a malfunction arises and should be spelled out in the performance contract.

Firm Actions, Damage

Contracts proffered by ESCOs will discuss actions the customer might take that could have a negative effect on savings. The management needs to determine if these conditions are reasonable and determine to what extent the organization should have the same protections.

Consideration should also be given to the impact the ESCO's redress may have on the organization. If termination provisions provide for equipment removal, the conditions of the facility after that removal and the consequences for facility operations need to be stated and understood.

Equipment Selection and Installation

The customer should reserve approval rights on selected equipment provided approval is not "unreasonably" withheld.

ESCOs must retain some rights if they are to guarantee the savings. Under some bid procedures, the ESCO may take on the role of a general contractor; i.e., writing specifications, monitoring bid procedures and overseeing installation. These may be services a given organization needs, but they also serve to protect the ESCO's position on guarantees. In any case, with guarantees involved, an ESCO must retain sufficient control of the specifications as well as equipment and installer selection to assure guarantees can be met.

To effectively meet the concerns of both parties, equipment specifications are generally developed cooperatively. Often, one party then narrows the selections to two or three bidders and the other party makes the final selection.

Contractual conditions used in any construction project; i.e., liability, Occupational Safety & Health Administration (OSHA) compliance, clean up, performance bonds, etc., should apply.

Provisions for Early Termination

From the customer's point of view, contract language regarding termination should clearly set the parameters for equipment removal. These provisions should include length of time the removal will require and provide sufficient details as to restoration of the facility after the removal.

ESCOs incur major exposure early in the contract, for they incur the major expenses at this time and must depend on eventual savings to cover these costs. Buy-out provisions must provide for ESCO recovery of costs incurred and a proportionate profit. Frequently, buy-outs are not offered as an option until a specified period, as long as two years, has elapsed.

For further protection, ESCOs, or their financiers, frequently specify that organizations which rely on outside support; i.e., government, use non-appropriations language in which the owners agree not to replace the equipment with equivalent equipment within a specified time frame. In this way, the non-appropriations provision cannot serve as a "dodge" to avoid making required payments. The investor, in such a scenario, may seek further protection by making sure the equipment in question is vital to the organization's operation.

Conditions Beyond the Control of the Parties

The usual contract language absolves the ESCO of certain contract responsibilities under force majeure, or acts of God. These conditions should be examined, and the merits of similar provisions for the organization should also be weighed. Increasingly, the language is written to absolve both parties equally.

Default Language

Language frequently limits the conditions of default and remedies for the ESCO, but may leave it wide open for the customer. When the financial burden is carried by the ESCO, this is not necessarily inappropriate. Similar language for the organization should be considered, especially if the organization holds the debt service contract on the equipment.

The owner's attorney should carefully consider the default and remedies it would impose on the ESCO and accept for itself.

Indemnification

Both the ESCO and the customer should be indemnified. Some ESCOs attempt to secure indemnification from indirect and contingency damages. These are frequently too broad and should be analyzed carefully by the organization's attorney.

Assignment

The customer should insist on prior approval for any assignment, including any changes of service responsibility, or key personnel. Prior approval of subcontractors is also desirable.

Applicable Law

The ESCO typically presents a printed contract as the basis for agreement. The ESCO is apt to specify the applicable laws of the state in which it is incorporated. Should court action be necessary, the ESCO has a cost advantage and possibly a legal advantage. This places an additional burden on the customer if the organization is located in a state different from the one specified in the contract

Since applicable law provisions may just as easily specify the customer's state, this provision may become "trading stock" in the negotiation process.

Savings Calculations Formulas

This procedure is frequently made far more complex than it needs to be. The reduction in units of fuel and/or electricity multiplied by the current cost of energy by unit is the standard procedure for calculating the cost of saved energy. The first step, of course, is to reliably ascertain, using a broadly accepted measurement and verification (M&V) procedure, the number of units which have been reduced. Attribution of electrical demand charge savings also needs to be evaluated, negotiated and specified in the contract language.

Weather or occupancy changes, added computers, etc., can effect savings. Extensive contract language, however, which tries to anticipate every contingency that might happen only benefits the legal profession. The clearest way to address this problem is to have a broad-based base year and annual adjustment provisions related to anticipated variables, which both parties approve. Then the parties should agree on a re-open provision, which in effect says, "If the annual baseline after adjustment differs from the base year by more than + 'X' percent, the base year provision can be re-opened to negotiate a new base year reflective of current conditions." It is important that it be stipulated that only this one section can be re-opened unless opening a can of worms has appeal.

Calculation of Base Year and Adjusted Baseline

In order to avoid confusion in this discussion, "base year" is used to describe the historical consumption and "baseline" is used to address the annual reference point. Provisions for calculating a base year should be clearly presented. Historically, base years were often established simply by averaging multiple years of consumption. That is not enough! In all cases, conditions that have caused that consumption should be noted and any potential changes, which could have a significant impact on consumption, should be clearly identified (the closure of a hospital wing).

Base year consideration should include (1) mild or severe weather in recent years; (2) recent changes in the structure, building function, occupancy, etc.; (3) recent O&M work which could affect consumption; and (4) any recent renovation which could also affect energy consumption. Reopen language should provide for some adjustment beyond the agreed upon variations, so neither party pays for unexpected contingencies. No owner wants to pay the ESCO for the savings realized from the closure of a wing of the facility. Nor does the ESCO want to foot the bill for added computer labs.

After these adjustments are made in the historical consumption pattern, a baseline is established which serves as a reference point for making savings calculations. It is recommended that both customers and ESCOs refer to the procedures established in the International Performance Measurement and Verification Protocol (IPMVP) (see web site *evo-world.org* for information on IPMVP).

The share of the savings will vary with the length of payback, the services delivered, the financing scheme selected, the risks assigned to the ESCO, the length of contract and the like. The interrelationship of these factors needs to be considered in negotiating each party's share of the savings.

Measurement and Verification

Procedures for M&V of savings will vary with the EE measures installed, the size of the project and a number of other factors. In the ESA, the contract language typically states that the M&V procedures will be decided jointly following the IGA report and according to the IPMVP. Further discussion of key M&V considerations are presented in Chapter 7.

Energy Prices

Price volatility needs to be given careful thought. How the "burden" of falling prices or the "benefit" of rising prices is to be shared should be clearly addressed in the contract. Since no one should ever be foolish enough to predict energy prices two, five or twenty years out, ESCOs should insist on a floor price. In other words, an ESCO should not only be able to guarantee the amount of energy saved, but should also guarantee the value of the energy saved will be sufficient to meet the customer's debt service obligations *provided the price of energy does not go below a set floor price.* If the ESCO insists on a price floor, however, the customer should enjoy the benefits associated with increased energy prices. Further, a ceiling price may be requested for financial energy savings calculations.

Comfort Standards

The greatest fear employees associate with EE, and more particularly performance contracting, is the loss of control of the work environment, particularly comfort factors. The frequently voiced supposition that an ESCO will control the building operation is simply not warranted—unless management abrogates its responsibility and gives the control away.

The management can, and should, establish contractually acceptable comfort parameters for temperature, lighting levels and air exchange as well as the degree of building level control needed (and override required) to assure a quality environment. Wherever possible, these conditions should be expressed in acceptable ranges, not specific set points. The more rigid a customer's expectations are; the greater the ESCO's risks become. These risks need to be hedged. As a result, the more control the customer keeps, the greater the ESCO risks and the lower the customer's share of the savings will be. Indirectly, customers always pay for actions which increase ESCO risks.

Projected Compensation and Guarantees

The most attractive part of performance contracting is the idea that there is an entity out there that will make sure the organization has new capital equipment that works, and can assure that the savings will occur and all this will happen without any initial capital cost to the organization. The manner in which the energy savings are guaranteed to cover debt service payments, and the risks associated with meeting that guarantee, are key components of a performance contract and deserve careful consideration.

Since the quality of maintenance on energy-consuming equipment affects savings, most ESCOs require specified maintenance provisions and may ask for related maintenance contracts. They may, however, not guarantee that energy savings will cover the required maintenance fee. If an organization regularly contracts for maintenance and the ESCO's fee is not greater than the existing fee, this may not pose a problem.

A major reason for a contract is to identify and assign risks and provide appropriate recompense. The "guarantees" are the bottom line in making sure a contract works in the organization's favor. However, the more risks accepted by the ESCO, the lower the savings benefits will be to the customer. To repeat: the underlying concept prevails as *money always follows risk.*

As with any contract, the organization's attorney should review the ESA before it is signed. Through all the negotiations, frustrations and delays, it is well to remember that a good contract, which both parties feel is fair, is essential to a successful project.

Model contracts are frequently requested of the authors. Experience has proved, however, that the use of a model ESA can be dangerous, for a contract must be carefully modified to meet the unique conditions of state laws, local ordinances and customer conditions.

THE SCHEDULES

The schedules attached to the general ESA serve to make the contract specific to the project, or a phase of the project. They establish the details of the work to be done and the conditions under which the work will be accomplished. They include the manner in which savings will be calculated, measured and verified, and services that will be provided for specific measures.

The content and number of schedules may vary. Usually, the following types of questions must be addressed, each of which may be the subject of negotiation:

- **Equipment**. What equipment exists? What equipment will be installed? What is the projected cost of the equipment? Who will install it? Who will maintain the equipment? Building modifications are usually addressed in this schedule as well.
- **Warranties**. How and when will the manufacturer's warranties be conveyed to the owner? How will the maintenance tasks be monitored? Some contracts couple guarantees with warranties, a practice that can obfuscate the exact nature of the guarantee.
- **Savings formula**. What are the assumptions and formulae that are the basis for the energy savings calculations? Allocation of demand charge savings and positive cash flow are treated here. Changes prompted by utility restructuring; i.e., real time pricing, also need to be addressed.
- **Measurement and verification**. How will the savings be measured and verified? Actual procedures, measurement devices and assigned responsibilities should be set forth varying with the measures to be installed. A complete M&V plan should be included for use and reference during the whole duration of the contract, as discussed in Chapter 7.
- **Guarantees**. What are the guaranteed savings per year? What are the payments to the ESCO from the savings per year? What are the guarantees by the ESCO to the customer? What are the guarantees, if any, by the customer to the ESCO? What are the procedures to adjust the baseline for reconciliation? When will the reconciliation occur? *[In organizations which operate on a fiscal year, it saves a lot of budgetary grief if the reconciliation is done just before the end of each fiscal year.]*
- **Price variation**. What happens when costs vary (especially energy prices) due to inflation or other factors? Is there a floor price? How are price increases shared?
- **Performance standards**. What customer operating performance standards must be met by the improvements installed by the

ESCO; e.g., lighting conditions, acceptable temperature ranges, steam flow, etc.? What are the equipment installation schedules? And what standards of service must be met?

- **Ownership**. Who will own the equipment during the life of the project? If the customer wants to purchase the equipment during the project earlier than planned, what will be the terms and conditions of the purchase?

NEGOTIATIONS

Effective negotiations lead to effective contracts. And more importantly, effective contracts make for good projects. Ideally, when the negotiations are over, all parties should walk away from the table feeling they have laid the foundation for a strong partnership—one which will serve everyone well for many years.

In any contract procedure, the customer need not, in fact should not, "lose control." Every client can, *and should*, first determine the key elements the organization must have in an energy service contract and the latitude within which it is willing to negotiate. Understandably, an energy service company and the financier have to have some assurance that they can protect their investments and that the savings can be reasonably guaranteed.

Before negotiations start, each party should take stock of its own operation and what strengths it brings to the table. Careful consideration should be given to what conditions are negotiable and how much latitude can be allowed on the negotiable items.

It is also important to learn what one can about the other party. Not just what is known that brought you to this point, but what is the other party's negotiating history and behavior at the table. A couple of phone calls in advance can often prevent some surprises.

Both parties will profit from a little self-interrogation. What do "they" offer that you must have? What does your organization offer that is particularly attractive to them? What is the best way to position your strengths in the discussion?

Know the process. Ignorance can weigh heavily against you. Negotiations have some uniformity regardless of the topic. Recall previous negotiations, even union negotiations.

Negotiations are not as mysterious as they are frequently made out to be. Consider what your strategies will be, and anticipate what kind of strategies can be expected from the other party. Common sense serves as a good guide. Never agree to something that you cannot live with on a day-to-day basis over the contract years.

NEGOTIATING STRATEGIES

Careful thought regarding some very basic negotiating strategies can make you and your organization feel more comfortable, as you head for the table.

The suggestions offered below have been roughly drawn from *Roger Dawson's Secrets of Power Negotiating*. For those with limited negotiation experience, this book is strongly recommended for use as a guide.

- The customer should carefully review the sample contract submitted with the proposal before the ESCO is initially selected.

- The customer's attorney should meet with those negotiating the contract prior to negotiations and go over the draft contract submitted by the ESCO. Then, the group can:
  - set-aside the parts which are acceptable;
  - note those parts that need slight modifications;
  - note those parts that might be key to the ESCO, but not necessarily to the customer;
  - identify the parts which are unacceptable and what needs to be changed—and how—to make it acceptable; and
  - decide just how much latitude there is on each item and what other parts have some "give" to be sure the key parts are developed to your liking.

- Never accept their first offer. Even "printed" contracts can, and are, revised.
- Ask for more than you expect to get. The other party assumes you will. Starting where you wish to end up too often leads to getting less than you wanted—or should have.
- Avoid confrontational negotiation. The other party will be your partner for many years, so start as you mean to go.

- Display some traits of the reluctant buyer/seller as part of your strategy. Eagerness has its place, but seldom at the negotiating table.
- Reserve the right to defer to a higher authority; i.e., the boss or the attorney. Generally, attorneys complicate things and too often want to get into legalese. Whenever possible, it pays to keep them out of the negotiation's room. (Remember, attorneys will not be living daily with the project.) Attorneys, however, can be very useful out of the room as the "higher authority."
- Remove *their* resorting to a higher authority by appealing to their egos or pressing for them to commit to making a recommendation of a certain position to that authority.
- Be on the lookout for their "problem," which you can help solve. Recognize it as a "hot potato" and test its validity.
- Never, ever, offer to split the differences, but you might encourage them to do so.
- A critical point, which has been noted by Mr. Dawson, is particularly important for performance contracting: perceived values during negotiations go up for materials and *down for services.* Protracted negotiations can, therefore, diminish the perceived value of services. Considering that performance contracting has a strong service focus, the negotiation process can work for or against a party depending on whether you are buying or selling those services.
- There are two basic rules on making concessions:
  — always get something in return; and
  — start big and taper off. If your concessions get bigger as you go, the rewards for the other party to continue negotiations are obvious.
- If you truly reach an impasse, consider setting it aside to deal with later.
- Should you reach deadlock on a key issue, consider intervention or mediation.
- Position a point for easy acceptance by leaving something on the table.
- Watch out for the "Oh, by the way" when it seems the negotiations are over and everyone is smiling and shaking hands. This last little "nibble" could be bigger than it seems.
- Never lose sight of the fact that a good contract is one in which both parties feel they have a fair and workable deal.

IF I WERE ON "THEIR" SIDE OF THE TABLE

To balance the scales, the customer should picture himself/herself on the ESCO's side of the table. When guarantees are part of the picture and performance is tied to the guarantees, there are some items that are virtually non-negotiable for the ESCO. A county once put out an RFP that glibly stated that the ESCO would carry the financing, make the guarantees and the county would select the equipment. Surprisingly, they got several responses—not so surprisingly, most were from very new ESCOs. Happily, several conditions were changed before any contracts were signed.

In order to make a guarantee on the savings from the project, an ESCO will expect to:

- write the specifications in cooperation with the owner and participate in the final selection; or
- select the equipment with the owner's final approval;
- select the subcontractors who will install the equipment with the owner's tacit approval; and
- decree the level of maintenance and tasks to be performed by the owner with some key maintenance provisions reserved to the ESCO.

If the customer feels a strong need to have control of any of these items, they can expect the ESCO to hold back a large financial cushion to cover the risks. This step, in turn, leaves the customer with a smaller project, less savings and fewer overall benefits for the same level of investment. Occasionally, the alternative is to remove the guarantee provision related to this item, or entirely from the contract.

If the guarantee, which states that savings will cover the debt service obligation, is removed, some performance conditions can still be maintained by developing a shared-savings model for the excess savings. In this scenario, there is no assurance that the savings will cover the debt service obligation, but the owner is somewhat assured of the ESCO's continued interest in the project's performance by splitting any savings over and above the debt service payment.

Section II

Implementation Aspects and Trends

One sale does not make a successful business any more than a grain of sand makes a beach. Any business model, including the performance contracting model developed in this book, requires effective marketing strategies and sales techniques for long-term growth.

To gain the most from a project, the customer is also very dependent on the energy service company (ESCO) having a successful business. The ESCO must be there to back the guarantee as well as to provide services and guidance through the life of a project. In fact, due to the expected longevity of a performance contracting agreement, partnership quality is a critical criterion in customer and ESCO selection.

Energy efficiency (EE) has many obvious benefits. In addition to preserving our limited fossil fuels, it reduces the client's operational costs and pollution emissions. Even better, it can be done through self-funded work. EE offers a truly unique opportunity to make money while improving the environment. Further, EE savings can help pay for a range of other services, including renewable energy, which ESCOs can provide. Since ESCOs invest considerable time and resources in acquiring customers, it is good business to make the most of a new customer once acquired. ESCOs are, therefore, increasingly looking for related services that they can offer which benefit the end-user.

The self-funding aspect of performance contracting means that the needed investment capital can come from money already in the client's budget. It is hard to overstate to the client the benefits of creating a more efficient operation with money already in the budget—money which is being paid to the utility for wasted energy. The proof of these benefits, as outlined in Chapter 7, relies on clear and concise means of measuring and verifying energy savings.

In countries around the world—transitional economies, industrial, or developing countries—people in the EE field have trouble making the sale. Be it an energy policy maker in a government, an energy manager in a corporation or sales people of an ESCO, every energy professional shares these frustrations. The sale to businesses, especially industry, has been particularly difficult. Since business is the backbone of the economy, the ultimate source of new revenue and a major consumer of energy, it is a sale that we really need to make. Making the business case for EE, as addressed in Chapter 5, lays the foundation for much of our energy performance contracting (EPC) growth.

Pressing business needs have prompted ESCOs to share concerns with colleagues, and ultimately it frequently leads to the formation of an association. Once in place, associations have been able to perform a range of services. ESCOs around the world can benefit from learning what other ESCOs and their associations are doing. Chapter 8 offers interesting information regarding today's ESCO associations while Appendix B presents a list of active associations known to the authors.

Chapter 5

Markets And Marketing

If someone claims energy performance contracting (EPC) is an easy sale, it is evident that this person does not know what they are marketing.

In their book, *Quality Selling through Quality Proposals*, Kantin and Hardwick list five criteria and then emphasize that satisfying any one of the criterion constitutes a complex sale. To paraphrase, these "complex sale" criteria from the customer perspective are:

- an innovative and sophisticated business approach;
- financial and non-financial benefits, which are not readily apparent;
- multi-level buying approval is needed in the prospect's organization;
- the products or services, which may solve a complicated business problem; and
- the process requires team selling techniques.

Those familiar with the EPC model will quickly recognize that it meets each and every one of these criteria. Thanks to the implicit energy savings, a self-funded project is a simple concept. When layered over a complex process, it creates a huge marketing challenge. EPC represents such a challenge. Marketing and selling performance contracting is a very difficult and complex sequence of events.

MARKETING vs. SALES

Before we explore effective marketing, it is important to differentiate between marketing and sales. Effective marketing will help design the most efficient use of sales time and talent. Marketing is a staff function. Sales has a direct relationship to profit and loss (P&L). This connection to P&L carries demonstrable responsibilities and makes sales a line function.

There is an old adage, "Nothing happens until somebody sells something." "*Sales*" is the engine that drives an organization. Without it, there is no revenue stream. The key function of marketing, therefore, is to make sure that sales happen as smoothly and cost-effectively as possible. A "sales-driven" organization, however, can lose money. Sales must lead to net profit.

In the EPC world, the marketing challenge embraces market selection, picking the right potential customers, and orchestrating the sale.

MARKETS

The natural inclination in selecting a market is to look around for opportunities and to pursue at random those that look most promising. That, however, is seldom the best use of sales time.

A more sophisticated, and typically more profitable, approach is to first examine what an organization has to offer. This requires a hard, objective look at the organization's strengths—and weaknesses. The fundamental question is how an organization stacks up when compared to its competition in a given market.

To make this analysis, it helps to look at an organization's skills in relation to various markets being considered, including:

- sales techniques and experience in working with a particular market segment—every market has its peculiarities and jargon;
- design capabilities;
- construction management;
- project management;
- measurement & savings verification;
- financial acumen; and
- legal expertise.

It is easy, however, to be lulled into thinking that these criteria apply evenly across all markets. A closer look at construction management in hospitals, as an example, will dissuade that notion. Project implementation in healthcare facilities does not have the luxury of installing the new boiler *off hours*. Facilities that operate 24/7 create unique challenges. Further, while under construction, management must be ever mindful of critical infection control procedures. Positive and negative pressures as well as iso rooms must be respected every day work progresses. If an organization cannot deliver a quality project under such conditions, it should not seek to work in hospitals.

After a careful analysis, management needs to figure how its organization can capitalize on its strengths and how they will play best in certain markets. At the same time, management needs to identify weaknesses, also with particular markets in mind, and assess ways through which those weaknesses can be bolstered. In the face of these weaknesses, the question must also be posed as to where the organization gets the expertise to know if it is getting the right support.

Addressing an organization's weaknesses relative to a particular market also has its economic side—new hires, subcontracted services, costs. Further analysis requires weighing such costs against the opportunities being offered by the new market.

But the analysis does not stop there. There is a further need to consider what the various markets offer in relation to other criteria. For example, what is the economic viability in a given market? If the opportunity is large, then the obvious questions are: Why aren't other ESCOs serving this market? Are there hidden risks in serving this market? Are the risks manageable? What will it take to penetrate the market and establish an ESCO under prevailing circumstances?

An honest assessment of how an ESCO compares to its competition in a given market is a critical step, but very hard to do. If an objective analysis cannot be performed by the in-house staff, it may pay to get an outside consultant, who understands the ESCO industry and knows what other firms bring to the table.

Once the potential markets have been narrowed down, further consideration needs to be given as to how the selected market(s) can be accessed. Are there suppliers, trade allies who regularly work with the market? Can sales be channeled to the more attractive markets? Does the market have definable leadership? Can a business relationship be established with these leaders? A successful project, which involves the leaders in a particular market, can become a huge sales advantage. Imagine the benefits that can come from those leaders describing the project to colleagues at a state or national convention. The sales implications are tremendous.

Market selection criteria will, of course, be swayed by the opportunity for an ESCO to make money. But caution is warranted. Basing marketing strategies on the size of a sale can be misleading. It is quite possible to have a strong cash flow and very little profit. Too many ESCOs commission their sales people on the size of a sale. This can push an organization to become sales driven and ignore the serious implications regarding profit. Further, the focus on sales can provide an incredible incentive for sales people to "cook the books"—to distort sales potential in order to make the sale and get a commission.

To avoid misrepresentations, it is preferable to provide the sales force with only a portion of the commission upon contract signing, another portion after the project has proven itself (maybe one year later), and the remaining commission as additional payments in the out years. This commissioning structure not only restrains initial misrepresentations, but provides an incentive for sales people to stay involved with the customer and preserve the savings level. This approach also provides fertile ground for resale opportunities.

All downstream commissions should be based upon the energy savings actually realized and the actual net profit yielded by the project. In these calculations, it is important to adjust for the changes in energy supply pricing. The sales force should neither be penalized nor rewarded for energy price changes.

The true measures of project value are actual energy savings and net profits over time. If an organization, or its management, says it is "sales driven," then it does not fully understand the business proposition under which it operates.

Once an organization has performed a careful analysis and selected its market(s), identifying potential customers is clearly the next step. But before this can happen, it really pays to establish some customer selection criteria. The value of adhering to those criteria and not veering away without clearly documented reason(s) cannot be overstated. The industry is rife with horror stories of money lost pursuing a potential customer only to realize the customer did not measure up in another key area and the project could not go forward.

PICKING A CUSTOMER

One of the most dramatic differences between direct sales and energy performance contracting is the duration of the vender/customer relationship. In direct sales, the relationship is typically sporadic and contact is often limited to days. In EPC, a close relationship that extends for years is common.

Since the whole concept of EPC rests on guaranteed results, the process is basically risk management. A fundamental consideration must be the risks that can prevent a firm from achieving the predicted savings. One of the biggest risks is the customer itself, so great care needs to be taken in selecting a customer. Unlike direct purchase, getting someone eager to buy a product is not enough. Finding a customer that is willing to work with the ESCO's team to achieve energy savings over time is far more important.

A scoping audit, which should be guided by the ESCO's customer criteria, is the ultimate customer selection tool and should determine if a project should be pursued. Whether or not established criteria are being adhered to should be an oversight function of the ESCO marketing group.

In analyzing prospects, ESCOs are inclined to go with what they know; e.g., focus on energy savings potential and facility-related issues. This is good, but not enough. Financiers insist that the ESCO look at the creditworthiness of the potential customer. This includes organizational and business aspects that are not always an energy auditor's strong suit. Having a financial person on the team, who can assess the current financial condition of the end-user and its business prospects, is crucial.

The weakest, and frequently the most dangerous, part of assessing customer potential is the absence of attention to the people factor. If you ask an experienced auditor to compare the potential savings of identical measures in similar facilities, the auditor will invariably acknowledge that the savings will not be the same. When asked why that conclusion has been reached, the auditor is apt to shrug his/her shoulders and look decidedly uncomfortable. The problem is that they are being asked to make a subjective judgment and there is nothing in the engineering handbooks to cover it.

When the dust settles, it becomes apparent that what is really being asked of the auditor is what obstacles stand in the way of achieving the predicted savings. In short, the auditor is being asked to make a risk assessment of the customer and every recommended energy conservation measure. Further, the auditor needs to also weigh the risk management/mitigation strategies available and their associated costs. This process has given rise to the investment grade audit (IGA). (For more information, see *Investment Grade Energy Audits: Making Smart Choices* by Shirley J. Hansen and James W. Brown.) A well-executed IGA is a fundamental aspect of customer selection. EPC is basically a financial transaction, but the project's integrity rests on the quality of the IGA.

A successful firm depends on a healthy net profit. In EPC, those profits derive directly from the ESCO's ability to manage risks. The first step in an energy efficiency (EE) project is to identify those obstacles that might prevent an ESCO from realizing the predicted savings. This necessitates a process for identifying the risks and determining how to cost-effectively manage/mitigate those risks. The biggest risk is the customer itself, and closely behind are the risks associated with each contemplated measure.

DEVELOPING MARKETING STRATEGIES

Marketing is not an isolated event. It does not belong to a designated division. Peter Drucker summed it up well in *The Changing World of the Executive* when he observed:

> *"Marketing is the whole business seen from the point of view of its final result, that is, from the customer's point of view. Concern and responsibility for marketing must, therefore, permeate all areas of the enterprise."*

Many business decisions are fundamental to developing effective marketing strategies. It can also be said that developing effective marketing strategies is fundamental to key business decisions. A good case in point is an analysis of the cost of customer acquisition.

The Cost of Customer Acquisition

Sales people frequently have potential customers that they love to call on. A chat over a cup of coffee will have great appeal even when a buy is not in sight. Sales people typically spend valuable company resources on entertainment and other efforts to influence a buyer.

Unfortunately, few companies take the time to fully analyze the resources it takes to put a sales person in the field. Or, to decide in advance how much time and money the firm can afford to expend in an effort to acquire a certain customer.

This is particularly true in EPC, as long, long sales cycles are the bane of the ESCO industry. The cost of customer acquisition requires the gathering of key data and effectively using that information to make critical marketing decisions. Procedures also need to be established to make sure those decisions are followed.

For every hour a sales person is in the field, the costs mount up. It is important to quantify those costs, so management knows exactly what it is spending to acquire a particular customer—and when a sales person could more profitably spend his or her time somewhere else.

Rough calculations can be made to approximate the costs behind each hour of sales time. If the time spent serving existing customers is subtracted from the company's total operating costs, the remainder establishes approximately how much of an ESCO's resources are behind the sales effort. That figure can then be divided by the total sales force hours, and the extent to which the firm's resources are used to back up each hour of sales time can be determined. Once that figure is known, the number of sales hours to penetrate a certain market and/or acquire a given customer can be set.

To guard against a sales person "over indulging" in acquiring a given customer, two safeguards are recommended: 1) establish the "affordability" criteria; and 2) establish a marketing/sales committee, which will periodically review the sales efforts in relation to each potential customer.

Affordability Criteria

Some "hard-nosed" objective thinking needs to go into the level of investment a firm can commit to acquiring a certain customer. These criteria might include:

- potential partnership quality;
- perceived risks associated with the customer's enterprise;
- the administrative burden entailed in relation to project size;
- the net profit potential;
- project size and energy savings possibilities; and
- the possibility that this customer might influence others in the same market.

Marketing/Sales Committee

After establishing the criteria and providing some scale of time/customer, the committee should be charged with periodic reviews of the progress of individual sales people and the marketing strategies being used with particular customers. This same group should be in a position to make "Go/No Go" decisions regarding the amount of time sales can afford to spend seeking a certain customer.

As each prospect is considered, a "Go/No Go" decision may vary from a full "cease and desist" order to enthusiastic encouragement to proceed. More typically, the committee operates in the grey area; e.g., two more weeks, four more calls, or a decision to put the potential customer on the "back burner" and revisit an opportunity in about six months.

The committee should also be charged with a periodic review of the criteria and time allocations.

Perhaps the most critical function of the committee is to make sure that the criteria are being followed. A classic example of a procedure gone wrong involved an ESCO working with the management of a 22-story building in San Francisco. The management exhibited all the partnership characteristics an ESCO could possibly want and the building virtually dripped with energy saving potential. The ESCO casually asked about the customer's financials and was just as casually put off. The ESCO proceeded to spend over USD 200,000 auditing the facility before it eventually learned that the building belonged to a foreign company that was going bankrupt. There was absolutely no way the ESCO could get financing to do the project. Customer creditworthiness was, of course, one of the ESCO's key customer selection criteria, but other factors had over-shadowed the financial concerns. It was a painful lesson.

GUIDING SALES

One basic tenet should undergird all marketing policy decisions: *everybody in an ESCO should be regarded as a sales person*. At some point, each individual in the firm is going to be asked the infamous passenger-to-passenger airline question: "What do you do?" There is always a possibility that the questioner just might have a huge project in his pocket, so the answer could pay great dividends. Everyone should have the 30-second "elevator speech," which captures the essence of what an ESCO does, ready to go.

At the same time, marketing strategies should recognize the limits of various people in the firm to actually "make the sale." Zealous engineers, who eagerly try to sell energy efficiency and EPC to top management, typically find themselves relegated to the boiler room. Once a technical professional see the exciting energy saving opportunities in a facility or process, he or she can get so caught up in all the technical benefits that the bigger picture sometimes escapes them.

The horrible truth is that top management is not interested in ENERGY! They don't want to hear about gigajoules or British thermal units. In fact, that stuff really turns them off. Try talking "energy" to a CEO or CFO and you are almost guaranteed a one-way ticket to the catacombs where the heating and cooling systems are housed. The guiding principle? Remember: management does not buy "energy," it buys what it can do. An effort to sell the idea of the more efficient use of a product that management is not conscious of buying just doesn't work.

We now know that if you want to get management's attention, even when you are selling energy, talk money.

In selling EPC, sales people should be trained to regard themselves as resource managers. If we revisit Kantin and Hardwick's criteria for a complex sale, we will recall that EPC is a multi-level sale and a multi-level buy. Orchestrating who in the ESCO firm should call on the various levels in the potential customer's organization is an imperative. Engineers seldom talk the language of the CFO, or CEO. An ESCO's financial officer is typically ill-equipped to talk to facility people. Sometimes, the CEO will only listen to the ESCO's CEO. Reading the needs and the character of various individuals and of a prospect's organization, performing the duties of a matchmaker and doing it all with timing that will prompt swift closure is not easy. A manual on EPC sales can make Machiavelli's *The Prince* seem like a first grade reading.

The business case for EE and EPC can be made ***IF*** marketing strategies are designed to learn what the *customer's* concerns are and how to talk *their* language. Addressing their concerns in their language truly comes into focus when responding to a request for proposal (RFP), or tender.

THE PROPOSAL

Particularly in hard economic times, the customer truly wants to see what you can do for them *in writing*. Since ESCOs rely on proposals for much of their work, especially in the public sector, attention to proposal preparation is a key marketing strategy.

Some guiding thoughts:

- EPC is customer driven, so successful ESCOs develop a clear understanding of their customer's business and current needs;
- a proposal should identify and confirm the customer's needs and how the proposed work will respond to those needs;
- stress that your organization plans to work WITH the customer's people;
- offer action alternatives for solving the customer's business problems;
- give them the quality they have a right to expect, go the extra mile—meet or exceed the customer's expectations; and
- establish a tone of honesty and integrity and back it up with all sales activities.

In preparing a proposal, consider what Steve Jobs, founder and guiding light of Apple Computer, once observed about quality:

> *"Be a yard stick of quality. Some people are not used to an environment where excellence is expected."*

Communicating with management, whether in a sales call or in a proposal, three approaches can effectively respond to the management's point of view. First, for public and/or private customers, it helps to urge them to view energy efficiency and conservation as a very cost-effective delivery system for meeting environmental mandates—a way to make money while reducing emissions. For the private sector, two other approaches often work: a) positioning energy savings as a percentage of the bottom line; and b) providing an effective cost/benefit analysis procedure, which compares the net benefits of EE and energy conservation to increased production. Implementing these approaches is discussed in a paper prepared for the 2007 World Energy Engineering Congress Proceedings, "Making the Business Case for Energy Efficiency" by Shirley J. Hansen.

Finally, we need to remind ourselves and top management as forcefully as possible that EE can be a self-funding endeavor. CEOs and CFOs have a tendency to compare energy investments to other business investments and fail to appreciate that frequently no new money is required to carry out EE work. When CEOs and CFOs start talking internal rates of return (IRRs), hurdle rates, and return on investment (ROI) with a two-year ceiling, it's a cinch they are trying to fit the EE investment into their regular investment model. Then, it is time to remind them once again that the money needed for energy investments is already in the budget—and being spent on wasted energy.

A basic tenet of any EE sales message should be:

Energy efficiency is an investment; not an expense.

Even better, EE is a very sound investment. In the ESCO industry, we talk rather glibly about two-year payback periods. That is a 50 percent ROI. It is hard to find another source that can give customers a 50 percent ROI. Yet, ESCO potential clients walk right by the opportunity on a daily basis. The heart of marketing EPC is to make sure they do not.

Chapter 6

Related Services ESCOs Offer

The energy service company (ESCO) industry is plagued with slow sales cycles. The cost of customer acquisition is high. The drain on ESCO resources can be debilitating. Ways to shorten the sales cycle, as discussed in the previous chapter, are important. It also pays to examine ways to expand customer services once a customer is acquired.

To maximize income opportunities, sales people and project managers need to maintain vigilance as to other opportunities to serve existing customers in a meaningful way.

The first step in examining potentially related services is to consider those services that can cost-effectively enhance the quality of the project. Near the top of the list is a key service, which can help manage risk and improve a project: training.

TRAINING

Energy-efficient equipment does not generate optimum savings by itself. To gain the most savings from such equipment, the latter needs to be correctly installed, regularly maintained and operated efficiently. Knowledgeable experts are needed at all stages of project development and implementation. Operators of the various systems need to be knowledgeable about every aspect so that they will be able to maintain and optimize the savings. Expertise can be, and should be, mobilized so that a project will be sustainable even after the ESCO has completed its work.

Training, therefore, is a key component of a good EPC project. Good ESCOs will develop and offer extensive training programs to the different stakeholders in an energy efficiency project. These may include:

- **Energy purchasing.** ESCOs can play a valued role in a range of administrative aspects in energy purchase contracts. As an added service, an ESCO employee, who is knowledgeable about a project to be implemented, can assess its impact on the use of energy to be purchased and help control supply-side costs.

- **Design team**. Those responsible for any new construction and improvement project can usually benefit from consultation with an ESCO. When new projects are as energy efficient as is economically feasible, they are more apt to maintain energy efficiency and be cheaper to run over their life span.

- **Mechanical and electrical maintenance team**. To the extent that operations and maintenance (O&M) functions are the responsibility of the facility team, training will give the team a head start in maintaining savings. Research has shown that up to 80 percent of the savings in an effective energy management program can be attributed to the energy-efficient practices of the O&M personnel. Training needs to be a strong focus of any project. ESCOs will benefit from making sure that the technicians are as knowledgeable as possible on key aspects of energy efficiency. Training can be offered in an impressive number of fields, including the maintenance and operation of specific equipment (chillers, boilers, motors, etc.), the optimization of global systems (ventilation, cooling, heating, etc.) and the utilization of simple and complex control systems.

- **Operations**. The operators of different industrial systems are often key in deciding how the different lines of production will be used. When they are more knowledgeable about how energy is used in their systems and how it is charged by utilities, they are in a better position to take energy efficiency factors into consideration in the decision process.

Special training can also be offered by the ESCO to specific groups, such as the people working in kitchens, in shipping, etc., who are often in direct control on how equipment is used. The specialized training offered by the ESCOs should not be limited only to the first year of the EPC contract. Such training should be offered on a regular basis to make sure that knowledge is well mastered and maintained by all personnel throughout the life of the project. Manpower turnover in any of the facilities that are benefiting from an EPC project is always a factor. Training newcomers is fundamental to achieving predicted savings in every year of the project as well as delivering post-contract success to the owner. As new approaches and technologies evolve, training becomes even more critical to maintaining the energy efficiency of the facility.

AWARENESS

Energy efficiency awareness is certainly one of the most basic aspects relative to the success of an EPC project. Even if the price of electrical and mechanical equipment and systems have been optimized, if operators do not use the systems efficiently, the ESCO is apt to fall short of its savings goal.

It is not an easy task to have all occupants in a facility be supportive of an energy efficiency project. Many people have misconceptions about the subject; e.g., energy efficiency causes poor air quality, the project will reduce work place comfort and the project is designed to solely benefit the owner of the business. Developing awareness and providing correct information will challenge these misconceptions, motivate energy users to avoid wasting energy resources and reduce operating costs of the enterprise.

Effective ESCOs develop techniques to increase awareness, which include:

- designation of an annual energy efficiency week, where conferences and different activities are held in facilities;
- posters, stickers and different information tools installed in strategic places, such as sub-meter read-outs, which depict results of specific user actions; and
- idea boxes and prizes for best ideas of the month.

An effective awareness campaign must be a constant effort. New approaches and techniques to get the best of occupant practices is apt to increase the performance of a project and make it sustainable. Effective ESCOs with quality awareness programs will increase the likelihood of post-project savings.

ASSOCIATED SERVICES

In addition to those which can enhance the energy saving aspects of a project, there are many associated services ESCOs can consider, including:

- Indoor air quality. In a walk-through (scoping) audit for energy purposes, an educated eye can identify up to 80 percent of the indoor air quality (IAQ) problems in a facility. Simple remedies, such as drainage of condensate pans or changing inadequate air movement, often cost little and can avoid major IAQ problems. (See *Managing Indoor Air Quality* by H.E. Burroughs and Shirley J. Hansen, Fifth Edition.) An ESCO, however, should NEVER guarantee IAQ results. Inability to accurately measure contaminants (or the high cost of doing so) plus the number of variables the ESCO cannot control make guarantees impossible.

- Waste removal. One ESCO realized a million dollars per year by installing an incinerator to burn hazardous waste at a hospital site; thus, avoiding the high disposal costs previously incurred.

- Automated meter reading. As utilities change their modes of operation, there are an increasing number of services that an ESCO can provide. Sub-metering can also play a big role in increasing the possibilities to optimize energy consumption in the different systems.

- Risk management. When risk management is viewed as a cost center in the customer's operation, there are many things an ESCO can do to help manage those risks. Since EPC is risk management, and successful ESCOs are risk management experts, it is an expertise ESCOs can sell.

WATER MANAGEMENT

As ESCO managers become increasingly aware of the opportunities that exist throughout the customer's operation, it becomes clear that there are other opportunities on the horizon—beyond energy per se. An excellent example is water management/conservation. Some ESCOs have added this service to their arsenal when the economics favored it. A huge revenue potential awaits those who see our emerging global water crisis as an opportunity.

Today water is more in demand and shorter in supply than ever. It has become critically important in more and more countries. As an example, in the US, Lake Mead, the water source of Las Vegas, Nevada, is dropping dramatically, or in San Diego where the new desalination plant is costing billions. Singapore is also a good example of such situation where the water is actually imported.

Over the years, "alternative energy" has become an accepted phenomenon. Finding "alternative" water will not be so easy.

It is very difficult to seriously address yet another major problem when economic conditions are so uncertain around the world, but it must be addressed. The longer we wait; the more costly it will become. It is possible to cut back on water, use it more efficiently and process grey water for non-potable (and maybe potable) purposes, but it is not possible to do without it! It is absolutely essential to life. The economics and the growing need have all the markings of an ESCO opportunity in the making. The list of water crises which threaten us is long and getting longer. Figure 6-1 depicts the relative water consumption of some major countries. An impressive number of countries around the world are in great need; e.g., Africa, where desertification is still a huge and unsolved problem. Even countries that have been perceived as rich in water are now facing problems. We now know that rain is not sufficient to replenish these depleting resources.

We hear a lot about our infrastructure needs, namely roads and bridges. According to a recent report from the Urban Land Institute and Ernst & Young, there is no greater challenge facing us today than our deteriorating water systems.

Sadly, typically low water bills (or even free water) as well as the absence of needed funds for improvements too often work against taking care of this critical need.

A decade long drought in Australia, however, has prompted huge investments in reducing water use and developing new water sources. Recent statistics show that national consumption has been brought down 20 percent, water recycling is targeted to hit 30 percent by 2015 and six desalination plants are scheduled for start-up next year.

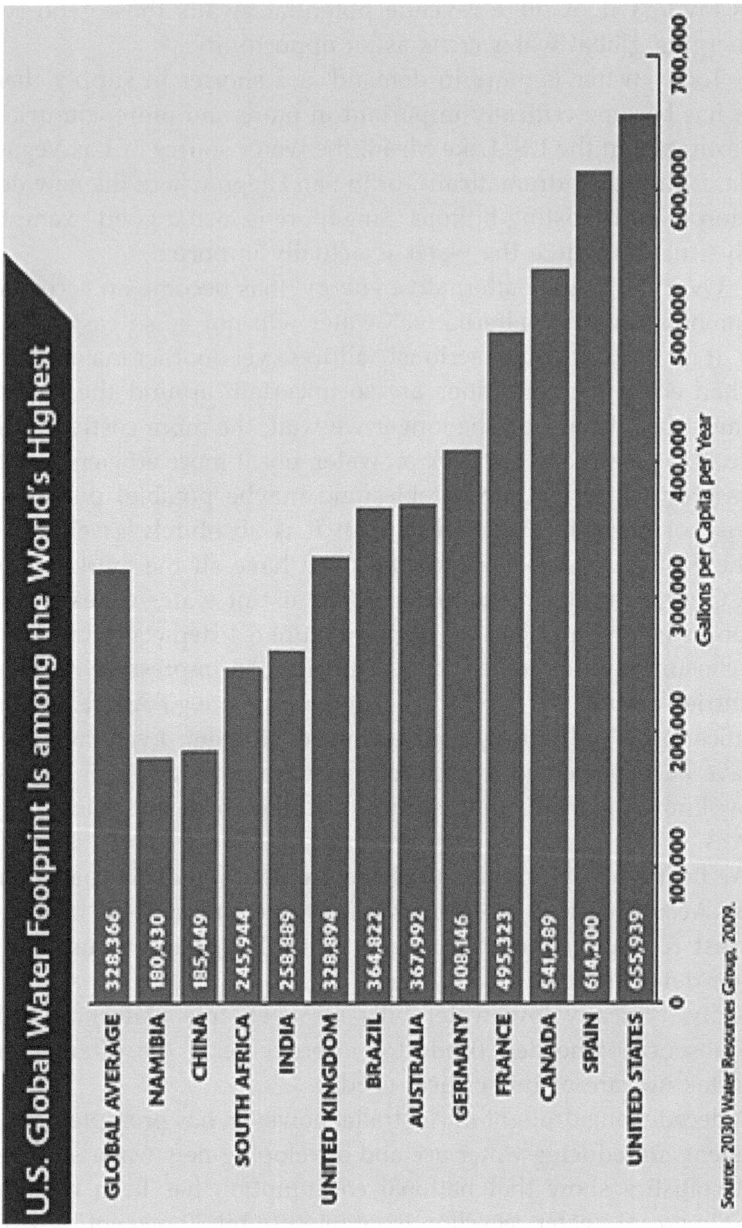

Figure 6-1: Relative Water Consumption per Capita

Singapore again offers a global model of efficiency, as it operates a nearly closed-loop water management system, which supplies nearly 5 million people with about 60 percent of their water needs. Currently, the island nation is dependent on Malaysia for its water, but its goal is to be self-sustaining within 50 years.

WATER ABSORBS ENERGY

As we focus on water needs, a related factor has pretty much escaped us. Most cities around the world get their water upstream, then use the water, pollute it, clean it and put it back in the river far downstream. This process described as a linear paradigm, uses a lot of energy. The desalination plants being constructed in Australia and the US as well as those already present in many Middle Eastern countries use a lot of energy. This is energy that many areas can ill-afford.

Earlier this year, the ACEEE and the Alliance for Water Efficiency published a white paper that described the co-dependence of water and energy. They highlighted the following facts:

- Sourcing, moving, treating, heating, collecting, re-treating and disposing of water consumes 19 percent of California's electricity, 30 percent of its natural gas and 88 billion gallons of diesel fuel annually, according to a 2005 California Energy Commission report.

- The River Network in 2009 found that energy use for water services accounted for 13 percent of US electricity consumption, at least 520 million megawatt-hours annually. Thermoelectric power accounted for an estimated 49 percent of US water withdrawals and 53 percent of fresh surface-water withdrawals in 2005.

The opportunities for ESCO financial gains through water management are obvious. The fit to energy efficiency and the ESCO model are equally obvious. Performance contracting is positioned to play an important role in the solution of this worldwide problem if ESCOs can just seize the opportunity.

Considering conditions similar to the water management example, ESCOs have before them a number of related services they can incorporate into their offerings. Doing so will afford them more mileage from existing customers, enable them to present a broader offering to potential customers, make better use of staff expertise and add a richness to their range of services.

Chapter 7

The Critical Role of M&V

When money changes hands based on the level of savings achieved, all parties should be comfortable with how those savings are measured, verified and correctly attributed to those who performed the work. This issue is becoming one of the most critical concerns in the industry

Under the financiers' general guidance, the energy service company (ESCO) and owner should jointly decide on the level of verification and attribution necessary. It is basically a case of cost vs. accuracy, and it is possible to reach the point of diminishing returns rather quickly. By following the good concepts and recognized principles of measurement and verification (M&V), it is possible to find the right balance that will satisfy all stakeholders in the project.

It is also essential that the financier get a good sense that the project benefits are measurable and they are measured through broadly accepted and recognized protocols. In the final analysis, a bankable project is one in which you, as an individual, would wish to invest. If we take a few steps back and view it as a banker would, then, an economically viable, bankable project is simply one which demonstrates good business sense.

As essential as M&V is to energy performance contracting (EPC), it can add value to the global project in many different ways. Indeed, while an EPC project will reduce energy consumption and cost in a facility, a thorough M&V can help ensure that savings will persist over time. During the timeframe of the EPC contract, and more importantly, after the ESCO is gone, the end-user can directly and fully benefit from the savings generated.

The following section briefly explains M&V concepts and applications in some detail. It is based on the world reference on M&V, the International Performance Measurement and Verification Protocol (IPMVP). Developed and maintained by the Efficiency Valuation Organization (EVO), the IPMVP can be downloaded in different languages for free at www.evo-world.org.

ENERGY EFFICIENCY CANNOT BE DIRECTLY MEASURED

Technically, the results of energy efficiency (EE) projects cannot be directly measured, as they can only be defined by the absence of energy consumption. In the context of EE initiatives, the adage that "what cannot be measured cannot be managed" is frighteningly true.

According to EVO, "M&V is the process of using measurement to reliably determine actual savings created within an individual facility by an energy management, energy conservation or energy efficiency project or program. As savings cannot be directly measured, the savings can be determined by comparing measured use before and after implementation of a project, making appropriate adjustments for changes in conditions."

THE PRINCIPLES OF M&V

According to the IPMVP, M&V is a science, which should follow some fundamental principles. M&V work should be:

Accurate: M&V reports should be as accurate as the M&V budget will allow, and the level of accuracy should be addressed wherever possible. M&V costs should normally be small relative to the monetary value of the savings being evaluated. M&V expenditures should also be consistent with the financial implications of over- or under-reporting a project's performance. Accuracy trade-offs should be accompanied by a conservative approach in any estimates and judgments.

Complete: The reporting of energy savings should consider all major effects of a project. M&V activities should use measurements to quantify all the significant measurable effects while estimating significant results where feasible.

Conservative: Where judgments are made about uncertain savings quantities, M&V procedures should be designed to underestimate savings.

Consistent: The reporting of a project's energy effectiveness should be consistent among:

- different types of energy efficiency projects;
- different energy management professionals for any one project;
- different periods of time for the same project; and
- energy efficiency projects and new energy supply projects.

"Consistent" does not mean "identical," since it is recognized that any empirically derived report involves judgments which may not be made identically by all reporters. By identifying key areas of judgment, the IPMVP helps avoid inconsistencies arising from lack of consideration of important dimensions

Relevant: The determination of savings should measure the performance parameters of concern, or least well known, while other less critical or predictable parameters may be estimated.

Transparent: All M&V activities should be clearly and fully disclosed. Full disclosure should include presentation in the M&V plan and M&V savings reports of all of the elements defined in Chapters 4 and 5 of the IPMVP.

The balance of these principles enables an M&V expert to present a flexible framework of basic procedures for achieving M&V for energy efficiency projects.

In the development and negotiations of an M&V approach for an EPC project, the parties should refer to these principles in order to find the right balance between cost and accuracy. Further, both parties should find these principles acceptable in order to define the savings on which the contractual arrangements are based.

IPMVP: THE DIFFERENT OPTIONS

Based on these principles, the IPMVP provides four acceptable approaches (called options) for measuring and verifying savings. All four options use the following fundamental formula.

Savings = (Baseline Energy – Reporting Period Energy)
± Routine Adjustments ± Non-Routine Adjustments

ESCOs may all have their own methodology for calculating baseline energy use. The reporting energy period represents the actual energy use in a facility as determined by the results of a given period's M&V report.

The four options for determining savings—A, B, C and D—are described in Table 7-1. The choice among the Options involves many considerations, one of which is the definition of the measurement boundary. If the M&V approach calls for determining savings at facility level, Option C or D may be favored. However, if only the performance of an individual EE measure is of concern, a retrofit-isolation technique will be more suitable (Option A or B). More information on the Options and their applications can be found in the IPMVP, Volume I. (available at www. evo-world.org).

M&V PLAN

One of the most important aspects of M&V is the development of a plan, which is adapted for a specific project. Such a plan typically becomes an integral part of the EPC contract, even if it is provided by reference, and is expected to be implemented during the whole duration of the contract.

Based on EVO's IPMVP 2010 recommendations found in "The Preparation of an M&V Plan," a complete plan should include discussion of the 13 topics:

1. **Project Intent**: Presentation of the scope of the project and the way it will be implemented and commissioned.

2. **Presentation of the Selected IPMVP Option and Measurement Boundary of the Project**: Specification of which IPMVP Option will be used as well as the identification of the measurement boundary of the savings determination.

3. **Presentation of the Baseline**: Documentation of the facility's baseline conditions and energy data, within the measurement boundary. The baseline documentation should include:
    a. Identification of the baseline period;
    b. All baseline energy consumption and demand data;

c. All independent variable data coinciding with the energy data; e.g., production rate, ambient temperature;
d. All static factors coinciding with presented energy data, including;
   — Occupancy type, density and periods,
   — Operating conditions for each baseline operating period and season, other than the independent variables,
   — Description of any baseline conditions that fall short of required conditions,
   — Size, type and insulation of any relevant building envelope elements, such as walls, roofs, doors, windows,
   — Equipment inventory: nameplate data, location, condition. Photographs or videotapes are effective ways to record equipment conditions,
   — Equipment operating practices (schedules and set points, actual temperatures and pressures), and
   — Significant equipment problems or outages during the baseline period.

When whole-facility M&V Options (C or D) are used, all facility equipment and conditions should be documented.

4. **Reporting Period**: Identification of the reporting period (which is not necessarily the whole duration of the contract).

5. **Basis for Adjustment**: Declaration of the set of conditions to which all energy measurements will be adjusted. The conditions may be those of the reporting period or some other set of fixed conditions.

6. **Analysis Procedure**: Specification of the exact data analysis procedures, algorithms and assumptions to be used in each savings report. For each mathematical model used, it is important to report all of its terms and the range of independent variables over which it is valid.

7. **Energy Prices**: Specification of the energy prices that will be used to value the savings, and whether, and how, savings will be adjusted if prices change in the future.

Table 7-1: IPMVP Options and their Applications of

IPMVP Option	How Savings Are Calculated	Typical Applications
A. Retrofit Isolation: Key Parameter Measurement Savings are determined by field measurement of the key performance parameter(s) which define the energy use of the affected system(s) and/or the success of the project. The measurement frequency ranges from short-term to continuous, depending on the expected variations in the measured parameter and the length of the reporting period. Parameters not selected for field measurement are estimated. Estimates can be based on historical data, manufacturer specifications or engineering judgment. Documentation of the source or justification of the estimated parameter is required. The plausible savings error arising from estimation rather than measurement is evaluated.	Engineering calculation of baseline and reporting period energy from: o short-term or continuous measurements of key operating parameter(s); and o estimated values. Routine and non-routine adjustments as required[1].	A lighting retrofit where power draw is the key performance parameter that is measured periodically. Estimate of operating hours of the lights may be based on building schedules, occupant behavior and/or PDL readings.
B. Retrofit Isolation: All Parameter Measurement Savings are determined by field measurement of the energy use of the FIM-affected system. The measurement frequency ranges from short-term to continuous, depending on the expected variations in the savings and the length of the reporting period.	Short-term or continuous measurements of baseline and reporting period energy, and/or engineering computations using measurements of proxies of energy use. Routine and non-routine adjustments as required.	Application of a variable-speed drive and controls to a motor to adjust pump flow. Measure electric power with a kW meter installed on the electrical supply to the motor, which reads the power every minute. In the baseline period, this meter is in place for a week to verify constant loading. The meter is in place throughout the reporting period to track variations in power use.

[1] A routine adjustment is a baseline adjustment that would be done regularly, for example due to weather variations. A non-routine adjustment would occur once due to an isolated event such as the introduction of a new ventilation system.

Table 7-1 (*Cont'd*): IPMVP Options and their Applications of

IPMVP Option	How Savings Are Calculated	Typical Applications
C. Whole Facility Savings are determined by measuring energy use at the whole facility or sub-facility level. Continuous measurements of the entire facility's energy use are taken throughout the reporting period.	Analysis of whole facility baseline and reporting period (utility) meter data. Routine adjustments as required, using techniques such as simple comparison or regression analysis. Non-routine adjustments as required.	Multifaceted energy management program affecting many systems in a facility. Measure energy use with the gas and electric utility meters for a 12-month baseline period and throughout the reporting period.
D. Calibrated Simulation Savings are determined through computer simulation of the energy use of the whole facility, or of a sub-facility. Simulation routines are demonstrated to adequately model the actual energy performance measured in the facility. This Option usually requires considerable computer knowledge and skills in calibrated simulation.	Energy use simulation, calibrated with hourly or monthly utility billing data (energy end-use metering may be used to help refine input data).	Multifaceted energy management program affecting many systems in a facility but where no meter existed during the baseline period. Energy use measurements, after installation of gas and electric meters, are used to calibrate a simulation. Baseline energy use, determined using the calibrated simulation, is compared to a simulation of reporting period energy use.

8. **Meter Specifications**: Only in the case of the usage of Options A and B, specification of the metering points and period(s) if metering is not continuous. This should include the following elements: meter characteristics, meter reading and witnessing protocol, meter commissioning procedure, routine calibration process and method of dealing with lost data. This section can be omitted in the case of the use of Option C since a utility meter is used.

9. **Monitoring Responsibilities**: The definition of the responsibilities for reporting and recording the energy data, independent variables and static factors within the measurement boundary during the reporting period. This responsibility can be given to anyone (the ESCO, the client, a third party) as long as all parties agree to it.

10. **Expected Accuracy**: Evaluation of the expected accuracy associated with the measurement, data capture, sampling and data analysis. This assessment should include qualitative, and any feasible quantitative measures of the level of uncertainty, in the measurements and adjustments to be used in the planned savings report.

11. **Budget**: Definition of the budget and the resources required for the savings determination, both initial setup costs and ongoing costs throughout the reporting period.

12. **Report Format**: Specification on how results will be reported and documented. A sample of each report should be included.

13. **Quality Assurance**: Specification of the quality assurance procedures that will be used for savings reports and any interim steps in preparing the reports.

Depending upon the circumstances of each project, some additional specific topics should also be discussed in a complete M&V plan. For example:

For Option A:

- Justification of Estimates: Report on the values to be used for all estimated values. Explanation of the source of these estimated values. Presentation of the overall significance of these estimates to the total expected savings by reporting the range of the possible savings associated with the range of plausible values of the estimated parameters.

- Periodic Inspections: Definition of the periodic inspections that will be performed during the reporting period to verify that equipment is still in place and operating as assumed when determining the estimated values.

The lack of presence of a good M&V plan, or more dramatically, of any M&V plan, is one of the major causes of litigation in the EPC business. Indeed, if parties have not agreed precisely on how M&V of savings will be done prior to the implementation of projects, problems are apt to occur during the validation period. As a result, courts have huge difficulties defining what the parties have agreed to and the consequences of such debates often result in quite variable findings that penalize one or both parties.

Therefore, we cannot emphasize enough that a quality M&V approach must rely on a detailed M&V plan, which in turn rests on the principles and recommendations from a protocol like the IPMVP prior to project implementation.

THE COSTS OF M&V

M&V costs will vary depending upon the IPMVP options utilized in a project and upon many other factors inherent to the EPC project, including:

- The size of the project;
- Complexity of the project;
- Type and amount of equipment installed;
- Number of interactive effects among consuming systems;
- Level of uncertainty of savings;
- Risk allocation agreed to between the client and the ESCO;
- Other valued uses of M&V data; e.g., optimizing O&M, selling carbon credits; and
- Availability and capability of an energy management system.

There is absolutely no rule on the cost ratio of M&V in relation to project total value. Experience would tell us that the rules of thumb for a complete M&V process for a comprehensive seven-year EPC contract could range between 3 percent and 5 percent of the total investment cost of the project. But again, this information does not necessarily apply broadly. Each M&V process must be tailored to a specific project and its cost will have to be accepted by both parties based on the desired accuracy.

MEASUREMENT AND VERIFICATION: BY WHOM?

As M&V is now well recognized as one of the fundamental tools for the success of energy efficiency projects and programs. Increasingly, questions relate to who should develop and implement an M&V protocol. In theory, any of the parties involved in an EPC project can design and implement an M&V protocol. In reality, the qualifications of those designing the protocol are exceedingly important.

In the specific case of an EPC project, the most common approach is to have the ESCO develop and implement the M&V plan. This, however, can present a conflict of interest on the ESCO's part. On the other hand, the ESCO client frequently does not have the expertise to validate the work done by the ESCO. In such cases, a third party may be hired by the client to validate the work done by the ESCO. This validation may include development of the plan as well as monitoring of the regular M&V activities, which are conducted during the contractual period.

Chapter 8

ESCO Associations

Associations serving energy service companies (ESCOs) are almost as old as ESCOs themselves. The first ones were developed in North America: the Canadian Association of Energy Service Companies and the National Association of Energy Service Companies in the United States. These associations were designed to share the common interests of energy performance contracting (EPC) practitioners, market the concept and act as representatives of the industry to the different government entities regulating the market or interested in using EPC in their own facilities. A meeting of Asian ESCO associations at a recent ESCO Asia Conference in Beijing designed to explore association-to-association networking opportunities offers a testimony to the value of ESCO associations and the services they perform.

Since their inception in North America, an important number of associations have been created in many countries, with these same purposes and sometimes for additional ones. A list of ESCO associations known to the authors and their respective contact information is presented in Appendix B. As can be seen, most countries have only one association, but for varying reasons, some countries have more than one. For example, the US Energy Services Coalition has a membership with wide-ranging expertise working together "to increase energy efficiency and building upgrades through energy saving performance contracts." In some countries, ESCO associations are essentially a subdivision of larger associations interested in promoting energy efficiency (EE).

ESCO associations can play an important role in jump starting and/or maintaining an EPC market in a country. Some countries with active ESCO industries do not have such associations, implying that an association's role is not perceived as being essential to the development of a sustainable EPC market.

MEMBERSHIP

ESCO associations, of course, have ESCOs as main members. The membership, however, is typically not limited to ESCOs and may include:

- equipment suppliers interested in the EPC market;
- financial institutions;
- ESCO clients, mainly corporate ones;
- public sector entities interested in the EPC concept; and
- consultants and EPC advocates.

As with all associations, major goals of the association's management are to get and keep members.

FUNCTIONS

ESCO associations often play a central role in harmonizing the interrelated activities of their members, the consumer market, government bodies, the financial market, equipment suppliers, utilities and society as a whole. They often offer different services to their members, which typically include:

- gathering and disseminating information to the membership, including relevant government policies associated with EPC;
- disseminating activities to the market about the EPC concept through events, which bring together relevant players and discuss specific national/regional topics of interest to these institutions;
- holding sector-specific and region-specific meetings for the benefit of their members to exchange ideas as well as discuss finance- and market-related issues;
- offering technical assistance and consulting services, such as providing ESCOs and ESCO customers with training, technical assistance and consulting services;
- encouraging cooperation between
  — ESCOs and EE equipment suppliers

   - member ESCOs and foreign ESCOs
   - ESCOs and research institutes as well as other stakeholders;

- creating partnerships with a variety of organizations. The associations often see these partnerships as critical in broadening their reach, building capacity and strengthening the association's ability to exchange information with relevant players.

In order to meet their mission and achieve their goals, ESCO associations often establish certain functions, including:

- cooperation and communication
   - establishing and maintaining an association website, which provides an online platform
   - publishing newsletters
   - communicating with foreign ESCO associations
   - maintaining and strengthening communication with EE stakeholders, sector associations and other relevant institutions
   - standardizing member operations; e.g., contracts, and opening up fields for business
   - linking the EPC concept to current or upcoming EE national initiatives;

- capacity building
   - providing new member ESCOs with the needed knowledge/expertise to enter the EPC market
   - fostering relations with the banking sector and the financial community in order to attract them into more effective working relationships with ESCOs
   - striving to increase their own capacity to more effectively serve association members;

- undertaking policy and sector-specific research in order to provide additional information to its members and the market on EPC and EE issues; and

- developing representative case studies for distribution.

FINANCING SOURCES

In most developed countries, ESCO associations are essentially financed by their membership fees, benefits of mandates they perform, and events they organize.

In countries with transitional economies, ESCO associations may receive the support of international or bilateral development agencies, which see such associations as a good way to promote the EPC concept.

CHALLENGES

Associations typically aspire to do more than they can afford. The struggle, which ESCO associations usually have with financing, limits their activities and often their capacities to play their expected roles.

ESCO associations are frequently short of permanent staff, which poses another challenge, as it often causes them to be directly linked to a few individual volunteers interested in the development of the EPC concept for their own benefit.

One of the most difficult challenges ESCO associations face is the pressure to establish an ESCO certification scheme. The intent is to provide ESCOs with an avenue for accrediting their businesses.

Since ESCOs are service companies, their qualifications typically rest on individual expertise within the ESCO. Turn over within an ESCO and the perennial problem of being judged solely by the last project completed makes accrediting ESCOs a challenging act. An accreditation procedure also needs to be carried out in a manner that enables all member ESCOs—regardless of size—the chance to gain accreditation. Furthermore, as the accreditation process (and the maintenance of such accreditation) involves significant costs, designing a scheme that does not favor relatively wealthy ESCOs can pose a problem. The process also poses another question: should the association reach out and accredit ESCOs that are not members? Is the association's goal to create an industry standard or provide another service to its members?

In an industry that is still relatively young in every corner of the world, associations can play a vital role in fostering ESCO development. The reader is encouraged to explore the resources available in appendix B and consider ways to collectively support EPC.

Section III

Esco Reports: A To Z

Energy performance contracting (EPC) has been developing in many different countries and has taken various forms. It demonstrates the flexibility of the concept to meet the needs of a range of markets and cultures.

Analyzing how the concept works in different markets and various cultures affords us a unique way to understand how EPC can seize opportunities and address obstacles. By presenting analyses from recognized industrial leaders from over 50 countries in the following pages, the reader is given incredible insights as to how our growing industry can move forward.

These reports rely entirely on the expertise and the perceptions of the contributing authors. It is indeed fortunate to have individuals with such qualifications share their knowledge of the ESCO industry in their respective countries. It should be stressed, however, that each report is based on the author's perception. Others may see things differently. No attempt was made to verify the facts or descriptions offered in the reports.

As pioneers in the industry and as advocates for performance contracting around the world, gathering information for this book has been a very gratifying experience. Seeing the growth from its embryonic stages to the maturity now evident is truly a joy to behold. There is no question in our minds that EPC is alive, well and growing!

The unique contributions performance contracting makes to improved energy efficiency, economic growth and the conservation of our natural resources is widely recognized. Documentation is growing that energy efficiency (EE) is the most cost-effective way to reduce environmental pollution. Performance contracting has been recognized as a valuable tool in our efforts to control climate change and to achieve a healthier environment. In fact, it is becoming evident that energy efficiency is an extremely desirable way to finance our growing desire for more renewable energy.

With all this in mind, it is with pleasure then that we offer the final summary chapter building on the richness of the in-country stories, comparing the obstacles they have overcome and the lessons that they have learned. In so doing, we present the reader with a compendium of exceedingly valuable guidelines, which can help those striving to create and/or strengthen our ESCO industry.

We now leave the floor to these collaborators, who have used their experiences in their respective markets to develop the following reports, for they offer the most updated information available today on the use of EPC worldwide.

Chapter 9

Country Reports

The main authors are exceedingly pleased with the caliber of authors whose reports appear in this chapter. We sincerely appreciate the time they have taken to share their expertise regarding the ESCO industry in the countries where they live and work.

The main authors do not have the level of intimate knowledge of ESCO activities in each and every one of these countries. It was therefore our intent to rely on the most knowledgeable local stakeholders in order to present the reality of their own country. For these reasons, the authors cannot attest to the full accuracy of all the details presented. However, given the experience and knowledge of the contributing authors selected for providing these reports, we are confident that they have provided very valuable insights into the worldwide progress of energy performance contracting.

Argentina

Claudio Carpio

ACTIVITIES

Until 2000, there had been virtually no energy efficiency activity that could be linked or related to the participation of companies that met the definition of energy service companies (ESCOs) in Argentina.

In 2000, there were two programs in Argentina supported by the Global Environment Facility (GEF www.gefweb.org). One called Efficient Light Initiative (ELI): the original ELI program tested the quality certification and labeling concepts and focused on seven countries during the period 2000 through 2003. One of these countries was Argentina.

Another program called Argentina Street Lighting Program (ASLP) promoted mechanisms to encourage energy efficiency projects in public lighting. Among the options proposed, the program envisaged the participation of distribution companies in the collection of the levy of street lighting from users, thereby getting lower delinquency rates. With this increase in revenue, it was possible to finance the modernization of street lighting and improve the service. Once the way had been opened for the financing of the work, the distribution companies could act as ESCOs, making improvements to the street lighting system within their concession area. However, due to the socioeconomic and financial crisis from the end of 2001 to mid-2002, the system failed to meet its objectives.

The government of the province of Buenos Aires promulgated regulations under Decree No. 3570/00 SAPE (Efficient Public Lighting System—EPLS) for the municipalities to be members of the EPLS. Edesur, one of the most important distribution companies of the country, agreed with a firm named LESKO to the formation of a joint venture (JV) that was to provide the maintenance, expansion and conversion of the public lighting system, thus establishing a kind of ESCO in the country. This alternative was very well received in the municipalities of Esteban Echeverría, Lomas de Zamora and Florencio Varela, among others. In the municipality of Berazategui, the ESCO faced a similar process of modernization, which was successful.

The ELI program fostered the creation of other ESCOs for the redevelopment of energy efficiency projects in commercial and public buildings, as well as for street lighting.

As of 2002, an efficient lighting conversion project in Military Club buildings was set up in the form of an ESCO model with the company Construman (www.construman.com.ar). However, it did not meet the definition of a typical ESCO.

The ELI program has also allowed other companies to present themselves as ESCOs for the conversion of lighting systems. These companies included American Eco System Inc., Mainieri SRL and the Department of Electrical Engineering, Faculty of Engineering, University of Buenos Aires (UBA). It is important to clarify that none of them met the precise definition of an ESCO.

At the end of 2003, the Energy Efficiency Coordination Unit of the Energy Secretariat, with the assistance of a European consultant, began to design a series of energy efficiency projects. These projects

eventually became part of a program that would be subject to financial support from the GEF, which is affiliated with the World Bank. Then, it started, with the so called "PDF-B funds," six preliminary studies to define the projects that the Energy Secretariat would eventually support with GEF funding.

The following studies were conducted: 1) develop the Argentina Investment Fund for Energy Efficiency (FAEE) and Evaluation of Financial Institutions; 2) set up a Standards and Labeling Program for Energy Efficiency and Design a Development Program of Energy Service Companies (currently referred to as Empresas Prestadoras de Servicios Energeticos—EPSEs); 3) support the Design of Programs for Energy Efficiency in Electricity Distribution Companies; 4) study the energy market baseline, the alternative, the incremental cost of the project and expected emission reductions; and 5) study Regulations, Tariff Signs and Economic Incentives for Efficient Use of Energy.

Finally, in May 2008, after exchanging views with the World Bank/GEF, the continuation of the following three components was defined: a) development of the Argentina energy efficiency fund and of an energy efficiency portfolio to be funded by a specific donation; b) development of an energy efficiency program within electricity distribution companies designed to support the procurement and distribution of CFLs exploring the most appropriate mechanisms for dissemination and awareness for future users of CFLs; and c) a third component aimed at improving the management capacity of public and private energy efficiency projects.

Of particular interest to ESCO market development in Argentina, point a) includes:

- Development and strengthening of the capacity of ESCOs (or EPSEs); and
- Promotion of the development of contractual instruments (performance contracts) for the implementation of energy efficiency projects.

The new program called for an Expression of Interest (EOI) for 25 energy audits on a pilot basis. A total of 22 consulting groups (local, international and mixed) along with the Ministry of Energy responded. None of the organizations met the traditional ESCO definition. After a technical evaluation, a "short list" of 10 advisory groups were invited to submit their technical and financial offers to make such diagnoses in July of 2011. The consultancy was scheduled for late 2011, selection and diagnostics will follow.

CONTRACTS

At the suggestion of the principal of a mining company in Chile, a "working base" was established using a polynomial formula. This formula, presented in the box below, was based on the speed of perception by the mining company of the benefits of the consulting work in energy efficiency.

Consultant fees: Down payment (1) = 30% Savings by housekeeping (*)
+ 20% savings with repayment of less than 12 months (**)
+ 10% of savings over a 12-month repayment (***)

The initial fixed amount was designed to be equal to the costs incurred by the consultant in the first four months of work. This amount included a reasonable initial estimate for the future: it was assumed that at least 20 percent of the recommendations without investment would be implemented by the mining company.

(*) 30 percent savings verified within 30 days after the implementation of the measures of "housekeeping" (without investment).

(**) 20 percent counted since the moment of implementation of cost-saving measures, for 12 consecutive months.

(***) 10 percent of the savings for six months after the implementation of measures that have a simple payback period exceeding 12 months.

LEGAL FRAMEWORK

Currently, in Argentina there is no legal or regulatory framework clearly defined to promote energy efficiency. There is no law, although in 2003-2004 there were some efforts to reach a legal rule in Congress (House of Representatives) but they were unsuccessful.

What is valid is Decree No. 140/2007, by which was launched in December 2007 the NATIONAL RATIONAL AND EFFICIENT ENERGY PROGRAM (PRONUREE in Spanish), a legal range lower than a law. The program has some issues tangentially related to the development of ESCOs in Argentina, for example, for industry:

"Designing and developing cross-technology programs that encompass the various branches of industry and contribute to the development of a market for energy efficiency. These programs will cover—among others—**the development of Energy Service Providers** and the promotion of efficient technological applications." [underscore supplied]

Referring to the CHP technologies, the decree says:

"Promoting the creation and development in the country of new Energy Service Providers with the aim to develop cogeneration projects and provide services that are necessary for this purpose, involving a high degree of scientific and technological infrastructure available in the country as well as national engineering." [underscore supplied]

In terms of intentions (regarding the promotion of ESCOs), Decree 140/07 was adequate. Reality, however, has shown that the Decree has not moved with the same efficiency from paper to real things.

By Decree 140/07, no regulations have been achieved on promoting the development of ESCOs in Argentina. However, there was an attempt to trigger this issue through the GEF, as described above.

In the financial system, there are no regulations that support "performance contracts." It is still considered an "uncultivated land."

MARKETS

At the present time, there is no market in Argentina involving companies that are fully accountable to the definition of ESCO. It is not developed yet. Nevertheless, there are individual consultants, energy efficiency experts, universities, groups and companies that have this issue only as a secondary activity, almost marginal. These organizations typically work in this area only when there is a specific call for competition, which is currently happening with the call from the Energy Secretariat to make 25 diagnostics in industries.

FACILITATORS

There is no association of ESCOs in Argentina. Nor is there an agency of energy efficiency as such, independent of the government. What exists in institutional terms is the Energy Efficiency Coordination Unit, under the Department of Promotion, which is in turn dependent on the Undersecretary of Electricity of the Energy Secretariat (included in the Ministry of Planning). As noted, the institutional status of the dependence devoted to the promotion of energy efficiency (and by extension, the development of ESCOs) is placed in a quite low level of decision making.

GOVERNMENT ACTIONS

As noted above, the government, through the Energy Secretariat, has sought specific actions aimed at developing an ESCO type of business. Despite these efforts, ESCOs do not exist (as such) so far. Most of the existing companies are consultants or engineering firms that make the diagnoses, indicate savings recommendations and quantify necessary investments. They are able to do the engineering and installation, but do not do financial investments. Generally, they are paid through fixed fees, not on a future savings basis.

Difficulties in the development of energy efficiency and ESCOs, include (but are not limited to):

- subsidized energy prices—especially in the metropolitan area—due to government energy policies and decisions, which discourage investments because they become unprofitable;
- lack of experience (and therefore confidence) in the market for contracting energy services by using performance contracts;
- difficulties accessing finance for energy efficiency investments as the banking system is not technically prepared to evaluate energy efficiency projects, let alone finance ESCO services; as a result, there is high risk perceived by bankers; and
- lack of economic and financial guarantees to firms seeking to develop performance-type contracts.

Australia

Paul Bannister

ACTIVITIES

In recent years, the market for energy performance contracting (EPC) in Australia has been relatively buoyant due to a number of internal and external factors. Industry participants interviewed for the purpose of this report all identified the industry as expanding. The global financial crisis, which did not affect Australia as badly as much of the rest of the world, does not appear to have driven the market down. Indeed, it may have redirected some capital into the refurbishment market from the new construction sector.

There has been a limited amount of direct EPC activity in the private sector with the General Property Trust (GPT) undertaking major energy performance contracts at two of its sites.[1] There is potential for further sites. However, as a whole, EPC for refurbishment has had limited uptake in the private sector.

Outside the traditional contract-for-savings EPC area, there has been significant development of "chauffage" style contracts. This type of contract is used for the delivery of build-own-operate cogeneration and trigeneration systems in both the commercial and industrial sectors. In the commercial sector, these installations have been driven by green-rating systems such as NABERS [2] and GreenStar [3], both of which provide recognition of cogeneration and trigeneration impacts on building performance. Given Australia's high carbon intensity electricity supply (typically in the region of 0.9-1.35 kg/kWh for most major population centers), gas-fired cogeneration and trigeneration systems have a significant ability to reduce site emissions. The high capital costs and operational risks of these systems make outsourced ownership and operation highly attractive.

The construction industry continues to operate in a quasi-EPC style over the delivery of many new buildings. Most new building projects in the upper end of the office market are being required by developers and owners to achieve 4.5 stars or higher under the NABERS Energy Base Building rating. These requirements are being built into builders' contracts as a contractual guarantee. This continues to be a very high-value sector of the market with quite a different profile of participants from the conventional EPC market—mainly builders and some mechanical contractors.

There are a number of factors that are driving increased activity in the energy efficiency sector generally. Some of which are expected to provide impetus to the local EPC industry:

- growing awareness of the financial benefits of energy efficiency programs; i.e., energy efficiency is a financial efficiency measure rather than simply a "green" initiative;
- growing demand from high-quality office tenants (mainly government but also the high-end private sector) for energy-efficient properties (in particular the requirement for governments to have 4- or 4.5-star properties);
- reputational issues (media coverage of high-profile buildings with poor NABERS ratings);
- probable introduction of a carbon tax in 2012; and
- general increases in energy costs.

The uptake of EPC continues to be generally quite limited. It is possible to identify a number of possible reasons for this:

- Historically, Australian businesses have paid limited attention to energy efficiency in general and, therefore, demand for EPC as a tool to deliver energy efficiency has been low. Although interest is growing rapidly, it comes from a very low base.
- Limited familiarity with energy performance contracts reduces interest and raises transaction costs. The introduction of a model contract by the now-superseded Australian Energy Performance Contracting Association, with Federal government support, has underpinned the development of government EPC programs. Nevertheless, a simpler, more streamlined contract is probably required for the private sector.
- Procurement barriers working against the single-source nature of EPC. While Australian public sector contracting is strongly driven by competitive tendering models, energy performance contracts are not a good match to such an approach.
- Restrictions on loans and management of capital versus operational costs, particularly in the government sector.
- A continuing need for improved understanding of the EPC model and its benefits.
- Relatively short payback expectations limiting the scale and availability of projects.

There has been a notable increase in organizations offering EPC in the past three years. These organizations include significant historical players such as TES and Honeywell being joined by other controls and technology companies (including Siemens and Schneider Electric), retailers (Origin), mechanical contractors (AG Coombs) and smaller EPC start-ups such as Enesolve, Carboneti, Genesis Now and Ecosave. The cogeneration/trigeneration field is dominated by energy retailers: Origin Energy in the commercial sector (having acquired Cogent) and AGL in the industrial sector.

As noted earlier, the Australian Energy Performance Contracting Association has been superseded as an organization by the Energy Efficiency Council. This change has reflected a broadening of the agenda to the full range of energy efficiency issues and initiatives rather than the sole focus on EPC.

The pressure for increased efficiency is expected to continue over the next five years due to the following factors:

- increases in energy costs above the inflation rate, both naturally and through the introduction of the Carbon Tax [4];
- pressure for private sector efficiency improvement from government programs such as Energy Efficiency Opportunities[5], Low Carbon Australia[6] and Environmental Upgrade Agreements[7];
- continued pressure for the improvement of the efficiency of government operations;
- mandatory disclosure [8] of NABERS ratings for office buildings above 2,000 m^2 on sale or lease, driving increased visibility of poorly performing office buildings (this may expand to a wider range of building types in coming years); and
- mandatory disclosure or tenancy lighting power density on sale or lease, with the potential to drive pressure for improved lighting for large leases.

GOVERNMENT ACTIONS

Specific procurement of EPC from the government sector has been growing, although it is inconsistent between states and falls well short of its potential. The Victorian State government Department of Treasury and Finance[9] "Greener Government Buildings Program" is the only major new initiative in this area specifically targeting EPC. The Queensland State government Department of Climate Change [10] operates EPC programs in hospitals and other government buildings. However, the level of investment has dropped since the early 2000s.

Similarly, a number of local governments have used EPC to upgrade their municipal building stock, including the City of Sydney and the City of Melbourne. Local governments are also planning to use financing tools to drive energy efficiency upgrades in private buildings, such as the City of Melbourne's "1,200 buildings" program [11]. These may encourage EPC but have not yet come to fruition.

Austria

Monika Auer
Jan W. Bleyl

ACTIVITIES

The following activities—predominantly on the customer side of the ESCO market—are the cornerstones of ESCO market development in Austria. In the mid-1990s, the national Austrian Energy Agency—then called E.V.A.—initiated a variety of market facilitating activities. In 1997, a first and comprehensive study covered the general approach of EPC, its potentials in Austria and the legal framework. Subsequently, practical guidelines, text modules for contracts and M&V guidelines were disseminated and a web-based supplier directory was set up. The first two pools with some 50 federal "BIG" buildings were tendered for and signed in 1998 followed by regional projects in Styria, Salzburg and Tyrol, some of which were inter-communal.

During this early development phase, European-funded cooperation projects on the national and regional levels proved to be helpful for sharing know-how and lessons learned in particular, with other European ESCO market and project facilitators.

In March 2001, E.V.A. initiatives led to a ministerial order, which laid the foundation for a "Federal Contracting Campaign (Bundescontracting Offensive)" encompassing around 300 federal buildings. The program is managed by the federal building agency "BIG" in cooperation with the Economics and Environmental ministries. It is operationally supported by 22 so-called "Energiesonderbeauftragten." The program has gone through ups and downs but is still Austria's largest EPC program. To date, approximately 550 buildings, bundled in 17 pools of buildings, have been outsourced to ESCOs and further tenders are under preparation. Since then, while the Austrian Energy Agency has almost entirely left the ESCO market facilitator arena, some regional energy agencies are still very active.

In the state of Upper Austria, the regional energy agency "Oberösterreichischer EnergieSparVerband" (ESV) was an early mover and set up an "Energy-Contracting Program (ECP)," both for EPC and ESC projects. Up to now, 56 projects have been supported, with the majority for public building owners but also B2B projects.

Also the city of Vienna's MA 34 (Magistratsabteilung Bau-und Gebäudemanagement) has established "EPC as a financing model for energy savings in buildings" after having successfully tested eight pilot projects at the end of the 1990s. Its homepage lists 40 EPC projects currently running and 25 already terminated. In the last few years, the swimming pool department "Wiener Bäder (MA 44)" has outsourced comprehensive energy and water-saving projects to private ESCOs.

Predominantly in rural areas of Austria, a sizeable number of renewable heat supply systems are operated under ESC schemes. They often include local distribution networks and are often operated by agricultural cooperatives or other ESCOs. One example is "Regionalenergie Steiermark" (www.holzenergie.net), which recently commissioned its 230th heat supply project.

Around 2002, the "Gemeinde-Contracting" (Ge-Con) initiative tried to generate economies of scales for smaller public buildings in communes by setting up five pools of buildings in Salzburg, Tyrol and Styria with different levels of success.

Last but not least, Graz Energy Agency initiated a comprehensive EPC program under the brand name "Thermoprofit®" which introduced, among others, quality standards for ESCO projects, the foundation of a network of qualified ESCOs, their certification as well as project development and facilitation on behalf of potential ESCO customers.

Another important feature of Thermoprofit® was to extend the project scope of services to include the building shell, thus facilitating a number of comprehensive EPC projects. Despite the success of these pilot projects, a broader market diffusion has not been achieved yet.

In the last few years, state of Styria-owned real estate company "Landesimmobiliengesellschaft" has applied the newly developed Integrated Energy-Contracting (IEC) model to eight of its real estates. The IEC model combines renewable supply and energy conservation measures. Further projects are under preparation.

The Austrian website www.contracting-portal.at has the following entries in its ESCO data bank:

- ESCOs offering EPC: 16 entries with a diverse background including building automation manufacturers, plant engineering and construction companies, utilities and independent "Know-How" ESCOs.

- ESCOs offering ESC: 50 entries with a diverse background including utilities, agricultural cooperatives, fuel dealers, HVAC installers, plant engineering and construction companies as well as independent "Know-How" ESCOs. Four of the above ESC providers have been certified by the Austrian Ecolabel for Green Energy (www.umweltzeichen.at).

Another list of ESCOs (many of them overlapping) can be found at www.esv.or.at.

The Austrian ESCO association, DECA, founded in 2005, currently represents 12 members. Originally, it was limited to EPC providers but it has recently opened up to ESC providers and facilitators. Among other activities, DECA serves as an information hub, hosts a regular exchange of experiences for its members, organizes experts meetings, collects market data from its members and lobbies for the industry (www.deca.at).

CONTRACTS

In Austria, ESCO models are mostly labeled as "Energy-Contracting." Two basic business models can be distinguished in the Aus-

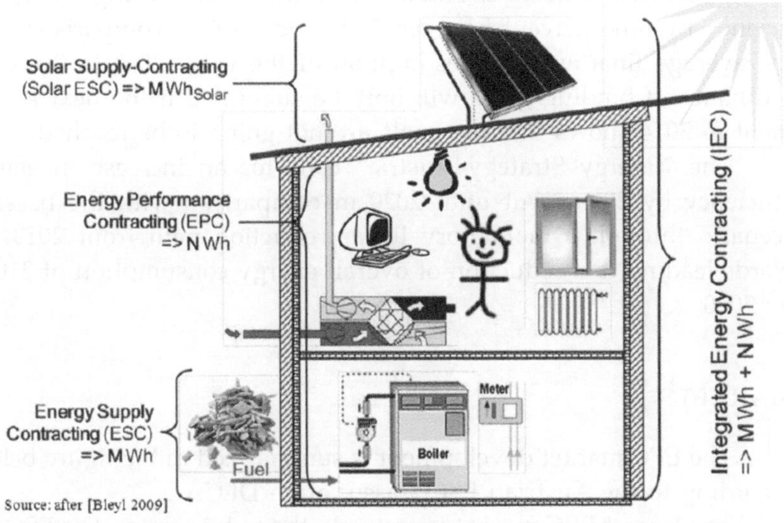

Figure 9-1: Scope of Services of Two Basis ESCO Models (left side) and IEC (right side)

trian market: ESC referring to a performance-based supply of useful energy and EPC, referring to a performance-based energy savings business model. The Figure below illustrates the typical scope of services of the above-mentioned Energy-Contracting models as well as the scope of the new IEC model.

The IEC model, which was developed as part of IEA DSM Task XVI[1], combines supply (preferably from renewable energy carriers) with energy conservation measures in the entire facility while simplifying M&V procedures (through quality assurance instruments).

Generally, ESCO services are seen as a modular service package where the scope of services can be adapted to the individual needs of a project. This implies that the building owner can (depending on resources) define what components of the energy service will be outsourced and which components will be carried out in-house (financing or ongoing on-site maintenance provided by a caretaker).

LEGAL FRAMEWORK

A major legal basis on the EU-level is Directive 2006/32/EG regarding final energy efficiency and energy services (EDL-RL, res. Energy Services Directive, ESD). It is aimed at increasing energy efficiency by nine percent between 2008 and 2016 in comparison with the average final energy consumption of the years 2001 to 2005. An ordinance of binding goals will only be suggested if the next assessment in 2014 shows that the goals are not going to be reached.

The "Energy Strategy Austria" calls for an increase in energy efficiency by 20 percent until 2020 in comparison with the baseline scenario through a mandatory linear reduction path from 2013 onwards leading to a reduction of overall energy consumption of 210 PJ by 2020.

MARKETS

The EPC market development is summarized in the figure below, according to the Austrian ESCO association DECA.

The largest EPC market is certainly the public sector. For ESC, no reliable market data are available, but there are indications that above 80 percent of the total ESCO market consists of ESC projects.

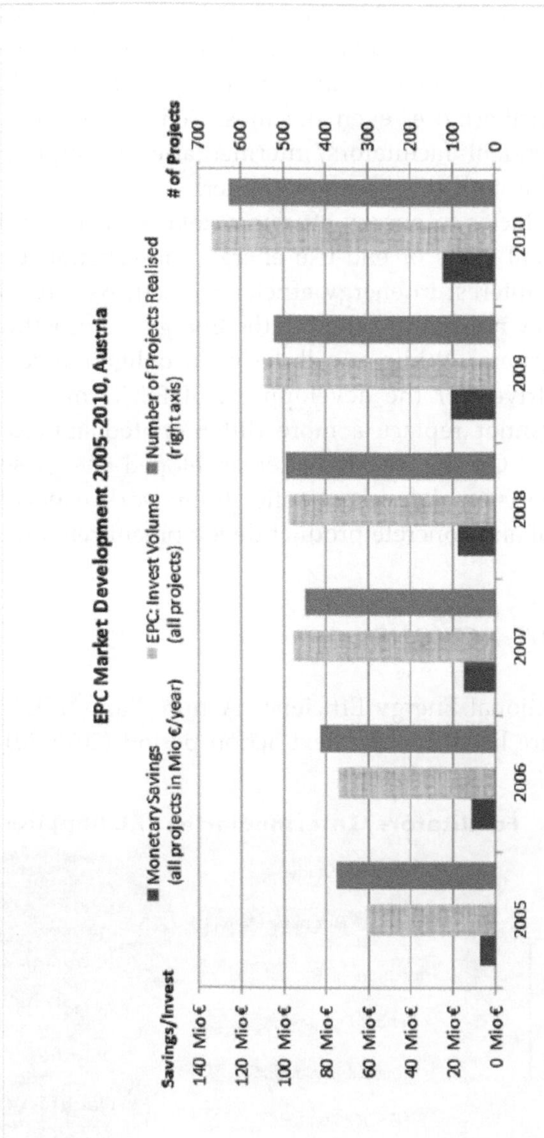

Figure 9-2: EPC Market Development

There are no easy or one-fits-all solutions as to how to implement energy efficiency projects. In any case, the decision of the building or business owner to tap into energy efficiency resources (either voluntarily or forced by regulations) remains a basic requirement—independent of the implementation model. In other words, efficiency markets need "educated" customers to demand energy efficiency (services) in the market. Furthermore, even the most "educated" customers will require independent facilitators/intermediaries to support them on their journey through this complex matter.

Many obstacles to energy efficiency take root in the fragmented nature and small units of end-use energy conservation potentials as well as in low interest in energy efficiency itself. As a result, it is important that they not be attributed to the Energy-Contracting approach or ESCOs in general. While a well-designed obligation scheme might be a helpful driver for the development of ESCO markets, it is not sufficient. It cannot replace a more differentiated approach in each market segment. On the way to better developed energy service markets, strong efforts on all levels of policy framework, capacity building, barrier removal and concrete product development remain to be done.

GOVERNMENT ACTIONS

In the National Energy Efficiency Action Plan (NEEAP), the following tasks are listed for the next action period (2012-2014):

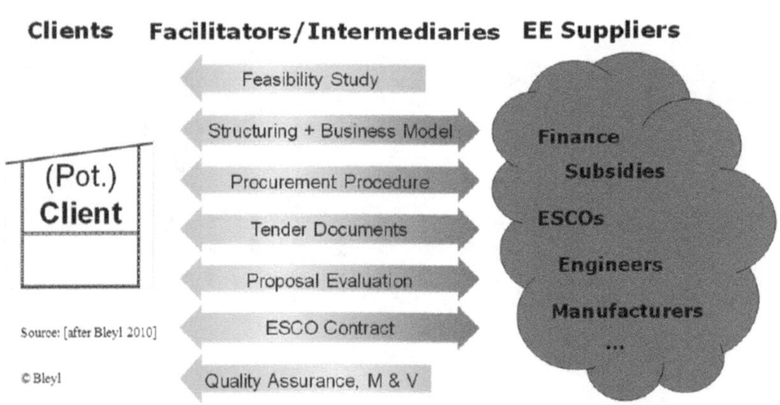

Figure 9-3: Facilitators

- increase of the refurbishment rate in the building sector to 3 percent;
- implementation/promotion of energy services (especially contracting) in the household sector.

The legal framework includes a "Climate Protection Regulation" and a so-called §15a agreement between the federal government and the federal provinces. It encompasses measures for the further development of the legal framework of the housing sector, the augmentation of the thermal refurbishment rate and the more frequent utilization of renewable energy technologies. Voluntary agreements with the respective umbrella association and an energy efficiency law (in definition, draft until July 2012) are planned.

Some of the key obstacles are:

- deficit of information on the customer side regarding functionalities and the range of offers for ESCO services;
- lack of know-how on the customer side regarding procurement (especially for small municipalities and enterprises);
- high resistance to change of affected parties (regarding household-related regulations, residential housing development promotion).

Strategies to address these obstacles are listed in the "Energy Strategy Austria" as follows:

1. incentive systems especially for the sector of non-residential buildings;
2. amendment of housing sector-related regulations for a more socially equal improvement of the thermal quality of residential buildings;
3. emphasis on new business models, such as contracting, new energy services, and ESCOs;
4. regarding residential buildings:
   — improvement of the financing situation through longer refinancing periods
   — incentives for highly comprehensive, high-quality refurbishments
   — possibilities for the increase of financial reserves in the Condominium Act

— possibilities for the increase of financial reserves in the Condominium Act
— reallocation of subsidies from housing subsidies to comprehensive thermal refurbishments and non-residential buildings
— minimum requirements for the allocation of subsidies for comprehensive energy-related refurbishments and subsidies for the renovation of heating systems
— tax deductibility and tax exempt amounts for thermal refurbishment costs
— doubling of subsidies for thermal refurbishments (one- and multiple-family houses);

5. regarding the services and industry sectors:
   — investment incentives and bonuses and/or early depreciation for wear and tear for energy-efficient engines, increase of the annual rate of lighting system renewals from three to five percent (current) to six to ten percent via incentive measures and energy efficiency solutions for air conditioning in service and residential buildings; and
   — adaptation of subsidy standards for service buildings to those of the residential housing development promotion.

The effect of the above-mentioned set of measures and instruments remains to be seen.

Belgium

Lieven Vanstraelen

ACTIVITIES

The earliest signs of ESCO business in Belgium appeared in the early 1990s. Some initiatives were taken by the federal and Brussels regional governments to promote the concept of third party investment.

The first projects were mainly carried out by large companies offering facility and building management using the "chauffage" model in the public sector (for the Ministry of Defense). New entrants included TPF-Econoler (a joint venture of the Canadian ESCO, Econoler), which started offering third-party financing (TPF) for buildings (for the city of Charleroi), and FINES, a Belgian ESCO specialized in relighting projects.

The public sector (mainly sports halls and schools), and the industrial sector to some extent, received attention from ESCOs in Belgium, in particular for supply contracts. Willingness by large consumers to outsource has been the main driving force for the provision of off-balance sheet solutions to energy efficiency investments. However, EPC was never a major factor.

As a consequence, the market for ESCO services did not really develop until the creation in 2005 of Fedesco. Set up by the federal government, Fedesco is a public ESCO that uses TPF and which is focused on energy-saving projects in federal public buildings. Fedesco, a limited liability company (NV/SA) under public law, is a subsidiary of the Federal Participation and Investment Company, a government-owned financial holding. It was started with a capital of EUR 1.5 million (USD 1.9 million) from the Kyoto Fund, raised to EUR 6.5 million (USD 8.5 million) in 2007.

Since January 2007, Fedesco has an exclusive right to apply TPF to federal buildings. Fedesco manages turnkey energy service projects on behalf of building occupants, in collaboration with the Federal Building Agency. It started the development of EPC projects in 2008. In October 2008, the federal government approved Fedesco's 22 percent CO_2 emissions savings objective to be reached by 2016, corresponding to a five-year gross investment plan of EUR 210 million (USD 276 million). The federal government agreed to Fedesco's proposal to apply the EPC model to at least half of the federal buildings, thus creating a potential market at federal level, ranging from 500 to 800 buildings in the following years.

The first project being initiated in that capacity is a pool of about 15 buildings part of the Fedimmo portfolio, owned jointly by Befimmo (90%) and the Federal state (10%). Fedesco basically turned the planned maintenance contract into a tender for an EPC project, including performance-based maintenance. The tender is to be launched in early 2012. Fedesco has announced new projects involving pools of 50 to 100 buildings, thereby bringing EPC in Belgium to the next level.

Fedesco also developed a national competence center on TPF and energy services (the "Knowledge Center"). Launched in November 2010, the Knowledge Center provides strategic and financial consultancy as well as project assistance to other regional, provincial and local public authorities and building owners. At the end of 2011, the Knowledge Center included the following customers: (i) the cities of Antwerp, Ghent and Ostend; (ii) the municipalities of Etterbeek and Watermael-Boitsfort; (iii) distribution network company Infrax; (iv) the regional development agency; (v) provincial distribution company Igretec; and (vi) public transportation companies De Lijn (Flanders region) and STIM/MIVB (Brussels region). Ostend will be the first city to launch an EPC tender for six municipal buildings, based on the Fedesco model contract. Fedesco and its Knowledge Center have put considerable efforts into developing a very well-designed and modular EPC model contract and manual, which includes performance-based maintenance following the Dutch standard NEN2767. As such, the Fedesco Knowledge Center acts as the main EPC project and market facilitator in Belgium. Other EPC project facilitators include EnergInvest, a highly specialized consultancy and project management company in energy services, TPF, and ESCO development.

Since 2007, there has been a growing interest in EPC, stimulated by Fedesco and growth opportunities from multinational companies in building automation and control (Siemens, Honeywell, Johnson Controls). They have realized a few very successful projects (in hospitals and schools). Furthermore, large multinational companies (Dalkia, Axima Services), offering facility and building management, have now started to offer EPC. Other European ESCOs like SPIE or Schneider are starting to show interest in the Belgian market as well.

The federal and regional governments have taken important steps towards increasing energy efficiency. In addition to transposing and implementing EU legislation, voluntary agreements, energy-saving obligations for utilities, green certificates and public sector obligations have been aimed at increasing energy conservation in Belgium.

Solar contracting is quite successful with the public and private sectors being targeted by companies like Enfinity, Ikaros Solar, Ecostream, Invictus and others that use private funds for Solar contracting based on TPF. Solar ESCOs install, operate and own the PV solar projects, offering rental prices (per m^2 of roof space or installed kWp power) as well as "green" electricity at reduced prices. They are targeted at large industrial, logistics and public buildings, mainly in

Flanders, typically fixing lower limits to roof surfaces, typically 1,000 to 3,000 m². Similar offerings exist for residential customers, mainly in Wallonia and Brussels, driven by Green Certificates schemes that are advantageous for smaller installations, typically 10 to 50 kWp.

The three regions in charge of energy efficiency and renewable energy policy (Flanders, Brussels and Wallony) have all expressed interest and announced some kind of initiative in the field of energy services. The most promising one is the creation in 2010 of the Flemish Energy Company (Vlaams Energiebedtijf or VEB) with EUR 200 million (USD 263 million) funding from the Flemish government. In addition to plans to invest in regional and local renewable energy projects, it will develop its own ESCO activities for financing savings in Flemish public buildings and schools. Operations are planned to start somewhere in 2012. The Flemish Energy Agency (Vlaams Energieagentschap or VEA) also started showing initial interest in EPC at the end of 2011. Specific plans by the Brussels and Walloon regions are still to be announced.

Regional distribution companies in the Flemish region (Eandis and Infrax) have launched their own ESCO services for municipalities and provinces that are shareholders in these companies. Funds that would normally return as dividends are used to finance energy-saving programs in municipal buildings. Through framework contracts with engineering companies and installers/building companies, they provide a turnkey solution including design, installation and financing. They do not provide performance guarantees or maintenance and they do not offer EPC.

One of the biggest barriers is the lack of awareness/knowledge about the concept and how to use it (public and private building sector and industrial sector). Fedesco's Knowledge Center and BELESCO play an important role in creating greater awareness and motivation to use the ESCO model.

Another barrier is the large amount of (long-term) existing maintenance contracts, including a growing number with guarantees on technical equipment. It is difficult to replace them with EPC contracts, although efforts are ongoing to integrate existing maintenance into EPC. The use of performance-based maintenance is another way to bring good practices into the area of EPC.

In the public sector, there is also an important number of owned buildings. The public sector may prefer alternative renovation projects with its own funds, manage technical measures individually or use traditional credit funding. At Fedesco, for historical reasons, approximately half of its investment plan may still not be based on EPC projects but rather on transversal measures (for boiler or chiller replacement, building controls, isolation, window films). Although the public sector uses third party investment principles from Fedesco itself, it does not require full energy service offerings or EPC contracts. Similarly, the ESCO services of regional distribution companies Eandis and Infrax are not currently based on outsourcing to ESCOs based on EPC. They are rather focused on the implementation of many individual measures by traditional auditors, specialized engineering companies and installers/entrepreneurs. This could turn out to be one of the major barriers to the successful development of EPC in the public sector. The role of those "regulated" public players in the ESCO market is thus increasingly being scrutinized.

It is expected that the ESCO market will continue to grow significantly in Belgium in the next five years. Fedesco's five-year investment plan including EPC and its federal competence center on TPF and energy services have provided key impulses to the market. Not to mention the major contribution from the creation of the Knowledge Center and BELESCO. Private ESCOs and third party investors will develop their EPC offerings and are expected to gain market share from more traditional players.

The rise of energy costs at world level is certainly a huge incentive for the development of the energy efficiency and ESCO markets in Belgium. The fact that the ESCO market is better structured today than it was in the 1990s will certainly play an important role in market growth in the near future. Nonetheless, the speed of development will strongly depend on continued support from regional and federal governments and on their willingness to support policy recommendations that have been made by existing facilitators.

CONTRACTS

In Belgium, the guaranteed savings model is the most commonly used model, although smaller companies like Factor4, REUS and Blue Energy Investments have developed shared savings models.

Some smaller third party investors and ESCOs have emerged

(Green Invest, Sophia Group, Terra Energy, Blue Energy Investments) often based on specific technology solutions or niche markets. One of Belgium's largest banks, Dexia, developed a specific TPF offer, called Energy Line, aimed at public authorities, Dexia's historical customer base. Energy Line includes solutions for photovoltaic (PV) solar panels and cogeneration (CHP). Dexia collaborates with specialized engineering companies for technical know-how. Energy Line will most likely be extended to include other offerings (full EPC projects). Both Dexia and BNP Paribas Fortis offer TPF to ESCOs. Other Belgian banks may follow.

Financing for ESCO projects is available but related knowledge is still not well spread. Customer financing, ESCO-based funding and TPF (mainly leasing) are all used in Belgium. ESCO-based funding is often preferred in order to limit participants and ensure only one responsible partner for the entire project.

Most ESCOs have access to private capital from banks or from their own organizations. ESCOs are able to structure the loan to be on or off balance sheet, depending on the tax situation and other considerations. Nevertheless, bank guarantees do add to credit exposure, thereby creating issues for smaller ESCOs. Cession of contract rates is expected to become a widespread technique.

Fedesco started with a EUR 5 million (USD 6.5 million) financing capacity, with state guarantee, which was increased to EUR 10 million (USD 13 million) in 2007 and EUR 100 million (USD 131 million) in 2009.

Private ESCOs use their own contracts with private customers. Fedesco has developed a model EPC tendering document and manual, which is being used both for internal projects as well as in consultancy projects for Knowledge Center customers. Agoria, the federation of the technical industry, has also developed a model contract for the private sector. Tendering procedures and contracts have been specifically adapted to Belgian public tendering law. It will be the basis of BELESCO's effort to develop a common model contract for the public sector.

Additionally, most utilities and regional energy efficiency agencies have a set of incentives targeting the public, commercial, residential and industrial sectors, which bring partial financing for eligible projects. The federal government provides tax reductions on energy-saving investments.

MARKETS

The ESCO industry in Belgium has been very concentrated in the public sector and in the commercial building sector as well. Projects include replacement of boilers and chillers, cogeneration, relighting and other technical measures. Fedesco also includes measures at building envelope level (insulation or spectrally selective window films).

At the present time, there are three public ESCOs as well as roughly six larger, and 12 smaller private ESCOs in Belgium.

ESCO specific activities are poorly documented as they are mainly operated by private companies that are under private agreements except in the case of public sector projects. BELESCO will provide more data in the future.

A small part of the industry is currently initiating activities in the residential sector. It is the aim of smaller energy consultancies to complement auditing services with the sale and direct installation of energy-efficient household equipment and lighting. Nevertheless, the residential sector is still a minor client for ESCOs. Some players are starting to target multi-dwelling apartment buildings or the social housing sector. Another federal government initiative in the residential sector is the FRGE (Fund for the Reduction of the Global Energy cost), a fund of EUR 100 million (USD 131 million) financed by a federal state obligation loan. Through local entities at city level, households can get low interest loans for energy-saving measures with a full service package for a target group of low-income households. Regional governments have recently created eco-renovation products in collaboration with banks.

GOVERNMENT ACTIONS

The Belgian ESCO industry benefits from support from the federal government. Through Fedesco, the federal government grants ESCOs immediate access to large public sector contracts for its own 1,650 buildings, although this number may decrease to 1,000 as a result of centralization of staff. While the credit rating of the government as a client makes financing projects relatively easy, the 2011 financial crisis and a very long political crisis have had a negative influence. Nevertheless, budgets for Fedesco have not been decreased so far.

On the other hand, energy efficiency programs are managed by regional governments. It is unclear what support they will provide for the development of the ESCO industry. This will depend on the results of the transposition of the EU Services Directive in Brussels, Wallonia and Flanders.

On the demand side, Fedesco has been asked to stimulate the development of the ESCO market through the creation of its Knowledge Center and, on the supply side, through the establishment of the Belgian ESCO Association, BELESCO. Set up in collaboration among private ESCOs, TPF companies and other market players, BELESCO is a non-profit organization that started its activities in 2008. It focuses on lobbying and stimulating the development of the ESCO industry as well as on training public and private customers. BELESCO is also building a database of EPC projects, setting up an accreditation program, developing International Performance Measurement and Verification Protocol (IPMVP) good practices and producing model contracts for the public sector. As such, it acts as a major EPC market facilitator.

Brazil

Alan Douglas Poole
Maria Cecília Amaral

ACTIVITIES

Brazil has had firms providing specialized energy efficiency project services since the early 1990s and an ESCO association since 1997. At that time, it was estimated that the total market for ESCO services was USD 16-17 million per year (BRL 16-17 million at the time). In 2009, the market was estimated at USD 65-90 million (BRL 130-178 million). Over the intervening period, and discounting inflation, this represents a growth rate of 7.5-10 percent per year.[1] While this is a significant rate of expansion, it hardly represents a take-off when one considers the small initial base and the large market potential.

Not only has long-term growth been relatively modest, it has also been erratic. For example, the survey for 2009 also covered 2008 and estimated that the ESCO market had expanded by 184 percent relative to 2008. This large increase represented a rebound from the recession in 2008, when most energy efficiency investments by businesses were suspended.

Until now, ESCOs in Brazil have been quintessential survivors. Does the apparent surge of growth from 2008 to 2009 presage a vigorous future expansion?

ESCOs are very diverse. The large majority are independent engineering consultant companies. Some ESCOs have chosen the model of representing manufacturers of energy efficient equipment, while a few are subsidiaries of larger companies that manufacture equipment (Johnson Controls) or subsidiaries of energy utilities (Light ESCO, Efficientia and Iqara Energy Services). Some ESCOs are joint ventures between larger firms from different sectors. Relatively few fit into the classic ESCO model, using performance contracts to finance relatively larger projects. These often face a shortage of capital. There has been some success in raising private equity to increase their capacity to invest.

Most ESCOs are small firms, many of which are very small. Table 9-1 shows the survey estimates of the energy efficiency project volume for three size categories of ESCOs:

- small = 5 employees or less;
- medium = 6-10 employees;
- large = more than 20 employees.

Most "large" ESCOs would qualify as small enterprises by most standard economic definitions.

Perhaps the most interesting point to emerge from this table is the very small average size of projects executed by Brazilian ESCOs. Projects carried out by small ESCOs are tiny, averaging less than USD 15,000. Even projects of larger ESCOs averaged only USD 42-44,000. This value is very low by international standards. Other evidence suggests that the payback period of projects is very short, at most a few months.[2] This pattern of small projects with short paybacks is probably due in large part to the limited access to financing by most ESCOs and the consequent need to use their own capital in projects. Another consequence is a predominance of lighting projects, which usually require smaller investments and have higher returns.

Table 9-2 shows the sources of capital for investments in ESCO projects for the year 2009. It also illustrates how they varied between the three size categories. Small ESCOs finance their projects almost

Table 9-1. Parameters of Energy Efficiency Project Volume of ESCOs, by Size Category in 2009 [a]

Category Firms	*# per Firm*	*# Project Firm USD 10^3*	*Investment per Category USD 10^6*	*EE Investment by Size USD*	*Average Project in Category*	*# Projects*
Small	14	15	215	3.0	14.348	210
Medium	8	28	1,201	9.6	42,900	224
Large	10	64	2,817	28.2	44,021	640
Total/Average	32	35	1,275	40.8	36,422	1,120

Source: (Gonçalves & Associados, 2010)
[a.] Surveyed firms only

Table 9-2. Sources of Capital for ESCO Investments in 2009, by Size Category

Source of Capital →	*ESCO*	*PROESCO*	*Utilities PEE/ANEEL*	*Private Equity, Venture Capital & Banks*	*End-User*	*Total Invested*
Small	94%	0%	6%	0%	0%	100%
Medium	60%	9%	9%	20%	3%	100%
Large	31%	13%	41%	16%	0%	100%
Average*	**42%**	**11%**	**31%**	**16%**	**1%**	**100%**

entirely with their own capital (94%). This value falls to only 31 percent among large ESCOs, which at the same time dominate the implementation of utility energy efficiency projects under the utility "Public Benefit Fund" (PEE/ANEEL) which is discussed below.

Curiously, medium-sized ESCOs had the highest share of financing from "private equity, venture capital & banks." They may be driven by poorer access to utility projects under the PEE/ANEEL. It is relevant to observe that, today, the largest ESCOs are the product of investments made by the PEE/ANEEL in earlier years, hence their big share of such projects. There is a bit of a chicken and egg question here. Did some ESCOs grow larger by accessing PEE/ANEEL resources or do utilities prefer to work with larger ESCOs? At the same time, some larger ESCOs avoid the PEE/ANEEL market. Some may have grown earlier with PEE projects but then shifted away and often added "adjacent services." It has been very difficult for "pure ESCOs" to grow offering only energy efficiency services.

Although the large majority of ESCOs (70%) have their headquarters in São Paulo or southern Brazil, ESCO services are available across the country. About two-thirds offer services at national level, including almost all of the medium and large ESCOs. Unsurprisingly, about three-quarters of small ESCOs operate at regional level only.

The development of the ESCO sector has been a "bottom-up" effort since the beginning. The incipient ESCO sector began to organize itself in the mid-1990s. The Brazilian ESCO association (ABESCO) was founded in 1997 with 15 members. Today, ABESCO has about 82 members, most of which are firms providing specialized energy efficiency services. The association (www.abesco.com.br) has been active in disseminating the energy efficiency and energy performance contracting concepts across the country. Furthermore, it has been involved in strengthening ESCO capabilities through training and other measures as well as in collaborating with the government and other agents for the development of initiatives with a view to overcoming barriers to sector growth. The latter has always been an uphill struggle given the minimal attention that the government has usually given to this approach to improving the economy's energy efficiency.

The government has just published a National Energy Efficiency Plan (PNEf). The PNEf is meant to define the means to achieve the energy efficiency objectives set out in the National Energy Plan published in 2007 and in the more recent Ten-Year Plans. These objectives are fairly ambitious. The most recent Ten-Year Plan explicitly assumes a reduction of 12 percent in the growth of electricity consumption until 2020. Slightly larger reductions are assumed for fuels in the non-transport sectors.

To achieve a reduction of this magnitude would be a major challenge, especially if it is beyond already established trends and reasonably verifiable. It implies strategic changes in the way Brazil now spends almost USD 400 million/year of public money for allegedly promoting energy efficiency. A rational energy efficiency policy would seek to leverage this substantial flow of public resources to catalyze larger private sector investments. In theory, it would be easy to increase the leverage of these public resources, since at least 80 percent are spent on donations of equipment. In practice, some well-placed people are happy with the political patronage that this approach to energy efficiency funding represents.

It is still hard to judge whether the new energy efficiency policy emerging from the PNEf will bring serious structural changes in the energy efficiency market. The document does not address the challenges of energy efficiency market transformation in any systematic way and ESCOs are barely considered. However, even though it can hardly be regarded as a plan at all, the PNEf may inaugurate a process that results in more far-reaching changes. If these occur, ESCOs could be major beneficiaries, since they are natural agents to leverage public resources. Even if the strategic changes are minor (which is more probable), the impacts on the small ESCO market could be significant.

Almost irrespective of the PNEf, there are several changes underway which could affect the ESCO market in the near future.

Regulated utility prices for electricity are unlikely to increase much relative to inflation and may even fall, as they have in recent years [3]. However, beginning in 2012 a new time-of-day tariff structure (associated with new "smart meters") will begin to be introduced for low voltage consumers—which include hundreds of thousands of business and public sector accounts. This will stimulate thousands of business consumers to seek professional guidance on how to exploit the opportunities which the new tariff structure opens. Something like this happened in the 1990s with regulations requiring high-voltage consumers to conform to power factor requirements (reducing their reactive demand).

Another factor is the expansion of the distribution network of natural gas. Consumers switching to this fuel represent an opportunity to identify broader energy rationalization measures. Unfortunately, few ESCOs until now seem to have been prepared to exploit this opportunity.

The new program to "transform the EE market for buildings" being implemented by the MMA with support from the IDB/GEF/UNDP could be a positive factor if the guarantee facility and the market promotion activities are well conceived and executed.

Last but not least, a certification program for specialists in measurement and verification (M&V) has begun in Brazil. This is the Certified Measurement and Verification Professional (CMVP) program, which is administered by the Efficiency Valuation Organization (EVO). The lack of certified specialists in M&V has been part of a broader problem of verifying the results of projects. Besides being of obvious importance for consolidating the credibility of performance contracts, this new capability should simplify issues related to public sector procurement and the Public Benefit Fund projects supervised by ANEEL.

Interestingly, the CMVP initiative was undertaken without any resources from official government programs. It was pure "bottom-up" and the experience brings an important lesson. The vigorous growth of the ESCO sector will depend in good part on the proactive vision of entrepreneurs and their association to overcome the barriers discussed above and to cultivate emerging opportunities. These include the public's growing concern about the environment and recognition that energy efficiency is an essential part of the response in addition to keener awareness in businesses of the importance of energy management (ISO 50.001). Hopefully, government policy will also become more supportive of the market-oriented approach represented by ESCOs. If there is even a modest move by government policy in this direction, the ESCO industry's growth could be quite dynamic in the coming years.

CONTRACTS

There is relatively little information available about the characteristics of the energy performance contracts used in Brazil. Most of those employed by ESCOs are "shared savings" contracts, under which the ESCO provides the financing usually with its own resources. The most common basis for remunerating energy performance contracts is the savings effectively achieved in monetary terms (Reals). However, savings in terms of physical units (kWh) are also common as a basis. Sometimes, the value of payments is set at a fixed level after the first few months in order to facilitate negotiations with banks. In general, one can say that Brazilian ESCOs are quite flexible in the methodology used to calculate remuneration.

LEGAL FRAMEWORK

Performance contracts as written today are not accepted by banks as a guarantee of future receivables, in particular by the *Banco Nacional de Desenvolvimento* (BNDES) (Brazilian Development Bank). The BNDES has an important role in project financing in Brazil (see discussion of PROESCO below). A work group was established to design a contract template that would be acceptable to the BNDES and other private banks. Furthermore, the contract template was to be feasible from the perspective of ESCOs; however, the work group was unable to define a satisfactory model. Another problem is that there is no established process of arbitration for resolving disputes involving performance contracts.

MARKETS

ESCOs do not earn all of their income from energy efficiency projects, although in the case of small and medium-sized firms, energy efficiency is the source of about 70 percent of the firms' receipts. In the case of large ESCOs, this falls to only 47 percent, which suggests that, in order to grow, an ESCO needs to enter the market for adjacent services. Common additional services provided include renegotiation of tariffs with electric utilities, power factor correction, electrical wiring, reduction of harmonics, standby generators and distributed power, facility maintenance, etc.

With regard to energy efficiency projects, there is an overwhelming dominance of measures to reduce or optimize electricity use. The frequency of projects to reduce fuel use is much lower, although about one half of ESCOs claim to provide services in this area. There are various reasons for such an emphasis on electricity. Government programs for electricity conservation are much larger (especially when the utility PEE program is considered). Investments for thermal processes in industry are larger, which complicates financing. Industrial firms also have more restrictions regarding the use of ESCOs than do building owners. Meanwhile, the consumption of fuels in Brazilian buildings is very small, especially if cooking is excluded. Space heating is almost non-existent.

Industries, services and businesses in the commerce/retail sector contribute in roughly equal proportions to the number of projects executed by ESCOs (Gonçalves & Associados, 2010). Smaller ESCOs tend to concentrate more on the services and commercial sectors, where projects are usually somewhat smaller and less complex. The residential sector has attracted little interest, while the public sector has been almost absent as a market due to difficulties in procurement.

Besides the normal breakdown by consumer sector, it is also useful to consider other market categories, which have had or could have an impact on the business volume. These categories tend to cut across consumer sectors. The most important is the Public Benefit Fund for Energy Efficiency, which is overseen by the power sector regulator—ANEEL (PEE/ANEEL). As discussed above, this program became a major source of income for ESCOs after 2000, although it has declined substantially since 2008.

GOVERNMENT ACTIONS

There have been several surveys conducted to pinpoint the perceived barriers to ESCO market growth. The most recent performed by Gonçalves & Associados (2010) found that lack of access to financing is considered to be the biggest barrier to expanding the business. This includes lack of credit at low interest rates for projects and/or equipment as well as difficulties in the liberation of PEE/ANEEL resources by utilities. [4]

A second major barrier is marketing energy efficiency as a concept. This begins with a lack of awareness and/or motivation on the part of many consumers outside energy-intensive sectors. Many of these do not regard energy as a strategic input and do not seek to measure or manage its use. Additionally, there is often skepticism regarding the claimed savings of energy efficiency projects.

The problem of marketing may be exacerbated by the patchy image of the ESCO sector. One frequently finds generic criticisms on the capability of ESCOs in the literature in Brazil, although no systematic review of the problem has been made. Certainly there are large differences in the technical and management capabilities of ESCOs. At the same time, there is limited use of formal standards and certifications to help provide a reference for the quality of this kind of service. The ESCO industry needs to invest in training and in developing firms' capabilities.

Although Brazil has long established energy efficiency programs and agencies, public policy over the years has given a low priority to promoting the development of the ESCO industry. Little has been tried and what has been attempted has not done much to mitigate the barriers facing the ESCO market. Below is a brief description of the major efforts and the consequences.

- *The Public Benefit Fund for Energy Efficiency (PEE/ANEEL)*, which mandates utility investments in energy efficiency, was opened in 2000 for "performance contracts" between the utility and consumers. Many utilities soon opted to contract ESCOs to execute these projects. This segment rapidly became the biggest single market for ESCOs in Brazil. However, while it produced revenue, it did little to transform the market. ESCOs were contracted by the utilities as installers, using standard cost-plus engineering services contracts, not performance contracts. Not a single project used commercial bank credit to complement the PEE resources of the utility. In recent years, utilities have greatly reduced this kind of project for diverse reasons related to the regulatory framework of the PEE/ANEEL.

- *PROESCO*: In 2006, the BNDES approved an innovative new credit line, called PROESCO [5]. It was expected to address the problem of guarantees for loans to energy efficiency projects and to encourage a move towards the acceptance of receivables in energy performance contracts. Most financing was to be done through cooperating financial intermediaries where the BNDES would assume up to 80 percent of the risk. However, this model has not worked in practice. Since the beginning of the program until early 2011, only 23 projects with total loans of BRL 87 million had been approved, none of which involved risk sharing with financial intermediaries. PROESCO is now being restructured. The risk sharing element with financial intermediaries, which was central to the original concept, has been eliminated. It will target firms which already have a well-structured relationship with the banks acting as financial intermediaries. There is no interest whatsoever in financing ESCOs.

- *Public sector buildings*: In 2000, there was a Presidential Decree which required that all federal buildings reduce their electricity consumption by 20 percent within two years. This raised hopes that a program for government buildings could emerge as in other countries where they have played a big role in consolidating the ESCO business. Unfortunately, after 11 years, the legal and budgetary issues related to tendering performance contracts for government buildings have still not been resolved and this market has remained closed for ESCOs.

Bulgaria

Boris Petkov

ACTIVITIES

There have been a number of market studies of the ESCO industry in Bulgaria. However, due to the very broad definition of ESCOs, it is difficult to assess how many companies are providing ESCO services in Bulgaria.

Some studies commissioned by the EU suggest that there are only five or six ESCOs active on the market while others identify over 20 companies working primarily in the public sector.

Many street lighting and public building rehabilitation ESCO-type projects have been completed by companies specially established for the purpose of implementing and guaranteeing energy savings. In 2002, CES (Company for Energy Saving) was set up by RWE, Stadtwerke Leipzig and the Municipality of Sofia to rehabilitate over 300 public buildings in Sofia. The project was successfully completed in 2009 and CES was acquired by Dalkia.

Most companies offering ESCO services have other core activities, usually consultancy services (Ivel Consult), manufacturing and/or distribution of energy-saving equipment (Erato Holding), contractors and engineering firms (Enemona) and energy suppliers (Dalkia).

MARKET

The ESCO market in Bulgaria is primarily stimulated by incentives available for:

- renewable energy projects funded by ratepayers through feed-in tariffs;
- home improvements and modernization of SME facilities and processes financed by energy efficiency credit lines and programs that encourage comprehensive retrofits in all end-uses of energy; and
- government programs for public buildings.

It is difficult to estimate the Bulgarian market for energy services because it is very fragmented, including such aspects as energy consultancy services, guaranteed savings contracts and heat supply (chauffage) contracts. Depending on the definition of ESCOs, it is estimated that ESCO industry revenues were in the region of EUR 10 million (USD 13 million) to EUR 250 million (USD 328 million) in 2010.

The main stream of ESCO industry revenues is formed by renewable generation projects that have been estimated to account for over 85 percent of ESCO industry revenues in the past few years. In terms of energy efficiency upgrades, the building sector has benefited most from ESCO services. Measures usually include the building envelope (window replacement and insulation) and building services (lighting and HVAC) improvements. A small number of street lighting projects have been completed. Not many ESCO projects have been undertaken in the industrial sector.

It is anticipated that the key contributing factors for further increases in deployment will be the availability of financing and publicly funded incentives that could be leveraged by ESCOs in Bulgaria. Additional key contributing factors are expected to include the growing trend of bundling renewable energy with energy efficiency improvements carried out by energy suppliers providing ESCO services.

Opportunities exist for providing ESCO services (street lighting, CHP and heat production) in various types of buildings, including administrative buildings, hospitals, schools, hotels, commercial buildings and premises of SMEs.

LEGAL FRAMEWORK

Over the last few years, the energy efficiency-related directives of the European Union (EU) have been successfully transposed into the Bulgarian legal framework. This led to the adoption of the Energy Efficiency Act and associated secondary legislation. Bulgaria has introduced a large number of policies and measures in support of energy efficiency for Small and Medium Enterprises (SMEs), industry, buildings and transport.

The new legislation established mandatory energy auditing as well as certification of buildings, appliances and heating equipment. This allowed opening up new opportunities for energy assessors and consultants.

More abundant sources of financing have become available primarily due to the steady economic growth in the country and financial support from the EU. In the 1990s, the ESCO market was shaped by funding coming from donors like the United States Agency for International Development (USAID) and the World Bank. However, in recent years, financing from local banks, national and municipal budgets as well as the EU structural and cohesion funds earmarked for Bulgaria have driven the growth of the ESCO business in the country.

The EU directives have also stimulated the improvement of the market for ESCO services and created an enabling environment for the development of renewable energy in Bulgaria. The country has put in place a comprehensive system of incentives based on feed-in tariffs, which attracted international ESCOs and developers. Currently, the preferential tariffs set by Bulgaria are comparable to feed-in tariffs in other EU countries like Germany and Spain.

In addition, efforts have been made by the national and local governments to support public-private partnerships (PPPs) and ESCOs in order to implement energy efficiency projects in the public sector. The whole was promoted by introducing more favorable procurement environments to stimulate third party financing (TPF), especially aimed at increasing the share of CHP in heat generation and the share of renewable energy such as hydro, wind and biomass.

Furthermore, the European Commission (EC) is pushing for bigger improvements in energy efficiency. It is also proposing a new directive known as the "Energy Efficiency Directive," which would replace the existing Energy End-Use Efficiency and Energy Services Directive as well as the Cogeneration Directive. The adoption of this new directive and its transposition into national legislation is expected to further stimulate the ESCO market.

GOVERNMENT ACTIONS

The Energy Efficiency Act (EEA) (as amended in 2009) sets the legal framework for how ESCOs perform their activities in Bulgaria.

The key provisions are set out in Section 4: Energy Services. For more details, the English version of the EEA can be downloaded from: http://www.seea.government.bg/documents/eng/LEE-2008.pdf.

In accordance with the EEA, Ordinance No. RD-16-347 of April 2, 2009 was adopted to spell out the procedures for assessment of savings and compensation for ESCO services performed on public buildings.

Apart from providing and enforcing the legal framework for sustainable energy in Bulgaria, the government supports a wide range of awareness raising campaigns. The Government of Bulgaria also sets energy-saving targets for energy traders, building owners (over 1,000 m2 of total internal area) and end-users with an annual consumption over 3,000 MWh. From 2013, targets will be set for owners of buildings with an area over 500 m2 and from 2016 for owners of buildings with an area exceeding 250 m2.

The government initiates a wide range of financing schemes and rebates for the rehabilitation of the building stock and the modernization of the industry in close cooperation with European institutions such as the EC, the European Bank for Reconstruction and Development (EBRD), the European Investment Bank and the Council of Europe Development Bank. As a result, credit lines for sustainable energy have been extended to local banks. Furthermore, the National Energy Efficiency Revolving Fund has been set up and the Kozloduy Decommissioning Fund/UNDP and state budget have provided funding for the refurbishment of public buildings and condominiums. The following consists of information about the key financing facilities available for sustainable energy projects:

Kozloduy International Decommissioning Support Fund (KIDSF)

It was set up in 2000 with contributions from the EC, EU member countries and Switzerland. KIDSF financially supports the early decommissioning of units 1-4 of the Kozloduy Nuclear Power Plant. KIDSF also supports energy sector initiatives associated with the decommissioning effort, such as improving energy efficiency in Bulgaria.

Bulgarian Energy Efficiency Fund

The Bulgarian Energy Efficiency Fund (BgEEF) was established in 2004 and initially capitalized entirely through grants, its major donors being the GEF through the International Bank for Reconstruction and Development (the World Bank), the Government of Austria, the Government of Bulgaria and several private Bulgarian companies. BgEEF is a revolving mechanism for development and financing of commercially viable projects and capacity building. BgEEF has the combined capacity of a lending institution, a credit guarantee facility and a consulting company.

European Union Energy Efficiency Finance Facility

In 2006, the European Commission decided to establish two multi-beneficiary programs on energy efficiency, together with the Council of Europe Development Bank in co-operation with Kreditanstalt für Wiederaufbau and the European Investment Bank. The projects covered Bulgaria, Romania, Croatia and Turkey. Both projects were aimed at providing financial assistance to acceding and candidate countries in increasing investments in energy efficiency in the industrial sector.

Residential Energy Efficiency Credit Line (REECL)

Launched in 2005 and available until 2014, it is aimed at helping Bulgarian households reduce their energy bills and consumption. It provides credit lines to reputable Bulgarian banks to make loans to householders for specific energy efficiency measures including (i) double glazing; (ii) wall, floor and roof insulation; (iii) efficient biomass stoves and boilers; (iv) solar water heaters; (v) efficient gas boilers; (vi) PV systems, (vii) heat recovery ventilation systems; and (viii) heat pump systems. Grant financing is provided in support of project development and incentive grants. Borrowers receive funds after verification by independent consultants that each eligible residential energy efficiency project has been completed.

Bulgarian Energy Efficiency and Renewable Energy Credit Line (BEERECL)

The BEERECL has been developed by the EBRD in 2004 in close cooperation with the Bulgarian government and the EU. The facility is available until 2013 and extends loans to participating banks for on-lending to private sector companies for industrial energy efficiency and small renewable energy projects. The BEERECL provides incentives for energy efficiency and renewable energy projects including cogeneration, heat and steam recovery, automation and control, upgrade/replacement of utilities, fuel switching (coal/fuel oil to gas), ground source heat pumps, and process optimization.

EU Operational Program "Development of the Competitiveness of the Bulgarian Economy 2007-2013"

Two such programs are:

- Grant scheme "Introduction of Renewable Energy Sources Satisfying the Needs of the Enterprises"
- Grant scheme "Introduction of Energy-Saving Technologies in Enterprises"

Eligible beneficiaries—enterprises (micro, small, medium and large) and cooperatives registered under the trade law or the law on cooperatives, having their headquarters in Bulgaria

EU Operational Program "Regional Development 2007-2013"

This EU program established a grant scheme "Rehabilitation of Multi-Family Buildings"

The eligible beneficiaries are associations of home owners.

In addition, while the government has provided some grants for energy audits in SMEs, it has been sporadic.

Canada

Peter Love
Pierre Langlois

ACTIVITIES

The rapid growth of EPC and ESCOs in Canada was unique in that it was a government-inspired solution, originating from the first oil price shock of the early 1970s and focused on inefficient use of energy as well as on cost savings expectations.

The first ESCO in Canada, and one of the first in the world, was created in the province of Quebec by Hydro-Québec and a local engineering firm. Econoler Inc. was founded in 1981 and developed a new concept for that time based on a unique shared savings approach with a first-out option. The first-out option consists of an open book approach with contract termination upon complete payment of all project costs, even if this occurs prior to reaching the maximum contractual agreement. That approach was very attractive for the market back then and all the more relevant at a time when the interest rates in Canada for the implementation of such projects were in the vicinity of 20 percent. This in turn significantly hampered investments, including those in the energy sector. Between 1981 and 1989, over 1,000 projects were implemented in all kinds of commercial, institutional and industrial establishments in the province of Quebec.

In the mid-1980s, many other companies began to offer a broad range of energy performance contracts as well as other forms of performance-based solutions. In addition to financing energy efficiency retrofits, they are now also being used to finance renewable energy installations as well as non-energy-related deferred maintenance projects.

Two major programs were launched in the 1990s to help the ESCO concept grow in the country:

1. The Federal Buildings Initiative (FBI), an energy sector initiative of the federal department of Natural Resources Canada. The FBI was officially launched in 1991, and two years later, the first project using EPC was implemented. The FBI provides fed-

eral organizations with the tools and services needed to make informed decisions about reducing energy, water consumption and greenhouse gas (GHG) emissions in their facilities. The FBI addresses three common barriers to improving energy efficiency: (i) inadequate capital budgets for energy efficiency projects; (ii) need for reliable information on current energy technologies and practices; and (iii) lack of required skills to manage retrofits. The FBI provides publications, case studies, sample tender documents and model contracts in addition to information on energy efficiency-related environmental, health and safety issues, employee awareness products and comprehensive training programs. In addition, the FBI maintains a qualified bidders list of ESCOs on its website. Qualified bidders are evaluated based on their financial capabilities and technical expertise to undertake major energy efficiency retrofits.

2. The Better Buildings Partnership (BBP) program began in June 1996 and focuses on curbing CO_2 emissions in Toronto through energy management firms' activities. BBP is a PPP between Enbridge Gas Distribution Inc., the Toronto Atmospheric Fund, Toronto Hydro and Ontario Hydro Energy, Inc. BBP comprises several programs including a residential energy awareness program, an office building and commercial building program as well as a loan recourse fund.

Unfortunately, Canadian ESCO activities suffer from the lack of continuous monitoring and reporting. The only real information available is based on data gathering from the FBI as noted above. It is expected that the ESCO market will grow significantly in Canada in the next five years. Many provinces are promoting the EPC approach for the implementation of energy efficiency projects in the public sector. The FBI program is also quite active and continues to promote its share of projects nationwide for federal buildings. The private sector is timidly starting to look at the ESCO concept as a way to eliminate the different barriers to the implementation of energy efficiency projects. It is likely that this market will grow at a steady pace over the next five years in Canada as there is still an important market to be developed.

CONTRACTS

In Canada, the guaranteed savings model represents an important part of current contracts being implemented. Canadian ESCOs are also using, to a lesser extent, the shared savings and the "chauffage" concepts.

In the case of the FBI program at the federal government level, efforts were made to overcome budget constraints in the government sector by promoting the use of private, third-party financing by ESCOs.

LEGAL FRAMEWORK

Generally, there are no legal barriers, which prevent the use of EPC in Canada. For federal institutions, the Federal Contracting Policy was amended to allow federal departments to enter into a service contract and acquire energy services with an energy management firm with a view to implementing energy efficiency improvements[1].

In some cases, there may be challenges on how EPC could be used more broadly and achieve its full potential in the public sector. Different organizations, including the newly formed Energy Services Association of Canada (ESAC) have been looking into the matter in the last few months and it is expected that they will make some recommendations about how to improve on the process at both the federal and provincial levels.

MARKETS

The performance-based solutions (PBS) industry in Canada is a CAD 450 million/year business (USD 450 million/year business), generating annual energy savings in Canadian dollars (CAD) of about CAD 45 million/year (USD 45 million/year). It is estimated that this industry is currently responsible for over 4,000 direct jobs and 5,000 indirect jobs. Up to 80 percent of the labor associated with these projects is local labor. Although there have been a few projects for commercial buildings and a small number of industrial projects, the vast majority of these projects are for public sector buildings. These types of buildings are commonly referred to as municipalities, universities, schools and hospitals (MUSH) and include all government offices as well as education and health care facilities.

In Canada, the early days of the industry took a different turn from that in the United States. In a typically Canadian way, the public sector got involved with the industry and this has continued to such an extent that a private-public "partnership" has developed.

On a more qualitative basis, the ESCO industry in Canada has been very concentrated in the public sector as well as in the commercial building sector. All levels of government are implementing energy efficiency improvements in their buildings using ESCOs.

Still, barriers exist in the marketplace. One is the internal competition of public and private sector organizations, which charge fees to manage buildings without operating under an EPC concept. Those fees are based on a percentage of the operational expenses. Therefore, an energy efficiency project is a source of fee reduction for them. In the public sector, some organizations own an important number of facilities and can obtain funding from existing capital budgets for building renovation without calling for the use of the ESCO concept. Even though the ESCO concept has been used in Canada for more than 30 years, there is still an important lack of awareness and knowledge about the concept and how to use it.

FACILITATORS

In 1987, the Canadian Association of Energy Service Companies (CAESCO) was established with the support of Ontario Hydro and the federal and Ontario provincial governments. CAESCO encouraged the orderly growth of the industry through accreditation, support and advice to both EPC contractors and customers. Membership in 1997 was over 50 and included, in addition to ESCOs, equipment suppliers, utilities, governments, lawyers and consultants. There are presently 13 accredited ESCOs in Canada. Unfortunately, CAESCO closed in 2001, for lack of support and interest on the part of the different stakeholders.

After nine years without an industry association, ESAC was formed in 2010. The founding members of the association are Ainsworth, Ameresco, Direct Energy, Honeywell, Johnson Controls, MCW

Custom Energy Solutions, Siemens and Trane. It is estimated that the eight founding members of ESAC represented about 90 percent of the CAD 450 million (USD 450 million) performance-based solutions business in Canada in 2009. The association's vision is that performance-based solutions become the premier choice for energy and infrastructure renewal initiatives, resulting in fiscally and environmentally responsible outcomes. Its mission is to actively promote government policies and regulatory support for greater use of guaranteed performance-based solutions to implement energy efficiency, renewable energy and infrastructure renewal initiatives.

GOVERNMENT ACTIONS

At the present time, there are a range of federal, provincial and municipal initiatives across Canada that relate to performance-based solutions including the following:

- **Federal Government**—The FBI has attracted CAD 320 million (USD 320 million) in private sector investments to date and generated over CAD 43 million (USD 43 million) in annual energy savings as well as in annual GHG emissions reductions (285 Kt). This was achieved through 87 projects in over 7,000 buildings or about one-third of federal buildings retrofitted between 1993 and 2011 under this program. In 2010, as part of its Federal Sustainable Development Strategy, the federal government set a target of reducing levels of GHG emissions by 17 percent below 2005 levels by 2020. This work was undertaken by the Office of Greening Government Operations and is expected to result in a major increase in the use of innovative financing and energy management services using energy performance contracts.

- **British Colombia**—BC's Energy Plan has set aggressive targets for GHG emissions reductions. As part of this plan, the government has estimated a requirement for about CAD 1.5 billion in energy efficiency upgrades to public buildings.

- **Nova Scotia**—The government has made a commitment to retrofit all schools in the province at "no cost" which is understood to imply the use of PBS. It has also selected ESCOs to undertake PBS for public buildings in four regions in the province.

- **Ontario**—Regulations under the new Green Energy Act require all public agencies to reduce energy consumption by 20 percent. The Ontario Power Authority has issued an overall requirement for utilities to reduce their electricity production and implement demand management programs as part of a mandate to achieve one of the most aggressive energy conservation targets in North America. To that effect, the Ontario Power Authority has launched a series of programs for the commercial/institutional sector. The Green Energy Act also included North America's first Feed-In-Tariff ("FIT") program that provides guaranteed premium prices for renewable energy.

  In Toronto, the city's BBP has completed more than 2,000 projects since its inception in 1996, improving over 20 million m2 of space. Over CAD 850 million (USD 850 million) has been invested with savings estimated to be over CAD 60 million/year (USD 60 million/year). Many of the larger projects used an energy performance contract.

- **Quebec**—The market in the province is basically focused on the public sector, and essentially in the health and education sector. Both ministries are favoring the use of EPC to attain the provincial reduction objectives of energy consumption set by the Ministry of Natural Resources. The program is implemented through an energy savings approach where the government uses its own financing capacities to finance projects that have to be paid back over a seven-year period. The ESCO has to guarantee the performance of the project during that period.

- **Saskatchewan**—SaskPower, the provincially owned integrated electricity utility, has been offering its own EPC services in the province for the last ten years through a joint venture with a national ESCO.

Information on specific projects in the different provinces can be found at the ESAC website at www.eenrgyservicesassociation.ca.

Chile

Pamela Mellado

ACTIVITIES

In Chile, the market for ESCOs is emerging. Only since 2005 has energy efficiency become part of public discussions, with the creation of the National Energy Efficiency Program (PPEE) in the Ministry of Economy. The ensuing institutional evolution resulted in the creation of the Chilean Agency for Energy Efficiency (AChEE) in late 2010. The Agency was established as a means to coordinate public-private initiatives in energy efficiency and implement the public policies that the Ministry of Energy had determined in this matter.

AChEE's mission is "to promote, strengthen and consolidate the efficient use of energy, coordinating the relevant actors, at global and local levels, and implementing public-private initiatives in the different energy consumption sectors, in order to contribute to the competitiveness and sustainable development of Chile." It is in this context that the development of the ESCO market emerges as one of the main issues to be considered to fulfill this mission.

In parallel, the principal initiative developed in Chile specifically oriented to promote the local ESCO market is the Inter-American Development Bank (IADB) FOMIN's "Promotion of Clean Energy Market Opportunities" (PCEMO) project, which is designed to "contribute to expanding market opportunities and improving the competitiveness of small and medium enterprises (SMEs) in Chile." In fact, one of its specific objectives consists in developing proposals for business models, which enables the development of an ESCO market in Chile and contributes to the development of the demand and supply of energy efficiency as well as non-conventional renewable energy (NCRE) projects.

Currently, there are about ten companies on the market that are associated under the National Association of ESCOs (ANESCO), which seeks to be the leader in the promotion of this business model.[1]

ANESCO emerged as a result of work through the PCEMO, where eight companies joined together through the "PROFO ESCO Association for Energy Service Companies Project" (Project Associative Energy Service Company), which enabled the first steps toward an ESCO association in Chile. From the eight ANESCO member companies, two are classified as medium-sized and six are small, measured by sales volume and declared capital.[2]

Regarding the size of that market, there have been preliminary approaches. First, and under the PCEMO, Fundación Chile developed a preliminary estimate, arriving at a figure of USD 187 million for the energy efficiency market. From this, it was determined that USD 136.7 million corresponded specifically to ESCOs. This figure excluded the residential sector.[3]

In 2009, the PPEE and the IADB commissioned a study entitled "Improving Chilean Financial Instruments to Support Energy Efficiency"[4] oriented to evaluate state-of-the-art options in terms of available financial instruments for promoting energy efficiency. Based on international experience, this study provides an estimate of the evolution of the Chilean energy efficiency market, in terms of three dimensions:

- i Size: In which the level of development of the energy services market is determined by the evolution of the ratio of the volume of energy efficiency investments carried out by EF/ESCOs to gross national income.
- ii Energy performance contracts: Participation in the energy efficiency market of EPC projects.
- iii Competition: The level of competition in the market is determined by the number of EF/ESCOs in operation.

Three scenarios—baselines, conservative and aggressive—were defined for each of these indicators, based on different variables.

Given that there are only a few companies operating in the market and that they are not all joined to ANESCO, Table 9-3 summarizes the predicted characteristics of the market in 2018 under each of these three scenarios.

Table 9-3: Predicted Characteristics of the Market in 2018

	Competency Size (USD million)		*EPC Agreements (number)*		*Competency (ESCOs present in the market)*	
	2010	2018	2010	2018	2010	2018
Baseline	1.5	27.6	0	33	3	15
Conservative	1.5	10.6	0	8	3	7
Aggressive	1.5	85.5	0	129	3	25

During 2011, AChEE developed eight demonstration projects through ESCOs, with five of them to be performed in public buildings and the remainder in the productive sector. These projects were intended to demonstrate the feasibility of financing energy efficiency investments through the lower energy consumption associated with them. For identifying these projects, two public tenders were carried out. The industrial sector was asked to present projects in a consortium between the end customer and the ESCO, based on a previous energy audit that allowed an appropriate analysis of the technical economic viability of the project. In the case of public buildings, energy audits were provided to serve as basis for the investment proposals of ESCOs. Subsequently, resources were allocated to proposals considering three main criteria: (i) the magnitude of achievable savings; (ii) the payback period of investment; and (iii) the quality of technical proposals.

Concerning financing, in the case of industry, the projects were being funded 40 percent by the Agency while the other resources were to be provided by a consortium between the end customer and the ESCO. The overall investment amount was to be recovered in a period shorter than four years through performance contracts in which the Agency would take part. Although the contest had been tested with the relevant actors and addressed the needs expressed by them, it was very difficult to find projects. In fact, despite having a budget of USD 500,000, it was only possible to identify projects for 50 percent of that amount. The main challenges identified were the lack of market confidence in the model, the ignorance of end-users, the limited experience of ESCOs to close this kind of negotiations, and the high risk aversion to address this type of project.

In the case of the five public buildings, the experience was different because 24 proposals were presented. Using criteria similar to the industrial case, seven projects were selected for a total of USD 700,000, where the Agency financed 100 percent of the investment with a payback period of less than four years. The decision to fund these sunken investments was based on fulfilling the purpose of measuring and verifying savings achieved to illustrate the viability of financing such investments through the savings generated. Parallel to this, a law firm was hired to offer legal alternatives for using performance contracts in public institutions in order to address barriers to that effect.

From the experience gained in 2005, it is possible to identify a number of barriers that underlie the operation of potential ESCOs in delivering the EPC model. Some of the most important barriers are detailed below.

At industry level, there is a wide variety of suppliers who daily come to offer new products or services that cannot be differentiated in terms of effectiveness due to the absence of any kind of labeling for industrial equipment. The next barrier appears when the staff responsible for energy and operation management consider that the existence of opportunities for improvement is a threat because they feel that it questions their own performance and therefore tend to impede scale-up to management levels. Finally, when both barriers are overcome and the client decides to take on an EPC project, new complexities associated with project implementation usually appear related to accounting and tax management of the assets. Among the barriers identified at this stage are, for example, how to justify the savings to pay expenses, what is the ownership of assets, who is responsible for ensuring operation and achieving the estimated savings and so on. At present, the only performance contracts that are being developed in industry are those that the Agency is carrying out as demonstrative experience. It seems that it is much easier for clients to directly finance the project making sure that they get adequate support during project lifetime. According to information obtained from ANESCO members, there is currently only one EPC project for this type of customer.

In the case of public institutions, the trust barrier has been reduced by the action of AChEE and the PPEE previously. This is because, since 2005, a sequence of actions were developed with public buildings, starting from the identification of opportunities through energy audits to implementation of projects directly or via ESCOs. There are new barriers associated with the rigidity of the public sector budget framework in Chile. This means that any implementation reflected in savings to an institution subsequently results in the resources being discounted for the next financial year, thereby eliminating incentives for the development of energy efficiency projects. On the other hand, budget formulation generates timelessness between the detection of the opportunity and the chance to implement it. Once detected, the opportunities and their execution must be considered for the next financial year, delaying implementation and undermining the interest of the ESCO. So far, projects in the public sector have been financed directly by AChEE or by each institution, and none of them have an associated performance contract.

In the residential sector, the volume of opportunities is less important since the climatic characteristics of Chile do not require the majority of households to have specific equipment for air conditioning. Nevertheless, the experience of the company "Energy Tracking" in the use of control systems for this customer segment has been quite successful (32 EPC projects during the last four years with a high growth potential). The method used is the guaranteed savings or shared savings approach, depending on the specific characteristics of each client.

CONTRACTS

In 2010, ANESCO hired a law firm to develop four model contracts (shared savings, guaranteed savings, "fast out" and "chauffage") that could be used by their partners in their future business. Until now, this has not been possible.

In practice, the most commonly used contracts are power sales contracts (chauffage), which offer a fixed price to maintain power under specific quality standards. The company, JHG Engineering, has developed such contracts for industrial customers.

As mentioned in the previous section, the company Energy Tracking has experience in performance contracts associated with control systems of power plants, which has had some success at residential customer level, but not in the industrial and public building sectors. The experience of this company combines the existence of contracts with different degrees of ESCO participation in financing, with repayment periods of three years and a baseline in line with the IPMVP.

On the other hand, regarding the Agency's experience in pilot projects in 2011, all of them were developed under the shared savings model, and the use of the IPMVP. The Agency fulfills both roles, as a financier and as an actor associated with M&V of savings.

FACILITATORS

As is clear from the background given above, the existing facilitators in the local market are ANESCO, Fundación Chile and AChEE.

For ANESCO, its legal establishment materialized in July 2011 and it is now in process of incorporating new members. ANESCO has been slowly setting up the energy efficiency market but still cannot position itself as a reference in the field. AChEE is also a new actor. The completion of eight pilot projects during 2011 constitutes practical experience in implementation of performance contracts in Chile. Developing those contracts appropriately was a major challenge in order to maintain this line of action within its strategic plan. It is important to mention that the Agency's budget for 2011 was 100 percent public and is likely to remain that way for a long period.

At the public level, the Ministry of Energy's Energy Efficiency division is responsible for establishing policies and regulations on energy issues, while AChEE is in charge of realizing those initiatives. Currently, there is the intention to address the issue of regulations aimed at strengthening the market, so it is expected that the actions taken will continue to be targeted at catalyzing market development through information, coordination, training and demonstration examples, such as those that were developed in 2011.

GOVERNMENT ACTIONS

During 2011, AChEE began implementing the project "Encouraging the Setting-up and Consolidation of an Energy Service Market in Chile." The latter fosters the participation of engineering companies and ESCOs in the promotion of energy efficiency projects in the industrial, commercial and public sectors. Among its components is the implementation of a line of partial credit guarantees given by the bank with the support of the Agency in order to overcome the barriers to financing faced by these companies. The goal is to promote the development of energy efficiency projects based on performance contracts. This project is funded by the GEF for USD 2.6 million for a period of eight years and has been developed with the IADB as implementing agency.

The development of the ESCO market represents a challenge for the Agency's work, especially because it is a specific way to promote the fact that energy efficiency investments are financed directly from the private sector through the savings generated. The role of the Agency is to catalyze the operation of this market, continuously detecting existing barriers and the way to address them.

China

Robert P. Taylor

ACTIVITIES

China's ESCO industry was launched as part of a deliberate plan by the Chinese government, with support from the World Bank, the GEF and several other international donors. The EPC concept was first piloted and adapted to the Chinese market through the China Energy Conservation Project, approved by the World Bank in March 1998. This project included GEF grants and World Bank loan funds to support the start-up of three new pilot Chinese ESCOs organized by Chinese investors.

Overcoming many initial hurdles, the three new Chinese ESCOs developed the first simple EPC projects in the country in a variety of industrial and building applications. Government support was especially critical in the first several years, as a plethora of legal, regulatory, accounting and taxation issues surfaced requiring resolution. Another important source of support was the World Bank line of credit which allowed the pilot companies to focus on making the business work in the early years with a funding source already in place. These factors helped the companies achieve success in demonstrating the business model in the Chinese market, with strong returns and corporate growth. Total EPC investments in the three companies rose to over USD 20 million per year during 2002-2004, and then over USD 30 million per year during 2005 2006[1]. Their success attracted the attention of others.[2]

At the end of 2001, about six small new ESCOs had been formed by interested independent groups and some 15-20 other companies were beginning to test the EPC concept. Many more were expressing interest. The Second China Energy Conservation Project was approved by the World Bank and Chinese government in September 2002, providing GEF grant financing for technical assistance to new companies through a new ESCO association, and helping backstop a national ESCO loan guarantee program. The ESCO Committee of the China Energy Conservation Association, commonly referred to as China's Energy Management Company Association (EMCA), was launched in April 2004 with 59 charter member companies. The ESCO Loan Guarantee Program also opened for business at the beginning of 2004. The program provided partial-risk guarantees to 12 different commercial Chinese banks for their loans to 42 different ESCOs, mostly privately owned. The loan guarantee program proved to be especially valuable as a means to introduce new ESCOs to commercial bank financing for the first time, after which many were able to further develop relationships with banks and other financiers by themselves.[3]

These and other efforts placed China's industry in an excellent position to respond to high demand in Chinese enterprises for energy conservation services and investment during China's aggressive national energy conservation drive of the 11th Five-Year Plan (2006-2010). Total EPC investments in China grew from USD 160 million in 2005 to USD 4.25 billion in 2010 (see Figure 1). ESCO members of EMCA grew from 47 in 2005 to 428 in 2010 (and increasing numbers of ESCOs have also been formed which have not become EMCA members).

Towards the end of the decade, the EPC industry began to attract increasing attention among top national leaders. With a vision to expand China's ESCO industry much further to become a mainstay in China's energy conservation project investment, China's State Council issued a major national government policy statement of support in April 2010.

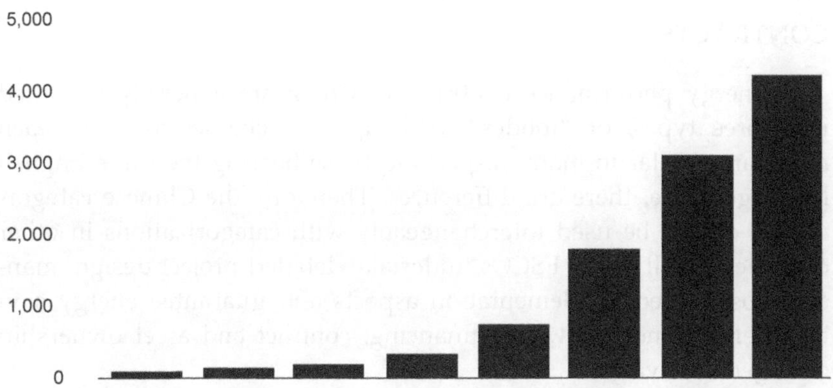

2003	2004	2005	2006	2007	2008	2009	2010
100	130	160	240	870	1710	2860	4250

Source: EMCA "2010 China ESCO Industry Summit Meeting Proceedings." January 2011. RMB converted to USD at annual average official rates of exchange.

Figure 9-4: Total Energy Performance Contracting Investment in China, 2003-2010 (million US Dollars)

The support includes specific new favorable taxation policies and post-project commissioning subsidies for energy performance contracted projects. In 2011, growth of ESCO companies and interest in EPC is reaching an all-time high. For example, by August 2011, over 1,700 companies had been accepted for registration on the national ESCO registration list, which is a requirement for application for investment subsidies. While many of these companies may be small and inexperienced, major Chinese corporations are also entering the ESCO business.

As development proceeds in the coming years, particular aspects that will require further progress to achieve the healthiest growth possible include:

- maturation of the many new companies entering the market and, accordingly, development of increasing trust between ESCOs and client enterprises;
- increasing project sophistication to include more multiple-technology, integrated projects;
- development of new ESCO markets, including the public institutional sector; and
- improving means for ESCOs to access formal financing sources.

CONTRACTS

Energy performance contracts in China are generally classified into three types, or "modes." Although the characteristics of each mode are similar in many respects to those bearing the same English language name, there are differences. Therefore, the Chinese categorization cannot be used interchangeably with categorizations in other countries. In all cases, ESCOs undertake detailed project design, manage most project implementation aspects and guarantee energy savings performance. However, financing, contract and asset ownership arrangements vary.

Shared Savings Contracts

In this mode, ESCOs provide the bulk of project financing and are compensated for their investment and services by their client from a portion of the energy cost savings resulting from the project. The assets created by the project are owned by the ESCO until contract completion, when they are transferred to the client, usually for no charge. The minimum energy cost savings stream from the project is estimated by the ESCO in the contract, usually conservatively, and acknowledged by the client. In most cases, contracts provide for payment streams to the ESCOs based on a decided percentage share of the agreed estimated minimum cost savings scenario, as long as project savings monitoring arrangements verify that at least the agreed level of energy savings has materialized with normal asset operation. Any additional savings are usually 'given' to the client. As long as the project is implemented with the basic results originally expected, these contracts typically result in a predictable payment stream. Although there are cases where payment streams vary every payment period, based on ongoing measurements of actual savings during the contract period, such cases are in the minority. Hence, most Chinese shared savings contracts are actually not the same as the traditional 'shared savings' contracts as defined in North America—they are probably closer in principle to the "ESCO-financed guaranteed energy savings contracts" typically used for federal government Energy Savings Performance contracts in the US.

Shared savings contracts currently are the only EPC mode recognized for the government's new EPC financial incentives. Hence, this mode of operations is likely to further expand and probably will become more standardized.

Guaranteed Energy Savings Contracts

In these contracts, clients provide the bulk of project financing themselves. Assets generated belong to the client. In addition to design and implementation services, ESCOs guarantee the energy savings levels from the project. To be considered proper EPC, failure to achieve the guaranteed energy savings amounts must have direct financing consequences to the ESCO.

Outsourcing Contracts

In this mode, ESCOs finance and develop energy savings assets within the client's facilities, and operate these assets over an extended period for agreed compensation, which is linked in one way or another to the energy savings achieved. The ESCO owns the assets and transfers them to the client at the end of the contract, which may last eight to ten years. One common example is the development of on-site "BOT" power generating facilities using waste heat or by-product gas from the plant. The ESCO erects and operates the plant, purchasing the energy resource for a small fee or no charge, and selling the electricity to the plant at a rate well below the plant's purchase price from the grid. Another case is where ESCOs develop or purchase local district heating assets, undertake energy efficiency renovations, operate the system and receive remuneration from the larger difference between heat sales revenue and fuel costs. In a final example, an ESCO develops, purchases or leases the lighting and/or space conditioning assets of a building, undertakes energy efficiency renovations, operates the systems, pays the building's electricity bills and charges the building owner or occupant fees for predefined lighting and/or space conditioning services, at costs lower than before the ESCO's involvement. This model is similar to chauffage practiced in other countries.

ESCOs and their clients have developed many variations on these modes, different types of financing arrangements and different types of risk-sharing regimes. Companies with leasing licenses are beginning to offer financial leasing contracts developed around energy efficiency projects. Some companies are beginning to look at the development of special purpose companies for large projects involving several ESCOs working together.

MARKETS

To date, Chinese ESCOs have developed their business primarily with corporate clients—especially industrial companies, but also commercial buildings, heat supply companies, residential building heat service companies and others. This is quite different from the ESCO industry in the US, for example, where public institutional clients (government facilities, schools, universities, hospitals) dominate the business. In China, the EPC business with many public institutional clients was stymied until 2010 by lack of enabling regulations. With new regulations now coming into force, however, this market will most likely open up in China.

Among EMCA-member ESCOs, about one half of the energy performance contract projects undertaken during 2007-2009 were in the industrial sector, while the remainder was primarily in the building sector. However, industrial projects averaged USD 1.6 million each, while building projects averaged USD 500,000 each. Therefore, industrial projects accounted for almost three-quarters of the total energy performance contract investment.

Over the last decade, Chinese ESCO projects have tended to be relatively simple—usually focusing on the application of a single technology, involving simple energy savings measurement and verification provisions, and often concluding in three to four years.

In recent years, Chinese-style shared savings contracts have accounted for about two-thirds of total energy performance contracts. Generally speaking, most Chinese clients prefer shared savings contracts if they can be provided at reasonable cost because the ESCO provides the financing off the client's balance sheet. This has been one the greatest attractions of industrial clients to EPC. Clients neither need to provide most of the upfront funds nor pay the ESCO if the energy efficiency guarantees of the project are not met. At the end of the contract period—usually three years or less—the client will receive the assets and future continued energy savings at no charge.

Dealing with predominantly commercial clients, the biggest risk for Chinese ESCOs is the risk of delays or failures in client payments of amounts due in contracts. Minimizing this risk is the single most important factor for decisions on client selection, project scope and technology as well as contract duration.

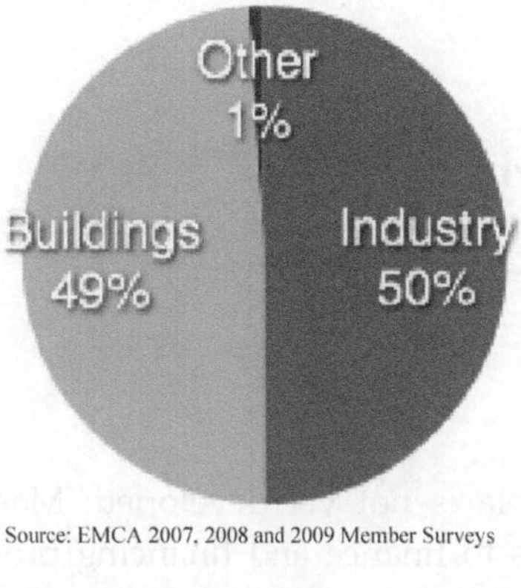

Source: EMCA 2007, 2008 and 2009 Member Surveys

Figure 9-5: Energy Performance Contracting Projects by Sector, 2007-9

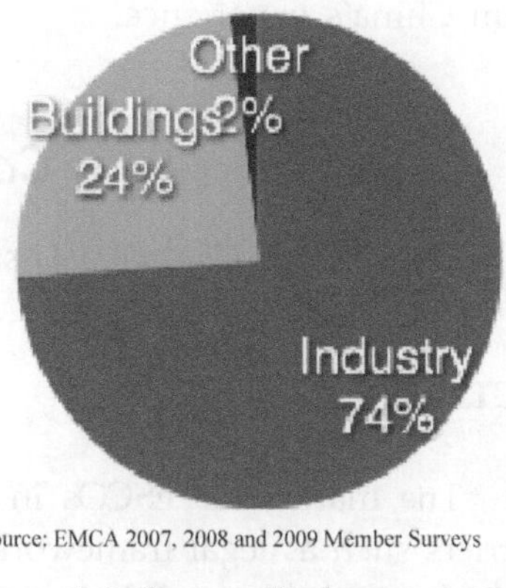

Source: EMCA 2007, 2008 and 2009 Member Surveys

Figure 9-6: Energy Performance Contracting Investment by Sector, 2007-9

FACILITATORS

Now in its eighth year of operation, EMCA has played a key role as a focal point for promoting the EPC concept to all parts of society, researching key problems and fostering solutions as well as assisting its members to expand business. EMCA's membership had increased to 560 members by the end of 2010, of which 428 have implemented EPC projects. The association has become a focal point for exchange with other countries on ESCO development.[4]

GOVERNMENT ACTIONS

The stable, strategic and steady support of the Chinese government for the development of the ESCO industry has been a major factor in the successful launch and rapid growth of the industry in China. Key areas of support have included (i) very active support for the initial piloting, development of cooperation with international donors for targeted support; (ii) assistance in legitimizing the industry in the eyes of the public; (iii) assistance in resolving various legal and taxation issues; and (iv) most recently, new policies providing incentives and favorable tax treatment. Especially in the Chinese context, it is highly unlikely that the achievements made to date could have been made without such support. The successful assistance of the government in fostering the development of the industry may be an area where other governments interested in promoting EPC might learn from China's experience.

Colombia

A. Gonzalez Hassig and C.A. Alvarez Diaz

ACTIVITIES

The market for ESCOs in Colombia is not yet developed. Many barriers such as legal framework, access to finance and financing products, demand-side confidence in local suppliers and lack of culture are all factors. A brief discussion is provided on the market history, key role players and actors, the legal framework and recent activities in the sector.

In the mid-1990s, the energy industry in Colombia went through major changes and restructuring which led the way to a more competitive sector with increased private participation. The primary objective was to increase the efficiency and competiveness of productive activities, which will eventually benefit industry, population life quality and economic growth.

These reforms in the energy sector along with economic market liberalization have provided the right environment for new actors and techniques, which have enabled energy users to select the best energy supply and sources, new conversion technologies and better industrial processes to lower energy consumption.

The use of these energy savings alternatives enabled an energy service market to emerge. This market is composed of a wide range of technical, financial and commercial services and products. This has allowed Colombian companies to adapt to a global economic reality.

In a country with such abundant natural and energy resources as Colombia, there is not enough attention to the economic opportunities presented by potential energy savings in industrial processes. The problem is compounded by the aggravating factor that industry's growth has been limited by social and political instability. Only in recent years has the political context changed, allowing envisioning an increase in industrial production.

Confronting the major industry changes and the existence of great energy-saving opportunities not yet taken, it is evident that there are certain barriers holding up the offer and demand of such services.

Energy service providers (ESPs) in Colombia are categorized as follows:

- energy marketers;
- power generators-IPP;
- power distributors;
- equipment suppliers;
- engineering consulting firms (with energy audits) and universities (consulting services and applied research);
- financial institutions (with complementary services).

Among these categories, just a couple of privately owned companies are operating under the ESCO model executing EPC projects in Colombia.

Since 2009, a program in Bogotá designed to promote energy

efficiency among industrial SMEs has been executed by the Bogota Chamber of Commerce, where more than 160 ESPs showed interest in EPC and in how to overcome market barriers.

CONTRACTS

It is necessary to differentiate ESCOs from the ESP market. Before 2002, ESCOs had no activity at all. In recent years, the number of ESPs has grown, leveraged by technology commercialization. The number of ESPs is close to 170 where 15 are the most relevant.

The demand side of energy services does not require long-term consulting and advisory services but instead short-term work agreements based on planned objectives.

Among the different types of performance contracts used in Colombia, the most commonly used is the "chauffage" type among IPP and steam, power and cogeneration equipment suppliers.

LEGAL FRAMEWORK

In 2001, an Efficient Energy Use and Renewable Energy law was issued to encourage education and research in these subjects, but not supporting or encouraging the development of these sources. Moreover, there is no political or regulatory support that enables investments or well-established goals to achieve a higher mix of renewable energy or energy efficiency increases. These limitations imply a legal gap for ESCO market development.

The 2001 legal act provided for the establishment of an Efficient Energy Use Committee (CIURE) which would define the policy guidelines for the development of the subsector. This committee belongs to the Ministry of Mines and Energy, the Ministry of Industry and Commerce, the IPSE, the CREG, the Mines and Energy Planning Unit (UPME by its acronym in Spanish), COLCIENCIAS and DNP. The Ministry of Transportation is also invited regularly.

Apart from the political aspects, there have been strong advances in technical standardization, labeling, biofuel market development and incandescent lighting substitution encouraged by legal acts.

In 2010, an efficient energy use indicative action plan (PROURE) was launched. This plan was intended to set energy savings targets by sector, raise awareness, strengthen institutional efforts and boost the energy efficiency market. However, without the proper tax incentives and the availability of capital resources, the targets will not be achieved.

MARKETS

In 2001, a study by UPME valued the energy services market in different sectors. This study took into consideration not only the energy demand size but also the investments required to achieve different energy saving objectives. Table 9-4 illustrates the market size.

Table 9-4: Energy Services Investment by Economic Sector

SECTOR	*USD million/year*
Industrial	210
Residential	109
Transport	123
Services	13
Total	572

As mentioned, there are different identified barriers to the full development of the ESCO market, as described in Table 9-5.

Table 9-5: Barriers to ESCO Market Development

Aspect	*Barrier*
Technical	-Technical deficiency of ESPs -Low price competence
Economic	-Instability in energy prices -Lack of financing mechanisms
Demand Side	-Client economic weakness -Lack of Culture -Neglecting to show processes
Local Context	-Political, social and economic uncertainty -Low continuity in institutional programs and planning -Lack of incentives -Excessive regulations and prices -Vague and burdensome tax regulations

FACILITATORS

UPME is an entity under the Ministry of Mines and Energy that has the function of comprehensive planning in the energy sector. UPME is the technical branch of the Ministry. UPME within its organization has an area of rational and efficient energy use and unconventional energy sources that supports with technical studies the planning of the subsector.

There is no public agency solely in charge of energy efficiency or renewable development, although recently, the ministry has expressed the need for such an institution.

The Colombian Energy Efficiency Council was created in 2010 to promote knowledge, awareness, rationality and improvement of technologies and processes that increase energy efficiency in Colombia. Its members are ESP private companies and consultants.

On the other hand, many capital funds, development agencies and multilateral banks have increased their presence in Colombia in recent years. These have brought forward different programs to overcome technical barriers, increase financing access and promote the ESCO business. Some of them are mentioned in Table 9-6.

Table 9-6: Presence of Capital Funds, Development Agencies and Multilateral Banks in Colombia

Institution	*Program*	*Activities*
IADB	OPEN	EE Promotion
IADB	Green Pyme	EE Promotion
IADB	IDEAS	Financing
IICA	AEA	Financing
IFC		Training
USAID	CCEP	EE promotion + financing
USTDA	Clean Energy	EE promotion + financing

GOVERNMENT ACTIONS

In countries with a fully developed ESCO market, ESPs have the opportunity to develop EPC in public buildings and lighting projects. In that way, the government becomes a facilitator for market development. This is not the case in Colombia, where government budgets cannot be obligated for more than the legal period, thus limiting the implementation of EPC in this sector. Furthermore, the economic savings produced from energy efficiency implementation cannot be used to pay the ESCO that makes the investment, thereby discouraging EPC with government and local authorities.

To encourage the ESCO market, the Colombian national government should lead by example and adapt regulations to allow the provision of energy efficiency in public buildings, which would result in lower power consumption and in more resources available for investments in facility renewal.

There is no access to financial resources to implement energy efficiency projects or ESCOs. Industries depend strictly on private banks to finance their growth and renewal of equipment.

The country is steadily growing and will continue to improve economically at high rates in the short term. This in turn will require the implementation of energy efficiency to seek business competitiveness locally and globally but, more importantly, to ensure a sustainable future.

Colombia will need to prepare a large number of professionals by increasing its academic capabilities in this subsector, promoting professional studies abroad and creating energy efficiency programs locally.

Croatia

Jasmina Fanjek

ACTIVITIES

Croatia launched the ESCO approach in 2003 under the initiative of the World Bank and the GEF with its first ESCO company, HEP ESCO Ltd. The importance and contributions of the company were huge, as it introduced energy efficiency in general in Croatia at a time when such a concept was at a very low level.

While to date energy efficiency has made remarkable progress at the legal, financial and administrative levels, nothing new has happened unfortunately regarding the establishment of new ESCOs and the removal of barriers to the widespread use of the ESCO model. There are only two active ESCOs in Croatia, which is the same situation as in 2003 when the first ESCO (HEP ESCO Ltd.) was launched along with private ESCO EETEK Ltd.

HEP ESCO Ltd. is a utility-based, state-owned company, which is 100 percent owned by the Croatian national electricity company. It was established on the initiative of the World Bank supported by the GEF grant and IBRD loan. The reason for such support was a high potential countered by very poor awareness for energy efficiency. The aim was to create the energy efficiency market and increase awareness for energy efficiency, especially in terms of the ESCO model. The other objective was to create an ESCO company that would show that the ESCO business could survive on the market while being profitable. The GEF financial support was used to increase international know-how and interest of Croatian financial institutions in financing energy efficiency projects.

Things started with simple projects such as lighting system retrofits in schools with very small investments. Subsequently, HEP ESCO built up expertise for the completion of comprehensive energy efficiency projects including cogeneration on biomass for a wood factory, which previously had to produce its own electricity.

At the end of 2010, the World Bank and GEF project was successfully finished and today HEP ESCO Ltd. has established a team of experts required for the preparation, financing and implementation of complex energy projects. Because of the changes in the budget law which almost entirely closed the public sector as an ESCO market, HEP ESCO had to give up projects that had already been prepared and focus on the private sector as the new market. It resulted in a significant number of new projects under preparation, which needed time to reach the implementation phase. As a result, the current gap of the projects under implementation is temporary and the potential of the new market does exist.

Private ESCO EETEK Ltd, a subsidiary of a Hungarian company which also provides energy services, is the second active ESCO on the Croatian market. Its first energy efficiency project in an industrial private company, EETEK, finished in 2004. Unfortunately, although the project was well done, with good results of implementation of several measures and a good payback period, EETEK was faced with financial risks (as is the case with most ESCOs). After financial problems, the client went bankrupt and EETEK lost the money it had invested in the project. Subsequently, EETEK was present on the Croatian market but without significant efforts and success in energy efficiency.

HEP ESCO is the leading and only really active ESCO in Croatia. It has invested USD 24 million in energy efficiency projects between 2003 and 2010. As depicted in Figure 9-7, 83 percent was invested in the public sector and 17 percent in the private sector, respectively.

From the total amount invested in the building sector, which mostly includes hospitals and schools, USD 11 million was invested (47%) along with USD 9 million in public lighting (37%) and USD 4 million in industry with cogeneration projects (16%) (Figure 9-8).

According to the data shown above, it is obvious that the public sector was the key market for HEP ESCO and for ESCOs in general. It also shows that there is a potential for the private sector.

Investments up to the end of 2010 resulted in yearly savings in the amount of USD 3 million and the average payback period was 7.5 years. Figure 9-9 shows how these savings are distributed by sector.

The same investments resulted in CO_2 emissions reductions in the amount of 11,000 tons/year. This amount is almost equally distributed among the main sectors of activity. Looking at the life period of the installed equipment, this amount cannot be neglected (Figure 9-10).

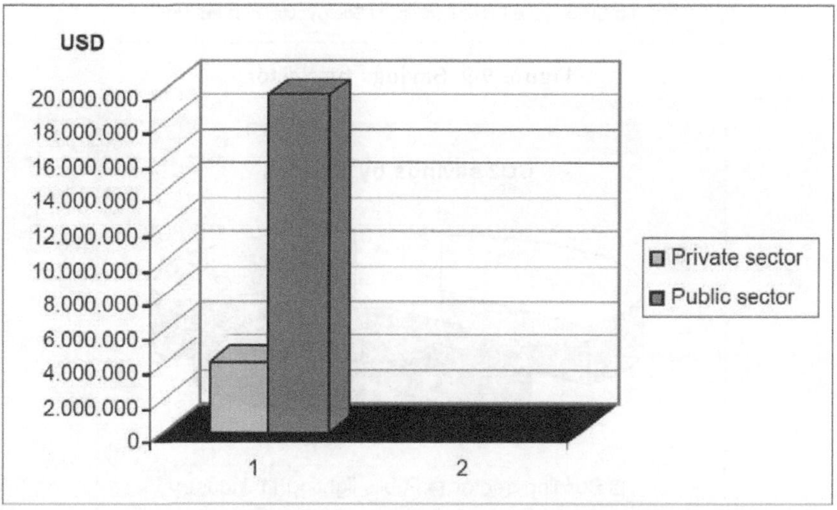

Figure 9-7: Investments in the Public and Private Sectors

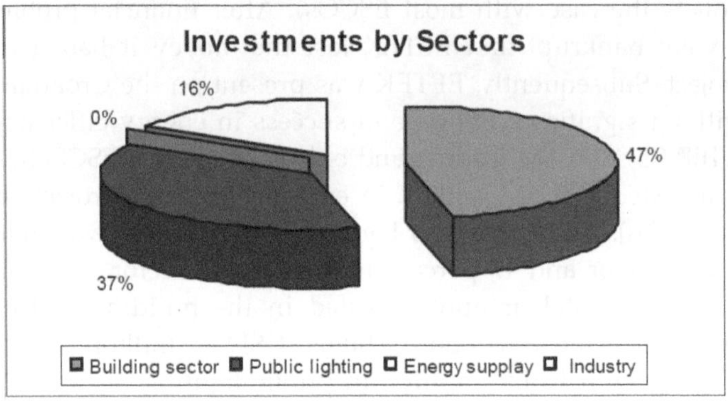

Figure 9-8: Investments by Sector

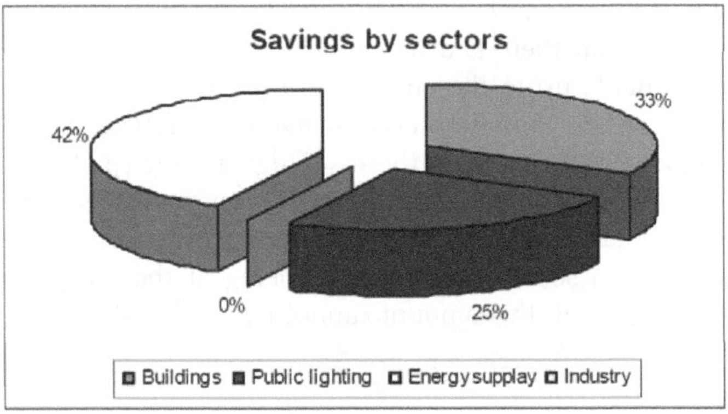

Figure 9-9: Savings by Sector

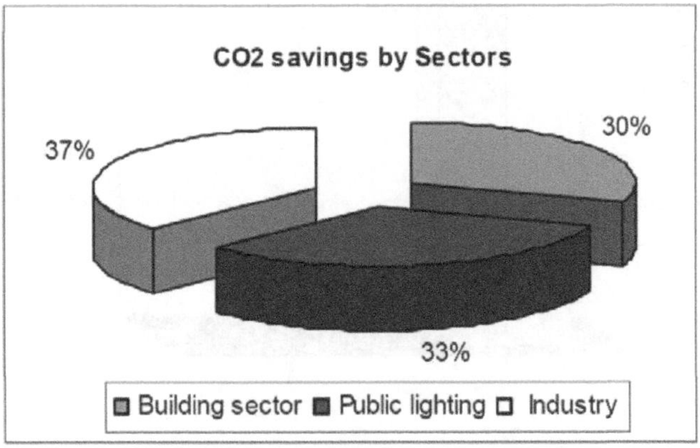

Figure 9-10: CO_2 Savings by Sector

MARKETS

Potential investments in energy efficiency in significant facilities based on a 2002 market study and projections for other sectors that were not included in the study are in the range of USD 1,600-2,400 million out of which, with very pessimistic projections, USD 1,000-1,500 million could be implemented through ESCOs.

These investments include institutional, commercial and residential buildings, municipal street lighting and water pumping, district heating distribution systems, cogeneration at end-user facilities. Investments exclude communities smaller than 40,000 people and small energy users that might need less than USD 25,000 in retrofit investments as they are not of interest to the ESCO market.

As a result, the potential for ESCO investments is huge in both the private and public sectors. However, the main problem and barrier which could and will appear (similar in all countries in transition and in the region) is the poor condition of the facilities that need investments for energy efficiency purposes. As the implementation of energy efficiency measures and investments in energy efficiency cannot be separated from other necessary investments for the successful development of energy efficiency projects, the question is—who will pay for these extra costs? This question always arises but, at present, the recession and the serious lack of funds for the ESCO model are real threats.

GOVERNMENT ACTIONS

Although energy efficiency plays an important role in Croatian energy policy, barriers to ESCO model implementation have not been removed yet. Among the barriers which ESCOs face all around the world, we will focus on explaining two of them.

The first barrier lies in the fact that the ESCO model is not recognized by government authorities as a business model providing services, but rather as a typical contract for delivering goods.

On January 1, 2009, the new budget law entered into force, which provides that budgetary users and municipalities can go into debt taking credit and loans (loans do not mean only money and also include other lenders, not just banks). The Ministry of Finance issued the opinion that contracts signed with ESCOs were commodity loans and should be treated as debt. Consequently, according to HEP ESCO's experience (shown in Figure 9-11), such treatment has a negative influence on the ESCO market bringing ESCO projects in competition with real investments needed to be done by budgetary users and municipalities, such as civil construction of new schools, hospitals, roads, public lighting systems or expansion of the latter. With such investments, energy efficiency could not compete in countries such as Croatia where the need for new facilities and institutions is crucial for normal life.

The second barrier is the public procurement law. Among the barriers, which this law brings to the ESCO model, we would like to place emphasis on the one which was experienced in Croatia with respect to the situation when multilateral institutions and some other financial institutions provided funds to ESCOs for financing ESCO projects. Such institutions demand public procurement procedures in using their funds. In such a case, two public procurement procedures appear. The first one is when the client hires an ESCO company through public procurement for the performance of an ESCO project and the second one is when an ESCO has to procure goods and works through public procurement to implement the project. This situation

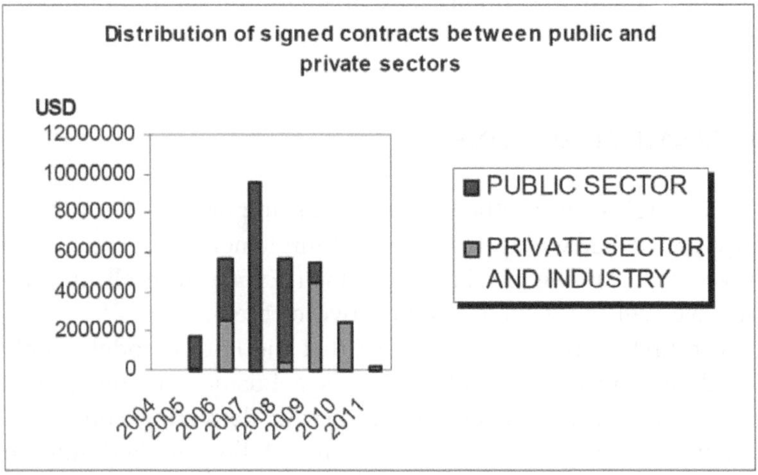

Figure 9-11: Distribution of Signed Contracts in the Public and Private Sectors Including Industry

actually disables ESCOs' ability to compete because it is difficult to establish the price to offer.

For utility-based ESCOs that can access such funds, it is another barrier to public sector projects.

It is proven that energy efficiency is an interdisciplinary category, so it should be clearly defined which organization is responsible for it. While energy efficiency must be regulated by just one entity, cooperation among the different ministries, agencies and other relevant parties must be assured. Government support and understanding of the issue is crucial. Good legislation which will take into account not one but all complex aspects of energy efficiency will give the results and the framework required for more widely using the ESCO model. Lack of understanding and cooperation on the part of key stakeholders and negligence of qualified experts can bring new barriers to the surface. The Croatian experience shows how government decisions and legislations can support, but also discourage good practice in ESCOs.

Czech Republic

Ivo Slavotínek

ACTIVITIES

The Czech ESCO market was initiated in the beginning of the 1990s. The first company, projects and general EPC know-how have been developed under the influence of US ESCOs. Therefore, the Czech EPC market is specific and differs from "western" European states like Germany, Austria and France.

Since its beginnings, the Czech Republic market has undergone a long evolution and today can be characterized as moderately developed. Complexity of completed projects, quality of services provided, project implementation procedures and structure of contracts and other related documentation are comparable, compatible and competitive with energy services available in other EU member states.

Nevertheless, the number of projects is still limited and clear rules for EPC implementation in the public sector, namely for state budgetary institutions, are missing.

Most of the projects have been completed in the public sector (specifically in education and health care facilities). About 150 to 200 EPC projects have already been realized, representing savings of approximately 800 TJ. Around one third of these projects have been in the private sector.

The EPC concept was introduced in 1994-1995 thanks to a EUR 3 million (USD 3.9 million) investment in the energy efficiency improvement of public healthcare institutions. Non-regulated, unsaturated and hungry markets looking for new approaches, solutions and technologies allowed for the rapid rise of EPC in both public and private sectors.

The dynamic start of EPC was followed by a deep recession in the late 1990s due to the instability and volatility of the local economy. Among the main factors were massive privatization, transition of the economy, convertibility of local currency, banking sector development and continuous changes of the business environment and legislation.

Until 2001, ESCO development was slow due to numerous barriers and obstacles. New hope was brought to the Czech market through a new energy efficiency law (obligatory energy audits for public and private entities) but targets under the law were not met.

Even if this decision looked like a strong push for energy efficiency investments, the final impact on EPC market development was negligible. All government legislative actions including the State Energy Policy adopted in 2004, the approved National Program for Energy Effective Management recognizing EPC, and the implementation of EU directives have not really promoted the ESCO business or opened public or government institutions for private ESCOs.

Since 2005, the market for energy services has experienced noticeable development. The number and size of completed projects are gradually increasing. The main reasons are the financial crisis and rising investment restrictions on public budgets.

The energy efficiency services target all energy utilities, including water. The majority of energy conservation measures include HVAC systems, electricity, heat and cooling generation equipment, control equipment modernization and expansion, efficient lighting, variable-speed drives, heat pumps, etc.

Czech ESCOs and energy efficiency facilitators are European leaders in their fields, as evidenced by a number of prestigious awards granted by the European Commission. For instance, SEVEn became the "Best European Energy Service Promoter of the year 2005," ENESA was recognized as "Best European Energy Service Provider of the year 2009" and Siemens was given an award for "Best European Energy Service Project in the Commercial Sector in the year 2009."

Currently about 8-10 active ESCOs operate on the Czech market offering EPC agreements or other types of energy efficiency services.

There are also approximately a dozen companies providing long-term energy delivery contracts (EC). In addition, another group of energy companies offer other types of energy service contracts, whereby savings are not guaranteed (energy consultancy, energy audits, installation of energy-efficient equipment).

ESCOs have both national and international ownership. The size, background and character of ESCOs are much diversified

In general, ESCOs offering EPC can be roughly divided into four groups:

- Independent ESCOs dedicated mainly to implementing EPC projects (ENESA, SUE);
- Subsidiaries, divisions or branches of energy equipment manufacturing companies (SIEMENS);
- Subsidiaries, divisions or branches of energy utilities (DALKIA, MVV Energie CZ, MARTIA);
- Subsidiaries, divisions or branches of facility management or engineering companies (AB Facility).

Mainly EPC remains a side business for the aforementioned subsidiaries of national or international corporations. Only a minority of local ESCOs are fully dedicated to providing EPC projects. The list of local ESCOs and completed projects are available at www.epc-ec.cz. The number of ESCOs has been increasing in the last years.

FINANCING

Long-term project financing is available in different forms and in reasonable conditions. In the public sector, the ESCO almost always provides project financing. The majority of public projects are financed through assignment of ESCO receivables to the local commercial banks. The rest of the projects are financed through TPF with a local commercial bank lending to an ESCO or through an ESCO's own corporate funding. Projects in industry or for private clients are financed through TPF with local banks lending directly to clients or through client funds.

CONTRACTS

Energy services are offered in different forms. The main concepts available for energy services are EPC and Energy Contracting (EC). Competition between EPC, EC, facility management and outsourcing type of projects still continue. However, principles of these types of schemes are clear and customers choose the one which is most suitable for their needs. EC agreements are more focused on energy supply modernization including DH systems, cogeneration, boilers etc. The BOOT contract scheme is also applied, primarily in mandates involving cogeneration and renewable energy projects.

Unlike EC, the EPC market is more structured. A standard contract for EPC has been developed by a group of ESCOs and independent consultants as well as legal and tax experts. This contract is now used in most public tenders. The adoption of the standardized contract lasted for years. Individually negotiated contracts via Eurocontract was adopted locally and became a widely accepted contract model.

The final version of the contract is, at the end, negotiated between the ESCO and the client usually under supervision of an independent consultant.

The most common contracting form is guaranteed savings. Under this contract, the ESCO guarantees a volume of savings achieved either in physical units (fixed prices) or in money. (A Czech contract that guarantees money savings does not fit the typical guaranteed savings model.) Excess savings (the savings over the guaranteed amount) are usually split between the ESCO and the customer according to a negotiated ratio.

A typical contract in the public sector lasts between 8 and 10 years. The longest contracts have been signed for 12 years.

Contracts in the private sector are much shorter, mostly because owners insist on a short payback of implemented conservation measures and, obviously, due to higher business and financial risks.

Recently, more attention has been given to the measurement and verification of savings. A task of the PERMANENT project sponsored by the EU is adjusting the IPMVP to the local environment and providing training for energy specialists and ESCOs.

LEGAL FRAMEWORK

State support to ESCOs and EPC is very limited, while energy efficiency is not a priority for the Czech government. Existing national legislation and documents forming energy policy and the business environment include:

- State Energy Policy;
- National Energy Efficiency Action Plan;
- Energy Act;
- Energy Management Act; and
- Act on Renewable Energy Sources.

None of the documents listed above specifies support to ESCOs and EPC.

Otherwise, the lack of state support to energy services is also visible from the fact that the Czech Energy Agency (CEA—the agency dedicated mainly to promoting energy efficiency and services) was closed by the end of 2007. Part of the agenda was moved to the Ministry of Industry and Trade. In any case, the role of the CEA had been limited over the last years of its existence.

The State Energy Policy (SEP), adopted in 2004, gave a general framework to "reliable and permanently safe supplies of energy at acceptable prices and for creating conditions for its efficient use that will not threaten the environment and will comply with the principles of sustainable development." It pointed out energy efficiency and use of renewable energy sources as the major ways of achieving these targets. Nevertheless, such statements have had a declaratory form only.

The National Energy Efficiency Action Plan (NEEAP) was prepared on the basis of the aforementioned Directive 2006/32/EC on energy end-use efficiency and energy services and is a main strategic document describing how to achieve the nine percent savings target. A similar document is prepared for the target on RES. The adoption of the Directive has not created any incentives for the development of new energy services and ESCOs.

The general legal framework governing the business environment, such as the Business Code, tax and accounting legislation, the public procurement law and other related legislation, does not limit the ESCO business.

The relatively stable legal framework, law enforcement, the availability of financing, the existing infrastructure and the status of the country's economy allow for further development of the ESCO business. Financial institutions, including most of the local banks, are ready to provide long-term financing for energy efficiency projects, particularly in the public sector.

MARKETS

At the beginning of the 1990s, EPC projects were equally spread over the public and private sectors. Now, the target market for ESCO projects remains the public sector, which is estimated to cover about 80 percent of the current market value. Projects in the public sector are mainly implemented in schools, hospitals, social care facilities as well as in culture and administrative buildings.

Industry accounts for ten percent of the market. Private non-residential buildings and district heating represent the remaining volume of the market. Recently, sophisticated and comprehensive EPC projects were implemented in cultural facilities of Prague (National Theatre, State Opera and Estate Theatre) as well as in several mental hospitals, 40 facilities of the Pardubice region and 31 schools of the Prague 13 District.

National Theatre Prague

Historically, the most active in EPC implementation have been middle-sized cities. In recent years, regional governments discovered the attractiveness of guaranteed energy services. The most resistant are still state institutions due to the existing barriers and resistance of the Ministry of Finance.

Apart from the public sector, the private sector (typically industry) is also on track regarding energy efficiency investments. However, the private sector is very specific not only for higher financial and business risks but also because ESCOs usually compete with energy efficiency equipment suppliers and face internal corporate rules that do not correspond to project implementation procedures specific to EPC.

The technical potential for savings in the Czech Republic reaches up to 300-400 PJ (35 50 percent of the final energy consumption or, in other units, around 80-110 TWh). The economic potential for savings is around 180-200 PJ. The achievable potential is then about half of this estimate. Looking closer at the potential for EPC, the highest potential lies in industry (about 100 PJ), but it is questionable to what extent this could realistically be achieved. The economic potential in other sectors (school, health care, administration, housing, etc.) is about 80 PJ. Typical average EPC projects achieve savings of about 5 TJ, so we are speaking about thousands of projects. It is clear that strong potential for EPC does exist.

In other words, the potential for energy savings through EPC, which are economically attractive, is about EUR 100 million (USD 131 million). Available estimations of the market potential vary, but they are in the range of EUR 10-15 million/year (USD 13-20 million/year). The current size of the ESCO market fluctuates between EUR 4-10 million (USD 5-13 million) annually.

The ESCO market has been growing in the last five years. The main attributes for such growth are marketing activities of local ESCOs and EPC promoters, rising energy prices and budget restrictions in the public sector. High operational costs, obsolete energy infrastructures, lack of experience in how to improve energy efficiency (and achieve energy cost reductions), lack of internal financing and increasing freedom of building and facility managers in renovation and construction decisions has also increased the interest in EPC.

FACILITATORS

The local market could not work efficiently without EPC promoters. The history of EPC is linked to SEVEn (the Energy Efficiency Centre) and also ENVIROS. New consultancy companies keen on offering similar services are appearing. EPC facilitators play a key role in government, public institution and customer education as well as in EPC marketing and public procurement management.

A working group supporting EPC development was established under an umbrella of the Chamber of Commerce with the purpose of exchanging experience and designing activities to tackle barriers in order to cultivate the EPC market. The working group reviews project implementation rules, prepares standard contracts and follows tendering procedures.

The ESCO community has not been organized until recently. Due to rapid changes in the market and the need to formulate and advocate the interest of local ESCOs, the Czech Association of Energy Service Companies (APES) was founded in autumn 2010.

GOVERNMENT ACTIONS

National programs promoting energy conservation have been continuously shrinking. Apart from EU-related programs, the state main national program on energy efficiency administrated by the Ministry of Industry and Trade is called EFEKT, which supports efficient use of energy and renewable energy sources. EFEKT's annual budget is about CZK 30 million/year (EUR 1.15 million) (USD 1.5 million).

There are two main EU programs on energy efficiency: operational program ECO-Energy and operational program Environment. Under the structural funds, energy efficiency projects (together with renewable energy sources) are to be supported in the budgetary period of 2007-2013. The major claimants are municipalities, although the private sector can also apply.

EU funding in the 2007-2013 period had a negative impact on EPC, until it was clear that using EPC did not disqualify the applicant from being granted the subsidy. Additionally, most of the public institutions wait for EU grants and put all energy efficiency projects, including EPC, on hold. Particularly, this happened in 2007 when new

EU funds were opened.

In general, the main bottleneck for higher deployment of EPC in the public sector is the lack of clear rules on how to apply EPC (registration and approval of the project, accounting about the project, etc.). These unclear rules definitely discourage many government institutions, particularly budgetary ones (military, police department, courts, prisons, etc.).

Despite the interest of several ministries in opening their institutions for EPC and intensive ESCO efforts, such restrictions given by the State Budget Act still exist. The Ministry of Finance prevents private business entrance to these entities.

Table 9-7: Summary of Basic Data on the Czech ESCO Market

Number of ESCOs	8-10 (EPC and guaranteed saving project providers)
Size of the Market	EUR 4-10 million annually (USD 5-13 million)
ESCO Association	APES (Czech Association of Energy Service Companies) since 2010
Types of ESCOs	Mix the independent ESCOs, engineering companies, subsidiaries of equipment manufacturers, energy utilities and facility management companies
Market Development	Stable with gradual increase in recent years
Sector for ESCO Projects	Dominantly public sector

Denmark

Jesper Ole Jensen
Susanne Balslev Nielsen and Jesper Rohr Hansen

ACTIVITIES

Until a few years ago, ESCO contracting in Denmark was very limited. There was a small market in the industry with only one supplier for the production industry (Energibranchen, 2008). However, in recent years, the ESCO market in Denmark has increased dramatically. Due to an increase in municipal ESCO projects, growth has been rapid compared to other EU countries (Marino et al., 2010).

In 2008, a handful of municipalities started ESCO projects in municipal buildings. 10 municipalities (of 98 municipalities in Denmark) have presently signed ESCO contracts, and at least another 10 municipalities are preparing to do so.

The first Danish municipality to sign an ESCO contract was the municipality of Kalundborg in 2006. It experienced delays, but restarted in 2009. Meanwhile, two other municipalities, Gribskov and Middelfart, had started ESCO projects. These three municipalities formed their own 'ESCO network' (ESCOmmuner). Close collaboration and knowledge sharing in the initial phases lead to several ideas on how the ESCO concept could be disseminated to other municipalities.

In 2009 and 2010, several other municipalities (at the moment 10 municipalities) signed ESCO contracts. The municipalities are at different stages; some are in the initial phase of preparing a tender, whereas others have finished retrofitting and have entered the operational phase.

In state-owned buildings, ESCO projects are emerging. The first ESCO project in ministry buildings at the buildings of the Danish Nature Agency was signed in 2011 with an investment of EUR 267,000 (USD 351,000) in 40 buildings designed to reduce energy use by 25 percent in 2013. In 2011, three ESCO projects in university buildings were sent to tender, and other projects are in preparation.

In social housing, the ESCO concept has a high priority for the Danish government, but in practice the initiatives have been few. The only project to date concerns the housing association Albertslund Nord, where a four-year ESCO contract has been signed with a guarantee to save 15 percent of the present energy consumption. While social housing is sometimes included in municipal ESCO projects, several uncertainties (financing, decision making, etc.) cause social housing to be generally avoided. it is hoped that the Albertslund Nord project will be a pathbreaker for introducing ESCO contracting in the broader social housing sector.

A large energy supplier, DONG (Danish Oil and Natural Gas), has developed an energy-saving scheme for house owners (DONG CleanTech) with consulting, loans and implementation of heat pumps, insulation and new windows in partnership with product suppliers. The energy suppliers have, as part of the National Energy Strategy from 2008, been instructed to save 6.1 PJ per year among their end-users, for the period 2010-2012. This motivates them to join partnerships with municipalities promoting energy retrofitting to local house owners, in partnerships also with local financial institutions and local enterprises. Although such initiatives are often labeled as 'ESCO light' these projects have little resemblance to the original ESCO concept, as no guarantees are involved and the private finance is provided only as loans.

Many municipalities have considered whether to carry out retrofitting of municipal buildings as an ESCO project or as an in-house project. A main reason for the municipality to enter into an ESCO contract is the possibility of financing the improvements of many buildings over a short time, and deliver instant energy savings. With in-house renovations, there is typically only room for gradual improvements, due to limitations in municipal budget and facilities management staff.

In Figure 9-13, the ESCO approach depicts a short analysis and implementation phase, meaning that the energy savings are reached relatively soon, although some adjustments in the operation period might be necessary to reach the guaranteed savings. The in-house approach normally requires a longer analysis period due to limited muni-

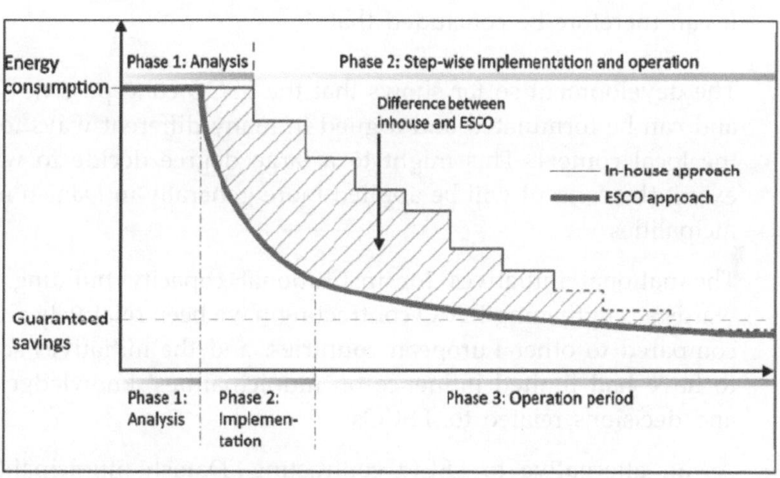

Figure 9-13: The Principal Difference between Energy Retrofitting with the Esco and In-house Approach

cipal staff and that the energy retrofitting is normally financed through annual budgets, making it a stepwise process where energy savings are reached gradually. This also makes the long-term goal more uncertain, in contrast to the ESCO approach where the end goal is defined in the contract between the municipality and the ESCO provider.

However, the ESCO model is often perceived by the municipalities as being too expensive, compared to an in-house approach. A survey from 2010 showed that 82 percent of municipal directors claim that in terms of economy, it is better to finance the improvements in other ways than as an ESCO contract. For municipalities, having signed an ESCO contract, a main argument is that the municipality does not have the administrative capacity, or the resources, to establish a volume necessary to carry out the building improvements. This is very important especially for the smaller municipalities for which it would be practically impossible to carry out such a task with an in-house approach. For the larger municipalities, choosing between the ESCO model and an in-house approach implies a question about the future role of the municipal administration: to what extent should municipal competencies concern hands-on knowledge of building management, including energy reductions, and to what extent should these competencies concern management as they relate to outsourcing, contracts and public-private partnerships? In Table 9-8 we have summarized the different characteristics of the in-house and ESCO strategies.

It can therefore be concluded that:

- The development so far shows that the ESCO concept is flexible and can be formulated and argued in many different ways to fit the local context. This might to a large degree decide to what extent the concept will be applied more generally in Danish municipalities.
- The national initiatives for institutional capacity building towards strengthening ESCO contracting have been relatively weak compared to other European countries, and the initiatives seem to have had limited influence on municipalities' knowledge of, and decisions related to, ESCOs.
- As an alternative to ESCO contracting, Danish municipalities often consider energy retrofitting with an in-house approach. Smaller municipalities have limited options for the latter solu-

Table 9-8: Main Differences between Municipal ESCO and In-house Strategy

	In-House	*ESCO*
Financing	• Step-wise renovation due to budget limitations • Long-term financing uncertain	• Guarantee for energy savings is politically attractive • Energy savings from 'day one'
Transaction	• Low transaction costs	• Transaction costs in partnership, require minimum building volume to make ESCO profitable
Capacity Building	• Keeps competencies in-house, more hands-on influence on solutions	• Learning and innovation from ESCO partnership (also depends on ESCO approach)
Fixation and Flexibility	• Flexible in relation to uncertainty on future building portfolio • Coordination between energy retrofitting and building maintenance easier	• ESCO lowers the risk of reductions in future investments in energy savings due to possible changes in political priorities

tion. As ESCOs increasingly are taken up by large municipalities, the choice between the ESCO and in-house approach might be accentuated. Therefore, a possible stagnation in ESCO contracting is not the same as a stagnation in energy retrofitting.

- It is a large challenge to develop an ESCO contract to include more ambitious renovation programs for municipal buildings, including more buildings and improvements in terms of building shell, indoor climate and renewable energy.
- ESCO contracting implies potential for public innovations, as learning taking place in the process can be exploited in different areas. We, therefore, see ESCOs not only as an option to pursue goals on energy reductions and maintenance back-log, but also as a way to develop municipal administration in more innovative directions.

CONTRACTS

The ESCO model used in Danish municipalities is similar to the European EPC model. The ESCO implements the retrofitting initiatives and guarantees a certain level of energy savings. If the reduction is not reached, the ESCO will pay the municipality the difference. If more than the guaranteed savings are reached, the municipality and the ESCO will share the surplus according to conditions defined in the contract. The ESCO contracts also typically include energy labeling of the municipal buildings and training of the municipal FM staff.

There is a large variation in the content of the ESCO contracts among the 10 municipalities, according to the number of buildings and the floor area included, as well as investments and guaranteed savings (see Table 9-9 for key figures).

Table 9-9: Variation Average in Key Figures for the Existing 10 Municipal ESCO Contracts

Number of Buildings included in Contract	*M^2 included in Contracts*	*Investments in Buildings*	*Guaranteed Savings on Existing Energy Consumption*
10-270 (av. 81)	20,000-270,000 (av. 135,000)	EUR 18/m^2 to EUR 89/m^2 (av. EUR 37/m^2	17% to 31% (av. 22%)

The goals are reached through improvements on regulation and control (introducing equipment for steering and monitoring with short payback periods) and, increasingly, physical improvements of the building shell (new windows and insulation as well as renewable energy sources—all with longer payback periods). The projects are typically divided into three stages: energy audit (incl. energy labeling), implementation and operation. Typically, the contracts give the municipality the option to cancel the collaboration after each stage. This often gives the municipal decision makers more confidence to sign the contract, as a collaboration lasting 10-12 years may be difficult to predict, especially when municipalities rarely know the ESCO supplier beforehand.

Municipal ESCO projects are different in relation to technologies, economy, ambitions etc., and subsequently in the way the process is organized. We have elsewhere identified three different approaches to

ESCO development (Jensen, Hansen & Nielsen, 2011):

- *Basic ESCO approach*: Includes replacement of installed energy systems and their regulation as well as services like CTS control, monitoring, light steering, heat regulation etc. Technologies are simple and relatively inexpensive but with high energy-saving potential and a short payback period. Only occasionally does this approach involve improvements of the building envelope.
- *Integrative ESCO approach*: Includes monitoring and regulation in combination with improvements on the building envelope and indoor climate, implementing renewable energy sources (solar panels, PVs, heat pumps, etc.). Here, the savings from the buildings with the highest savings potential are used as investments in a wider array of buildings. Such projects require major investments and have longer payback periods, but they typically include a wider segment of the municipal buildings.
- *Strategic ESCO approach*: Introducing the ESCO model might lead to new ways of thinking about the role of facility management in developing local competencies in energy retrofitting, and in public-private collaboration. Such examples are primarily seen among Danish spearheading of ESCO development, including the 'ESCOmmune' collaboration. The municipality of Middelfart has established initiatives in local partnerships to motivate local building owners to carry out energy retrofitting, as part of the 'Green Growth Strategy' in the municipality.

The three approaches typically have different levels of innovation and various forms of collaboration between the municipality and the ESCO supplier. A main challenge for the ESCO market in Danish municipalities is to go from the basic ESCO model to the integrative ESCO approach, to improve a larger proportion of public buildings.

MARKETS

As in other countries, the EU Directive on the Energy Performance of Buildings has been a main driver for governments to encourage the development of energy services. However, the development of the ESCO market in Denmark has, in practice, been a combination of legal framework and incentives, market development on the supply side and the municipalities' own ambitions on the demand side. The Danish market has been dominated by private companies having gained experiences from ESCO contracting in neighboring countries, primarily Sweden. Several consulting companies have developed ESCO competencies and taken on the role of consultants for municipalities in designing the tender and the contract for the ESCO collaboration.

LEGAL FRAMEWORK

At the national level, ESCOs have been promoted and encouraged in different policy papers on energy savings in the existing building stock. It is seen as an essential input for reaching international as well as national goals on energy savings and CO_2 reductions (including the European 20-20-20 goals). In 2005, the Danish Government signed a political agreement as part of the EU Directive on the Energy Performance of Buildings, where the main objective was the statutory energy labeling of both public and private buildings. To encourage energy savings, the municipalities were allowed to take loans for renovation. Normally, municipalities are not allowed to start building projects by taking up loans, but exceptions were made for energy retrofitting initiatives, giving municipalities an attractive financing opportunity.

Although energy labeling is a cornerstone for promoting ESCO contracting, many municipalities have not completed the energy labeling of their buildings yet. A 2009 survey showed that only 30 percent of municipalities had completed energy labeling. However, energy labeling of the municipal building portfolio is included in many ESCO contracts. Thus, the demand for energy labeling of municipal buildings might actually work as a driver for ESCOs, although it was not originally the intended way.

The climate agenda has been an important motivation for many municipalities, especially voluntary agreements. One is the Climate Municipality, a voluntary agreement between the municipality and the Danish Association for Nature Protection, which obliges the municipality to reduce energy consumption by 2 percent per year in the municipality as a whole; i.e., not just the municipal administration, but the municipality as a defined area, including private building owners. This includes not just energy for heating of buildings but all kinds of energy, including supply, transport, electricity, etc. At the moment, 70 of 98 Danish municipalities have signed such an agreement.

Another voluntary agreement is the Curb-Cutting Agreement with the Centre for Energy Savings in which the municipality promises to reduce electricity consumption in public buildings by 2 percent per year. In many cases, these agreements have motivated municipalities to enter ESCO contracts, as saving goals represent great challenges for the municipalities, and also a political acceptance to pursue energy savings. The COP15-meeting in 2009 undoubtedly helped promote the climate agenda and motivated many municipalities to set targets for CO_2 reductions.

FACILITATORS

Experience shows that besides formal regulations, institutional capacity building can be an important tool for developing an ESCO market. In Denmark, such institutional capacity building has not taken place as a coordinated effort, but different initiatives have contributed to it. Besides a general promotion as a tool for energy reductions in buildings, the ESCO model has also been promoted by the Ministry of the Interior and Social Affairs as a way to increase public-private partnerships (PPPs). This effort includes workshops with private and public partners to promote networks and to disseminate knowledge support to municipalities that consider ESCO contracting, action plans for public-private collaboration, collection of knowledge and best practices, etc. On another track, the Centre for Energy Savings, a publicly financed unit with the aim of promoting energy savings in general, has promoted the ESCO model, by disseminating knowledge of ESCOs, monitoring development and experience in Danish municipalities, information on best practices, etc. Moreover, the Energy Research Program has initiated a number of R&D projects related to ESCOs which, for instance, describe the elements involved in ESCO activities, experience from other countries, developing a standard ESCO contract, etc.

The development of ESCO contracting in Denmark can be characterized as mainly market driven. The development on the supply side has benefited from the development in other countries. Several ESCO suppliers have gained experience from projects, especially in Swedish municipalities, and some ESCOs have been very active in promoting

ESCO contracting. Municipalities have collected information on ESCOs from networks and other informal sources, including visits to Swedish municipalities that have helped them in their decision to take up ESCO contracting. Furthermore, the 'ESCOmmunes' have been an important source of information on ESCO contracting.

Finland

Sami Siltainsuu

ACTIVITIES

The ESCO business model was introduced to the market in the mid-1990s. During these years, the business was very small, with few players and little interest. The ESCO business expanded little by little, first mainly in the industrial sector and the first concrete results were seen at the beginning of this millennium.

In its beginnings, the ESCO business model was unknown to buyers both in the public and private sectors, creating long sales cycles and projects which were intensive to manage and administrate.

A great deal of time was spent during the sales process to explain and present the model at different stages within organizations' decision-making processes. When ESCOs gained more publicity and became more common in the private sector, minimum sized projects commonly had 1,000 MWh savings with a three- to five-year payback period. Typical customers were from the energy-intensive industry sector. ESCO projects were usually conducted at the end of a long energy efficiency cycle and founded on energy audits, often conducted by third-party consultancy companies.

In privately owned real estate, such as office and retail buildings, the ESCO model has not yet become popular. The few projects that have been implemented are mainly by long-term building owners of facilities, such as athletic training centers. The scarce amount of projects and low popularity can be explained by a changed business environment. Ownership of real estate has been concentrated in larger real estate development companies, many of them with international footprints. Quite often, these decision makers operate abroad and are therefore not easily accessed by Finnish ESCOs. In addition, behavior in ownership has shifted: long-term ownership has been replaced with volatile transactional real estate portfolios. Real estate is bought and sold frequently, which creates short-term ownerships, a condition that does not suit the ESCO business model well, given that payback periods quite often exceed five years.

In the public sector, the model was first used on single buildings. Actions were mostly concentrated around minor modernization and fine-tuning of existing HVAC technology. In addition, a few heat source changes (from oil or electricity to geothermal heat or wood chip fire burners) were implemented. In 2003, new players came into the market, presenting new ways of doing ESCO business in the public sector. In that business model, projects now included multiple buildings. Audits, detailed planning and implementation of energy conservation measures were done simultaneously in larger entities. This model facilitated the creation of projects that delivered a larger positive impact on municipalities' operational expenditures and carbon emission reduction targets. Municipalities were able to reduce carbon emissions by 10-11 percent, which exceeded the goals set by the EU Energy Performance Directive with one single project. The EU has set a goal to reduce CO_2 emissions by 9% by the end of 2016. By including multiple buildings in an ESCO project, municipalities are able to reduce carbon emissions by 10-11%, thus exceeding the EU directive with a single project.

As previously stated, when the ESCO market was new in Finland, the sales cycles were very long. In both the private and public sectors, new models were needed to be introduced and marketed. In particular, the intensive decision-making processes and levels in the public sector slowed down these new initiatives.

Today, the ESCO business has a solid position in the public sector market. There are several big international players and a couple of small domestic firms sharing the market. Typical project size today is between EUR 500,000 (USD 657,000) and EUR 3 million (USD 4million). A distinctive phenomenon in the market is that along with energy conservation measures, significant renovation investments and capital improvements are carried out. Replacement of an air handling unit is a good example. In order to add a heat recovery unit and a variable-frequency drive to a fan engine to improve energy efficiency, quite often, air ducts and fans have to be renovated as well.

Switching to an alternative heat source has become one of the standard energy conservation and carbon emission reduction measures. Finland has many sparsely inhabited areas where several schools and day-care centers are out of reach of district heating. Until recent years, the buildings were designed to use oil or electricity as a source of heating. Today, the heat source is switched to geothermal or wood chip fire burners.

Today, in the public sector, the ESCO model is seen as a good way to maintain real estate portfolios and upgrade them to meet today's requirements for both indoor air quality and energy efficiency.

In 2005, new International Financial Reporting Standards (IFRS) came into effect. In that procedure, capital assets acquired by leasing agreements were obligated to be recorded in profit and loss statements. Because of that, the ESCO business model was not seen as attractive as it had been. This slowed down expansion of the ESCO business in this sector significantly.

With EU carbon emission mandates and organizations creating sustainability goals, the ESCO business model can be a solution for many public and private entities, provided that the EPC provider is a reputable firm. By leveraging guaranteed energy savings and the energy efficiency expertise of an ESCO, a building owner can make capital improvements that can be completely or partially funded by energy conversation measures.

CONTRACTS

In the Finnish market, it has become more common to use the term EPC rather than ESCOs, due in part to the contract models used. This is true especially in the public sector, where the guaranteed savings model (and different variations of it) is virtually the only model used. In that model, the EPC company delivers turnkey energy projects for the customer. The company carries out audits, does detailed planning and implements energy conservation measures. It is customary that the customer finances the projects from its own budget or finds a financial institution to finance the project.

It is the guaranteed savings aspect of projects that make them EPC projects. Typically, 80-90 percent of total estimated savings are guaranteed. After installation, the guarantee period usually lasts the entire calculated payback period.

Realization of the savings is determined by readings of main utility meters. In the case of achieving savings that are lower than expected, the EPC company pays the difference back to the customer based on the conditions and prices agreed to. Alternatively, in the case of over-performance, the excess savings can be shared between the supplier and the customer. Usually, compensations are based on utility market prices. This "shared savings model" has not gained popularity in Finland, as it is riskier for the supplier and financing a project is seen to be less economically viable.

In private commercial buildings, energy efficiency projects are carried out with a model that is seen as a somewhat "lighter" version of guaranteed or shared savings. In this model, guaranteed savings are replaced with a performance assurance period. During that period of time (which is quite often five years), the supplier fine-tunes and monitors the customer's buildings. The fee is based on improved energy performance.

LEGAL FRAMEWORK

Common, standardized terms of contracts regulate agreements, projects and services countrywide. General terms of consultancy contracts are applied for auditing and design, while general terms of building contracts are the basis for installation and execution contracts.

Guaranteed savings and performance assurance are beyond the scope of general terms of contracts, and they are not applicable as such. Length of assurance period, comprehensiveness of guaranteed savings, definition of baseline, etc. have to be agreed upon separately. In addition, special circumstances (renovation or sale of a building) are dealt with on a case-by-case basis.

The law of public procurement regulates purchasing in the public sector and is applicable across the European Union. However, there may be different areas of priority depending on specific market conditions in respective countries. The main purpose of the law is to give equal opportunity to every vendor in the market to submit bids answering requests for proposals (RFPs) or requests for qualifications (RFQs).

The law allows municipalities to use quality criteria (in a RFQ) with different weighting percentages (instead of price criteria in a RFP) in order to identify the best possible supplier. In reality, indisputable quality criteria are very difficult to find. In worst cases, public biddings are contests whereby the vendor who bids with the lowest price and biggest savings wins. As a result, the ability to deliver a quality project and long-term performance assurance are neglected. When selecting a partner for EPC, a building owner should select based on qualifications, such as past performance of energy savings, references and savings calculations methodology, not solely on low price and highest (or sometimes inflated) energy savings.

In the last couple of years, public procurements have been exposed to disturbances. For example, some vendors have threatened municipalities with law suits if they use too many quality criteria. Often, vendors who have lost bids have gone on to prosecute municipalities in civil court. This behavior does not serve anyone's benefit; as a matter of fact, it has slowed down ESCO market growth significantly. Overall, the atmosphere in the market is nervous and sometimes almost hostile. The situation is expected to remain the same in the near future. One way out is to change public procurement laws and to find common denominators in the procurement process.

MARKETS

There are no thorough and updated statistics available about the ESCO market. This is due partly because ESCOs are reluctant to report projects to official databases. Government officials estimate that an equal amount of projects are carried out annually in the private and public sectors, but the amount of energy saved and the size of the projects are estimated to be larger in the private sector.

The ESCO business has grown steadily over the years. In just a decade, it has grown from near zero to over EUR 10 million (USD 13 million) annual turnover.

The total business potential and size of the market have been estimated by different organizations. Quantification is often done by estimating the energy savings potential and payback periods. In 2005, the entire market potential was estimated to be approximately EUR 600 million (USD 789 million). This estimate is based on assuming the total energy savings potential to be 2 TWh in both the private and public sectors.

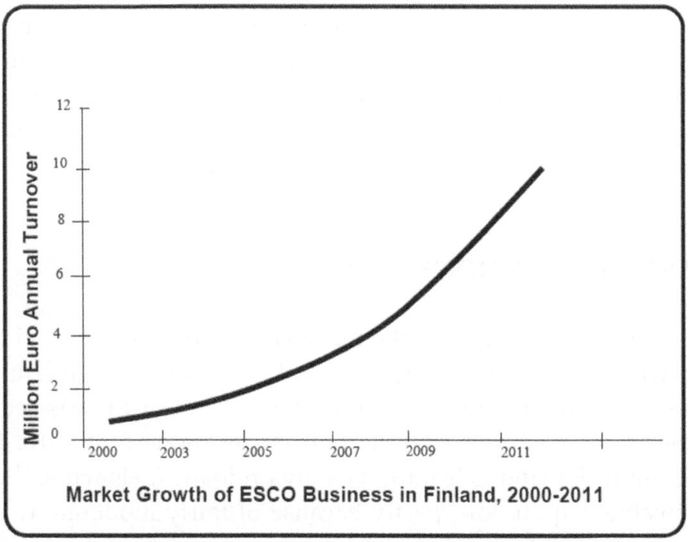

Figure 9-14: Market Growth of ESCO Business in Finland (2000-2011)

FACILITATORS

The Finnish government has founded an expert agency, Motiva, to encourage and promote energy efficiency activities in Finland. Motiva defines its tasks and objectives as:

- Motiva Oy is an expert company promoting efficient and sustainable use of energy and materials. Its services are utilized by the public administration, businesses, communities, and consumers.
- Motiva operates as an affiliated government agency (an in-house unit), and its functions will be developed as such. The company's entire share stock is in Finnish state ownership.
- Mission: We create well-being by promoting the use of energy and materials that are as harmless to the environment and as productive as possible.
- Vision: Motiva and its networks are decision makers' and end-users' best expert in issues of energy and materials efficiency and renewable energies. Motiva is the strategic partner of the main authorities in its field. Motiva is extensively known as a reliable and impartial expert.

Over the years, Motiva has strived to create a good and healthy ESCO market in Finland by creating agreement and procurement forms specifically for the public sector. Organizations and municipalities can enter into an energy efficiency agreement with Motiva. With an agreement, different entities set their own energy targets, plan how to achieve them and take action toward reaching their goals.

GOVERNMENT ACTIONS

Finland is committed to the European Union's energy and climate goals, which is one of the biggest motivators for the government to support the development of the ESCO market. The EU has set a goal to reduce CO_2 emissions by nine percent by the end of 2016. Energy production in Finland is leaning heavily on fossil fuels while the share of renewable is quite low, partly because of fairly moderate wind and hydropower conditions.

Therefore, energy efficiency is prioritized as a means to reach common goals. The government supports the ESCO business by subsidizing energy audits and energy efficiency investments. The amount of direct support to ESCO projects varies from 10 to 35 percent, depending on the scope and type of the investment. Additionally, if an organization has an energy efficiency agreement with Motiva, it is entitled to have five to ten percent more support.

France

Bruno Duplessis, Frédéric Rosenstein

ACTIVITIES

The current French Energy Service (ES) market has developed in three steps.

First Step: Evolution of a Historical Energy Service Model

The historical French energy services (ES) model; i.e., an energy performance contract based on operation and maintenance (O&M),

is the result of the public service culture in France. Within public procurement, O&M contracts were divided into different subcontracts (P1, P2, P3 and P4). This was done to ensure the indexing of prices, to apply different VAT rates, to distribute the invoice between the owner and tenants or occupants in accordance with the law and to apply them to public accounts that required an obligation of results. The same scheme has been extended to private buildings and industrial utilities (steam, cold and compressed air).

All the players of this market are federated into the French professional association FEDENE[1]. The market is largely dominated by two companies—Dalkia (a subsidiary of Veolia specialized in water, environment and energy) and Cofely (merger of Elyo and Cofathec, now a subsidiary of the former GDF-SUEZ).

Second Step:
Structuring the ES Market Due to the White Certificates Scheme

Energy suppliers or any obliged party contributes to increasing the awareness of energy users from any sector by promoting energy-efficient solutions and energy-saving measures.

The first three-year period of the White Certificates mechanism in France has just ended. It is not perfect but one of its principal benefits is that it allowed better structuring of the links between different professionals in the value chain of energy efficiency services. Moreover, EPC agreements are now eligible as a standard measure rewarded with a white certificates bonus. If the duration is more than 15 years and the energy savings guaranteed are, for example, 40 percent. This bonus may reach up to 50 percent of the energy savings amount achieved with standard conservation measures within the implemented EPC agreement.

Third Step: Anticipation of Future Evolution on the Energy Markets

This step really started with the spread of automation and communication technologies, whose progresses are huge in terms of O&M. Circuit breakers, transformers, turbines, motors/engines for example can now include self-diagnosis features and energy management systems that allow controlling and operating equipment as well as managing the energy flux of a site (building, factory, etc.).

The actors of this market are members of Gimelec, which is the association of providers of electrical equipment, automation and associated services. Three of these players (Siemens Building Technologies, Schneider Electric and Johnson Controls) became real ESCOs by offering EPC agreements based on the American model.

Thanks to new communication technologies, the centralization of all the alarms, operating parameters and energy/fluid consumptions and the possible transfer to outside allow (1) making tariff optimization (on the basis of historical tariffs); and/or (2) intervening just on time and then optimizing the efficiency of O&M tasks. New actors (Ergelis, Vizelia, Energie Systèmes) started to provide such contracts. These energy efficiency service providers anticipate the mass installation of smart meters to households, expected in France around 2017, as well as the development of associated services (optimization, load shedding, planned outage, etc.) and tariffs.

LEGAL FRAMEWORK

The French public policy framework promoting energy efficiency in buildings has been greatly enhanced in recent years following an extensive public process called the Grenelle Environnement (Environmental Round Table).

The Grenelle Environnement includes a major program to reduce energy consumption in all sectors of the construction industry and sets an overall goal of reducing primary energy consumption in buildings by 38 percent by 2020. Specific targets have been adopted for:

- new buildings with a maximal consumption of 50 kWh/m2/year after 2011 for public and tertiary buildings and 2012 for all other buildings;
- the renovation of public buildings, with a reduction target of at least 40 percent of energy consumption as well as at least 50 percent of GHG emissions for government buildings (50 million m^2) and its main public institutions (70 million m^2) within eight years, EPC is suggested to achieve these goals;
- the renovation of social housing buildings, private commercial buildings and private housing 800,000 of the most energy-inefficient social housing buildings are to be renovated by 2020;
- the development of EPC for condominiums.

Legal Framework in the Public Sector

The public procurement regulation does not allow global contracts, which include design, construction, operating and financing. The public procurement regulation requires that investment costs be paid separately from operating costs, which means that the savings cannot be used to pay investments in the same contract. The Grenelle law introduces the possibility of having global contracts including design, construction and operating only if energy savings are guaranteed. A next decree will implement this legislative modification and then allow EPC (but without third-party financing) in the public sector.

The Government Order of July 17, 2004 creates the possibility to draw up public private partnerships (PPP) contracts where a concession scheme is not available and where traditional procurement contracts cannot be implemented because of the legal restriction to have separate contracts for each phase of a project. The order also allows the public sector to pay the private company's remuneration periodically during the project. Additionally, it allows for payment based on performance indicators previously set out in a contract (instead of being purely revenue-based), especially regarding the definition of the energy consumption level to be attained. The common law procedure for placing contracts, except in the event of urgency, is the competitive dialogue, in accordance with community law concerning public contracts. The majority of EPC agreements were entered into within the PPP procedure. Nevertheless, this procedure is at the moment more adapted for large projects in the building sector (more than 10,000 m2).

Legal Framework in the Social Housing Sector

Social housing is owned and managed by local public companies (2.1 million units), public-private companies (0.3 million) and private companies (1.9 million). Energy retrofitting is currently financed through specific low interest loans (1.9 percent for 15 years from state bank Caisse des Dépôts) and some tax reductions up to 25 percent of investments. Energy savings certificates represent an additional source of income for energy retrofitting.

Rents are capped and cannot be raised after retrofitting, but since the law called "Molle," passed on March 25, 2009, energy savings can be partly recouped from tenants. Social housing operators (SHOs) can charge a flat amount representing 50 percent of energy savings for a 15-year period.

EPC agreements with third-party financing can be implemented as private contracts only by private SHOs. Public SHOs need to use PPP contracts, which entail the loss of fiscal benefits. Financing through EPC is therefore limited to private SHOs, which represent 42 percent of the social housing stock.

Legal Framework in the Co-Ownership Housing Sector

The stock represents 7.6 million units, from which 2.2 million units are under occupation or rented in residential buildings with central heating boilers. Rents cannot be raised (more than the inflation every year) after retrofitting. However, since the passage of the law called "Molle," on March 25, 2009, energy savings can be partly recouped from tenants. AS with SHOs, owners can charge a flat amount representing 50 percent of energy savings for a 15-year period.
The Grenelle laws introduce some obligations for residential buildings with central heating boilers, including:

- carrying out an energy performance certificate for co-ownership housing buildings with less than 50 units or an energy audit for those with more than 50 units;
- offers relative to energy performance contracting or improvement recommendations must be presented after the audit at the co-owners general meeting.

CONTRACTS

The most frequent ESCO contracts are still the first generation of contracts, based on operations contracting, with extensions. Operating contracts involve heating and air conditioning installations for which the service provider has a firm commitment undertaking (a temperature level to be guaranteed for the heating of premises).

Contracts may cover the following areas:

- *P1*—purchase of fuel, cost of heat energy (oil, gas, etc.);
- *P2*—daily operation, cost of labor and minor maintenance;
- *P3*—full maintenance, cost of major maintenance and of the total guarantee;
- *P4*—new equipment funding, investment depreciation.

The law specifies a framework for the duration of heating and air conditioning contracts in the public sector. Contracts which include P3 services have a maximum duration of 16 years. For contracts without P3 services, the maximum term is eight years in the case of fixed-price contracts. Five years is the permitted duration for all other contracts. The duration of French heating operations contracts is generally shorter than that of PPP contracts, such as performance or concession contracts.

However, very new contracts appeared with a second generation of actors that are only investment ESCOs and have no other role in the operation.

Various contractual formulations can then be envisaged, all of which have in common a guarantee on the value of the energy savings identified in a detailed audit carried out by the ESCO. The ESCO's remuneration is based on the value of the energy savings achieved and, if these are lower than the guaranteed savings, the ESCO pays the difference.

The factors which differentiate contracts include the following:

- The share of the investment assumed by the ESCO, which can range from 0 percent to 100 percent.
- The duration of the contract, depending on the value of the guaranteed energy savings, but also on the share of this amount from which the client wishes to benefit from the first year of the contract. If this share is zero, the energy savings pay off the investment and remunerate the ESCO; if the share is non-zero, the duration of the contract increases.

There is a third generation of new companies that try to add some "smart" controls to the usual operational control.

These third generation firms act as consultants but accept being paid from the savings. They do not take the place of the operational companies introduced first, but they request the right to set the control values from outside into the BMS.

An initial inventory of features of the operation in place is made and expressed in "customer-friendly" terms (ranges of temperatures, schedules of occupation, etc.). It will be respected in the new operation. However, the service is a "black box" for the owner and for the local operations company—the gains will disappear when the contract ends.

And finally, the state promotes a new type of public expenditure that may generate a new type of performance contracts.

The recent appearance in France of PPPs is changing the investment funding context. These special contractual arrangements should, in effect, allow performance targets to be introduced into invitations to tender, particularly with regard to defining the level of energy consumption to be attained.

The prior appraisal and the technical and economic analyses required in order to use this procedure have dampened the enthusiasm of local authorities, who wish to take up this new tool to fund their investments in energy efficiency. The definition of a functional program to open the dialogue with the future contractor requires total motivation on the part of the local authority that, nevertheless, wishes to have recourse to this instrument. This recent framework led the usual firms from the public works field (as Bouygues, Vinci or Eiffage) to enter the EPC provider community by providing huge and global refurbishment operations of social housing or public buildings (schools or colleges).

MARKET

As it has been described, the current ESCO market is segmented by historical reasons. The table below gives an estimation of the respective size of each identified segment. To that end, energy efficiency services and EPC activities have been identified as a fraction of the global turnover of each ESCO market segment.

FACILITATORS

Energy Efficiency Agency (ADEME)

ADEME is a public agency under the joint authority of the Ministry for Ecology, Sustainable Development, Transport and Housing, the Ministry for Higher Education and Research, and the Ministry for Economy, Finance and Industry.

Table 9-10: Estimated Respective Size of Each Segment

ESCO Type	*Examples*	*Number of ESCOs*	*Energy Efficiency Services and EPC Turnover (EUR million/year)*
Heating and cooling operators, facility managers, maintainers	Dalkia, Cofely	More than 250 companies	7.5
BEMS providers becoming ESCOs, namely by financing	Johnson Controls, Seimens, Schneider	3+	10
Optimizers of BEMS use or home automation devices	Ergelis, Vizelia, Energie Systèmes	5+	n/a
White certificates-related companies	Certeco + the obliged parties, subsidiaries/ partners	50+	6.3
Public works providing savings guarantees	Bouygues, Vinci, Eiffage or their respective subsidiaries, ETDE, VINCI Energie and Clemessy	10+	n/a

The missions of ADEME are encouraging, supervising, coordinating, facilitating and undertaking operations with the aim of protecting the environment and managing energy. ADEME supported the elaboration of guidelines on the implementation of EPC in different sectors: secondary schools, universities, social housing and public lighting. ADEME offers training courses on energy efficiency in buildings and on renewable energy. One of these training sessions is about energy performance contracts for public and private buildings.

The Building-Energy Research Foundation (Fondation De Recherche Bâtiment-Energie)

Created in 2005 by several French industrial companies (Arcelor, Gaz de France, EDF, Lafarge) with the support of ADEME and CSTB (French Scientific and Technical Centre for Construction), the Building-

Energy Research Foundation provides financial support for research carried out by public and private laboratories, and for the evaluation and development of this work. The last call for tenders of the foundation was about the development of an M&V methodology adapted for ambitious energy renovations of buildings.

PROFESSIONAL ASSOCIATIONS

The ESCOs are united within different associations: manufacturers of electrical and monitoring-control equipment (GIMELEC), professional federation of equipment providers and installers (SERCE, UCF/FFB), and association of heating and cooling operators (FEDENE). The different associations involved in energy efficiency services came together in an association formed in September 2005 with support from ADEME, under the name of Club S2E (Club for Energy Efficiency Services). This club wants to promote energy efficiency services and EPC through customers of its members, which in turn leads to the publication of methodological guidelines for the definition and implementation of energy efficiency services[2] and the measurement and verification[3] associated with energy efficiency services and EPC, which was deeply inspired by the IPMVP.

Germany

Jan W. Bleyl
Friedrich Seefeldt

ACTIVITIES

German ESCO market development started in the early 1990s. On the supply side, early market development activities were mainly driven by two groups of stakeholders. A few utilities (all of them still "bundled") started to implement least-cost-planning and demand-side management strategies, which may be considered as ESCO-type activities in some cases. At the same time, a variety of manufacturers of building technologies, automation and control equipment, as well as a number of plant engineering and construction companies started to extend their value-added chain and scope of services.

On the demand side of the ESCO market, a key driver has been the engagement of independent intermediaries (also labeled as market or project facilitators) since the mid-1990s. Mostly, these were energy agencies that spread the word and helped in particular the (potential) demand side of the ESCO market. The whole consisted in developing concrete projects, preparing tender documents (including model contracts) and putting the projects out on the market for ESCO bids. This holds particularly true for public sector EPC projects.

The first large statewide EPC program "Energy Saving Partnership (Energiesparpartnerschaft)" was initiated by the State of Berlin, grouping more than 100 public buildings into two building pools. This idea was successful and during the subsequent 15 years, more than 20 large EPC tenders were launched.

The most prominent facilitator example is BEA (Berlin Energy Agency) but likewise, a number of other regional energy agencies (EA-Northrine-Westphalia, KEA—Klimaschutz—und Energieagentur Baden-Württemberg or Bremer Energiekonsens as well as others). Also, the German Energy Agency deserves credit for developing the market, guidelines and innovative models. The State of Hessia, also, developed model contracts, which laid the basis for EPC contracts in federal, state and municipal buildings in many regions. Despite these enabling factors, the EPC market is still mostly limited to large projects in public buildings (pools), hospitals, leisure facilities and the like.

Energy supply contracting (ESC)—performance-based supply of useful energy—developed faster and succeeded in different end-use sectors, predominantly in the residential sector but also in industrial premises as well as in public facilities. Featured technologies range from standard boilers to CHP solutions (sometimes including distribution networks), but the measures are mostly limited to boiler rooms. The majority of projects run on natural gas but a variety of renewable heating systems and solar systems have been installed as well. The minimum energy cost line is about one order of magnitude below that of EPC projects.

The estimated number of ESCOs range from 250 to 500. A total of 250 up to 300 companies are continuously working with energy services (BEI/Prognos/Energetic Solutions, 2009), of which around 50 have more than one EPC reference (Berliner Energieagentur, 2009). More than 200 not very active companies may be regarded as "market observers" or "market entrants" (BEI/Prognos/Energetic Solutions, 2009).

The principal actors in the market are large national and international companies with a small dedicated ESCO unit as a supplementary business. SMEs with ESC as their main activity represent a smaller share of the market.

According to a Prognos market survey, service providers can be categorized as shown in Figure 9-15.

The German ESCO industry consists of national or super-regional utilities (17%), municipal utilities (25%), ESCOs which are typically branches of building technology or control manufacturers as well as building and metering service companies (37%), planning and engineering companies (5%) as well as others (18%) (Berliner Energieagentur, 2009/Prognos, 2010b).

The ESCO industry in Germany is represented by two associations: The "Verband für Wärmelieferung" (VfW) represents mainly SMEs that deliver energy supply services (ESC). However, VfW also hosts an EPC working group. The turnover of its 285 members is stated with EUR 1.8 billion (USD 2.4 billion), holding some 40,000 contracts.

The "ESCO Forum im ZVEI" currently lists 23 members who mainly serve industrial and real estate customers. Their accumulated

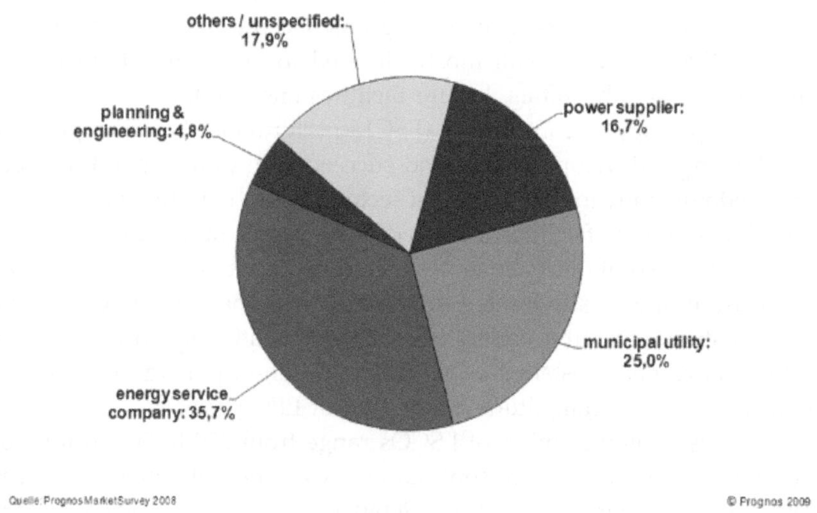

Figure 9-15: ESCO Markets

turnover is at above EUR 1 billion (USD 1.3 billion). The "ESCO Forum" resulted from a merger of the former "PECU" and "Contracting Forum im ZVEI" in 2007.

CONTRACTS

The German energy efficiency action plan (NEEAP) to the European Commission has adopted, to a large extent, the following definition of Energy-Contracting:

- **Energy Contracting**—also labeled as **ESCO**—or energy service—is a **comprehensive energy service concept** to execute **energy efficiency** and **renewable projects** in buildings or production facilities according to **minimized project cycle cost**.

- Typically an **Energy Service Company** (ESCO) acts as general contractor and implements a **customized efficiency service package** (consisting of e.g. design, building, (co-) financing, operation & maintenance, optimization, fuel purchase, user motivation).

- As key features, **the ESCO's remuneration is performance based**, it **guarantees the outcome** and **all inclusive cost** of the services and takes over commercial as well as **technical implementation** and **operation risks**. Over the whole project term of typically 5 to 15 years (after Bleyl 2008).

In Germany, ESCO models are mostly labeled as "Energy-Contracting" and in legislative texts also as "gewerbliche Wärmelieferung" (commercial heat supply). Two basic business models can be distinguished: ESC, referring to a performance-based supply of useful energy and EPC, referring to a performance-based energy savings business model.

Even though public attention on EPC indicates otherwise, ESC is by far the predominant business model in the German market with a share of more than 80 percent, according to the market data of the two leading ESCO associations as shown in Figure 9-16.

There is no institution for official market data; consequently, reliable market data are scarce or not publically available.

With a total revenue of around EUR 3 billion (USD 4 billion) (Prog-

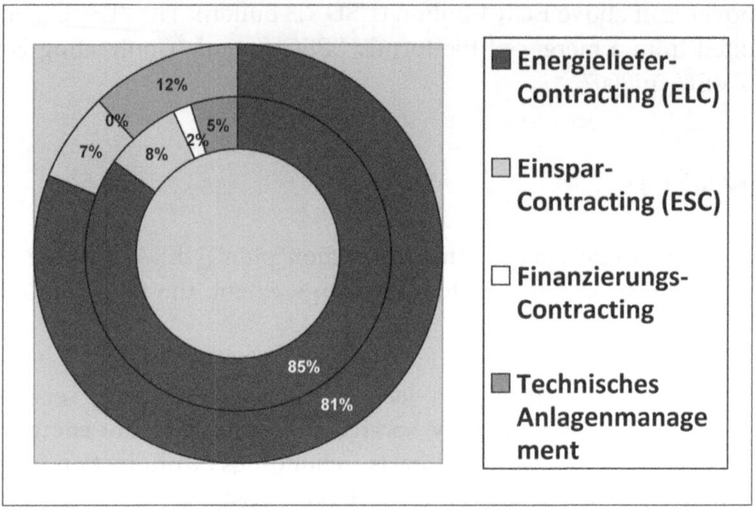

Figure 9-16: Business Models

nos estimate for 2011 for EPC and ESC), Germany is the leading market for Energy-Contracting in Europe. Market figures show a solid growth of 10 percent per annum compared with a total revenue of around EUR 2 billion (USD 2.6 billion) in 2008 (BEI/Prognos/Energetic Solutions, 2009). The market volume might be even higher if technical services for operation of the building equipment were taken into account.

The most important market segment is the residential sector, showing a share of more than 60 percent of the market in addition to a variety of small scale ESC projects and an increasing number of large housing companies sourcing out their technical facility services.

The public sector (comprised of schools, administration buildings and the health sector), with a total market share of 15 percent, is the most important market segment for EPC projects. A growing share of the market also goes to industrial and commercial projects, which may have a volume of 20 percent of the total market, showing considerable growth in the last few years.

Even though the market is experiencing considerable growth, transaction costs are hindering the market to move to smaller projects and explore larger potentials, especially the large segment of single-family houses. Indeed, the latter represents more than 60 percent of the surface area in residential buildings in Germany.

Due to the heterogeneity of small and large market participants coming from utilities, engineering companies and manufacturers of technical equipment, there is still high uncertainty regarding market figures. The newly founded Federal Office for Energy Efficiency and Energy Services (Bundesstelle für Energieeffizienz und Energiedienstleistungen—BfEE) has launched a market study to get a comprehensive and more systematic overview of the market.

MARKETS

German ESCO markets have experienced good market development, which was mainly driven by a variety of market actors in different regions. There was no centralized federal action for any obligation scheme or any coordinated federal support scheme. Market development, however, could have been more dynamic. Different policy measures, for example the newly established Federal Office for Energy Efficiency (BfEE) might give a valuable impulse for a more dynamic development.

We have learned the following key lessons for ESCO market development:

- Successful **market development**—in particular for EPC in the public sector—**was demand-side driven**. (Potential) ESCO customers defined their goals and needs for energy efficiency service packages and **put out requests for proposals on the market**. Studies and even investment grade audits (IGAs) are not sufficient to create projects.

- To foster market development, the role of **independent market and project facilitators as intermediaries between ESCOs and their (potential) clients** has proved to be a key success factor (as represented, e.g., by energy agencies).

The role and sample activities of project facilitators as intermediary between the demand and supply sides on the ESCO market are summarized in Figure 9-17.

This facilitator role requires more active and knowledgeable players as well as better funding.

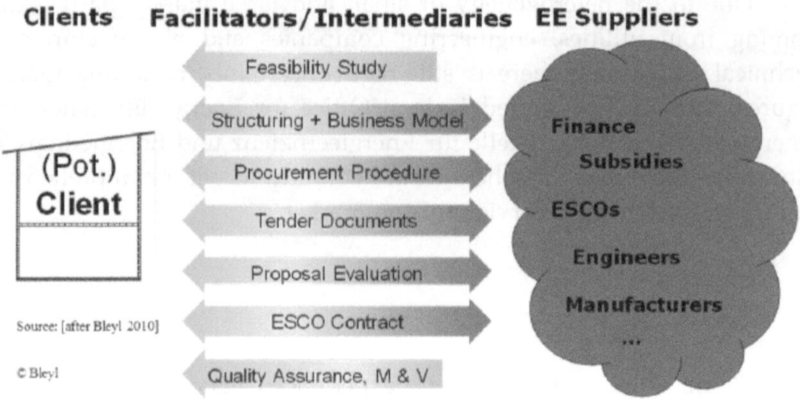

Figure 9-17: Facilitator Roles

- Contracting to an ESCO is a strategic "make or buy" decision of a (potential) client. Outsourcing to an ESCO competes with a standard in-house implementation and has substantial implications on the standard buying routines of the outsourcing institution. The decision also implies either entrusting one general contractor (ESCO) versus contracting to individual subcontractors for planning, construction, O&M as well as optimization.

- **Outsourcing requires new organizational routines on the customer side**; e.g., with regard to procurement practices (typically "negotiated procedures" are applied), interdisciplinary co-operations between different departments and project engineers or long-term cross-budgetary financial management.

- **Energy-contracting is a flexible and modular energy service package**. This also implies that the ESCO customer may define—depending on his own resources—what components of the energy service will be outsourced and which components he carries out himself.

- **Energy efficiency often is not the driving force and not a stand-alone business case** but a (beneficial) side effect. Better listening to the "real" needs expressed by customers and building strategic alliances with facility managers, security, automation and other building technology tasks to incorporate energy efficiency goals or minimum performance standards early on in project development is required.

- **Financing is not** necessarily **the core business of ESCOs**. Their core competence usually lies in technical, economic and organizational matters of an energy service package. ESCOs should serve as finance vehicle, not necessarily as financiers themselves. Nevertheless, of course, payments to ESCOs must be secured.

- The energy-contracting approach offers **integrated solutions for the project life cycle** (planning, construction, O&M and optimization) and is **interdisciplinary** (technical, economic, financial, organizational and legal aspects) in order to achieve guaranteed performance and results of the efficiency technologies deployed. The ESCO concept **opens up solutions, which are not achievable through standard, disintegrated implementation processes** (life-cycle cost optimization across investment and operation budgets, integrated planning or performance guarantees over the complete project cycle...). However, these opportunities also imply a **highly complex product**.

GOVERNMENT ACTIONS

Even though the German ESCO market has experienced a continuous growth over the last 20 years and can be regarded as having matured, there are still obstacles. Generally speaking, more and better substantiated information is required, especially on the demand side. Energy efficiency services still appear as low-interest-products, lacking clear guidance as well as incentives to make the external professional, and typically more efficient outsourcing solution to an ESCO, more desirable than the home-made version (or the status quo).

More concretely, barriers vary across end-use market segments. In the residential sector, the cost allocation between landlords and tenants (split incentive) needs to be resolved in Germany's tenancy laws. Furthermore, transaction cost logics are against smaller projects and hinder the development of a higher market potential, especially in single- and small multi-family houses.

A renewed and currently circulated ministry of justice draft[1] aims to resolve the cost allocation problem in the residential sector by an amendment to the tenancy law. The implications of the Renewable Feed-in Act (EEG), however, imposing a clear disadvantage for ESCOs with CHP plants compared to the in-house operation of the CHP plant, are still pending further treatment. Moreover, the transaction cost logic against smaller projects remains unsolved.

In the commercial sector, the need for flexibility and, hence, the need for a short payback period are the main barriers to deciding for high capital expenditures and life-cycle optimized energy infrastructures. In addition, both SME clients and ESCOs often lack a sufficient credit rating with a view to long-term investments. While price development is basically more and more in favor of CHP-based contracts, pricing and tariffs of the German Renewables Feed-in Act impose a clear disadvantage on ESCOs. The market is still moving forward against these barriers but the outlook could be much more positive without them.

In the public sector, public budget restrictions (treatment of ESCO contracts as debt and missing long-term commitment authorizations "Verpflichtungsermächtigung") along with the complexity of public procurement legislation are some of the major barriers in place. Although the majority of federal states have resolved such problems or offer help through market facilitators, there are remaining uncertainties at least among new clients.

EC directive 32/2006 for end-use efficiency and energy services demands a systematic promotion of the market for energy services. With respect to existing market traditions and cultures as well as market liberalization with its vertically disintegrated market actors, Germany refused to follow EC recommendations and did not impose any obligation on energy companies, as was the case in Italy, France, the UK or Denmark.

In order to observe and promote the markets for energy services and energy audits, Germany established the Federal Office for Energy Efficiency ("Bundesstelle für Energieeffizienz," BfEE) in 2010. The BfEE has launched several studies for market observation and is planning to take action to remove a series of barriers.

As a general conclusion, the German market for energy services has a long tradition of solid growth, especially over the last two decades. The majority of activities were market-driven. However, there are barriers remaining in each market segment. The implementation of the BfEE might be a first step to better observing the markets and systematically removing some of these barriers

According to German Market experts, financing typically has not been an unsolvable bottleneck for market development. Nevertheless, convincing potential customers to outsource comprehensive energy service packages to an ESCO still takes a long time, causes high transaction costs and requires solving a variety of obstacles. Not to mention that it is often times rooted in the "human factor" and the reluctance to change established organizational routines rather than in factual barriers.

Hong Kong

Dominic Yin

ACTIVITIES

Hong Kong is a sub-tropical city of seven million inhabitants, a part of China but operating under its own laws. Due to scarcity of land, almost all residences, offices and retail is housed in high-rise buildings. There is little manufacturing left in Hong Kong—most was moved across the border to China in the late 1980s and early 1990s. This situation has meant that most ESCO activity in Hong Kong involves air conditioning and lighting services in tall buildings.

Various parties introduced different ESCO models in Hong Kong in the early 1990s. One private energy management service provider, Energy Resources Management, began proposing both shared savings and guaranteed savings performance contracts to large property companies in 1992. The large US-headquartered BMS vendors also proposed ESCO model contracts to some select clients, but with limited commitment and success. At least one of the world's leading manufacturers of water chillers studied the market potential in depth but concluded that Hong Kong was not a suitable place to roll out its US ESCO offering.

The government-subsidized Hong Kong Productivity Council competed with private enterprises and their low overheads contributed to them winning a tiny number of contracts. However, the projects failed to create any forward momentum.

Many more years passed. Subsequently, in 1998, the government issued a specific policy to explore the concept of introducing ESCOs in order to enhance energy efficiency and energy conservation within the government itself. The government's Energy Efficiency Office set up an "ESCO Task Force" to gather experience and advice from local and overseas parties and to recommend guidelines for EPC. The Task Force recommended exploring the possibility of conducting pilot EPC projects for five buildings of similar characteristics. In the early 2000s, the Hong Kong government's engineering department, EMSD, awarded a number of projects, each around USD 70,000, under a guaranteed savings arrangement. However, again, no traction was derived from these and no widespread appreciation of the ESCO model came about, even among other government departments.

The three energy utilities operating in Hong Kong under basically monopoly conditions also offered ESCO-type services to a very select group of their customer base around that time, but with conflicting internal agendas.

Although there were exceptions, the ESCO approach remained largely unknown by most of the parties that would typically be decision makers in other jurisdictions; e.g., hospital, university and government facility senior management. The property sector is particularly important in Hong Kong, where services make up over 90 percent of the economy, and almost all of those services are housed in high-rise buildings. Nevertheless, real estate executives on both the owner and management sides remained largely unaware of the ESCO approach throughout the 1990s and well into the first decade of this century.

Finally, in the last few years, there has been growing awareness of shared savings and guaranteed savings EPC in Hong Kong. Product-specific suppliers dominate—vendors offering a shared savings or guaranteed savings clause in their sales contracts in order to sell their own particular products—as opposed to pure ESCOs, brand agnostic, as seen in mature ESCO markets.

Since 2007, the Clinton Climate Initiative (CCI) has been instrumental in the recent development of the ESCO model in Hong Kong, helping both potential facility groups as well as large multinational ESCOs represented in Hong Kong.

However, the number of contracts actually taken forward remains meager despite the CCI and major international ESCOs being present, and despite the sales presentations made along with the proposals submitted and follow-up conducted.

The major obstacles to ESCO success in Hong Kong have been the same as elsewhere: poor awareness, lack of experience in providers and prospects, a variety of complex legal and contractual issues, intransigence of the conventional procurement process and the host of challenges in any M&V attempts.

CONTRACTS

Hong Kong has seen shared savings, guaranteed savings and other hybrid formats. Statistically, the guaranteed savings approach dominates by far. Pay-from-savings contracts have been restricted to utilities. Full chauffage-style contracts have not been seen probably due to the quasi-monopoly position enjoyed by the three utilities. Nevertheless, some Hong Kong companies have found success with the "chauffage" approach serving clients across China.

Financing has been by both the ESCO and the client. Although attempts have been made to introduce TPF, create private equity funds and offer to securitize the positive cash flow from bundled smaller projects, no significant progress has ensued.

Moreover, from the very small number of performance contracts executed over the last 20 years, all have been based on extensive negotiation rather than on a formulaic approach. This has undoubtedly impacted true profitability to both the client and the contractor negatively. Expertise in M&V has not had an opportunity to develop.

LEGAL FRAMEWORK

There is no legal framework for ESCOs or the ESCO model in Hong Kong. Contracts are governed by normal commercial law. Hong Kong is particularly well regarded in Asia for its legal system and for the rule of law. Therefore, there is a good basis for considering Hong Kong as an ideal location to base an ESCO business and to specify in contracts and agreements that Hong Kong be the governing domain in case of legal dispute. However, this opportunity remains untapped.

Hong Kong follows common law, a heritage of its colonial past, and would typically look to Commonwealth countries such as the UK,

Canada and Australia for legal precedents when so required. American, Japanese and European case law in the area of energy services contracts might be more difficult to call upon.

The giant multinationals that have made up most of the ESCO work so far seen in Hong Kong offer only their standard worldwide contract term, again usually under US law.

Contracts have been known to take six months or more to negotiate, even for contracts of minimum size by international standards. The lack of a locally developed legal framework for ESCO contracts is seen as a major cause of such delays.

No performance contract dispute cases could be found in a review of Hong Kong court records. Known cases where penalties on energy savings shortfalls have occurred have been settled between parties of interest without getting into legal proceedings.

MARKETS

Given the very modest penetration of the ESCO model in Hong Kong, not too much inference should be drawn from any examination of current ESCO market data. However, it is worth noting that, as in other countries, the largest number and size of ESCO contracts in Hong Kong, up until now, have been the usual categories of hospitals and universities. The median project size is less than USD 1 million.

FACILITATORS

Associations, an EE branch of a government engineering department and other private facilitators are present in the Hong Kong market. Nonetheless, efforts, a strategic approach, government support and other catalytic elements seem to be lacking the ingredients that would get the ESCO business model beyond its current phase.

An association, Hong Kong Association of Energy Services Companies (HAESCO), was founded in 2008. It covered a very broad spectrum of energy-related topics and activities including biodiesel feedstock production, obstacles to nuclear energy proliferation and carbon trading. No particular initiatives related to promoting the ESCO concept were found. Only a small percentage of members actually offer shared savings, guaranteed savings or other performance contracts as an option, let alone as a primary business.

Little by the way of public education has been undertaken by any party, and none by HAESCO, to make the public aware of the ESCO concept. The Hong Kong Productivity Council (mentioned above) arranged one workshop on the topic in the early 1990s. A number of seminars have been held over the years which have included one or more speakers addressing the ESCO story from the speaker's employer perspective. No case could be found of in-depth training including worked examples, detailed study of the process of carrying out a contract under an ESCO business or even the basics of M&V.

The Hong Kong government's lead entity in the EE sphere is the Electrical and Mechanical Services Department. In addition to carrying out the ongoing repairs and maintenance of government buildings and facilities in Hong Kong, the EMSD has an Energy Efficiency Office, set up in the early 1990s. For many years, the Energy Efficiency Office's primary work product was a succession of appliance EE labeling schemes covering the domestic appliances that had contributed most to a typical residential electric bill—including five that are now mandatory: window air conditioners, refrigerators, compact fluorescent lamps, washing machines and dehumidifiers.

The clear leader among private facilitators is the Clinton Climate Initiative, mentioned above. In addition, the US Commercial Service sponsored the development of a P2E2 model (pollution prevention and energy efficiency) which, while targeted at end-users in China, specified that Hong Kong companies were to fill the ESCO component of the arrangement. Although a concept with multibillion dollar potential, this particular P2E2 initiative did not yield much in the way of outcomes during its active years.

GOVERNMENT ACTIONS

On the positive side, the Hong Kong government has recently enacted a Building Energy Code that came into force in September 2011. It can be hoped that this new law will oblige owners of existing buildings to upgrade their HVAC, electrical, lighting and vertical transportation systems when conducting renovation works of more than 500 m^2 (5,380 ft2). The potential exists for applying the ESCO model to fund such upgrades.

However, in terms of direct promotion of the ESCO industry, or even direct help to prospective users of ESCO services to understand and overcome the obstacles inherent in the ESCO model, or to provide any support when facing the giant corporations that end up getting most of the ESCO contracts in Hong Kong, the government has done little of substance.

Furthermore, the government continues its long-standing tradition of putting everything to do with EE through the very engineering-centric EMSD (described above) even though the ESCO model is largely a risk management approach. This involves financing taking the second most significant part, and engineering only comprising one of the minor components.

Other widely recognized barriers to EE, green and sustainable building renovation and new build endeavors, which are directly due to government action and inaction, apply to ESCO work as well. These include Buildings Department regulations grounded in concepts and technologies in place since the 1950s and conflicting permitting between different government departments.

Finally, the Hong Kong government has not supported the development of the ESCO business in a way that would be most tangible: executing a large-scale and full scope energy performance contract in one or more of its own vast property portfolio.

Hungary

Tamás László

ACTIVITIES

The concept of EPC, known in Hungary as TPF, was introduced in Hungary at the beginning of the 1990s. At the time, financing was a complicated process. EPC became an attractive opportunity to reconstruct existing infrastructures in public and private sector facilities. It also made it possible to improve the efficiency of energy production and energy consumer systems without end users having to use their own financial resources. Customers were also able to benefit from a relatively low risk based on ESCO performance contracts.

Prometheus later DALKIA

Prometheus was a Hungarian company before the change of economic and political system. Prometheus supplied companies with energy efficiency and chimney services. In the process of privatization, a French corporation bought this company using its network. Finally, the owner of the company changed the original name of the business to DALKIA.

By far, the biggest achievements were reached by Prometheus, contracting 150 facilities for TPF in hospitals (50), schools, and municipal institutions as well as smaller district heating companies (11). Generally, the old steam-based systems were converted to hot water systems with less wasted energy in the system, and secondary systems were renewed. The payback period ranged from 5 to 15 years depending on the condition of the old system.

The typical approach to TPF at that time focused mainly on the technical solution of projects through contracting. Subsequently, implementation resulted in energy savings.

Landis and Staeffa

Formerly Landis and Gyr, Landis and Staeffa began to deal with TPF in 1996. Within a few years, ESCO Landis and Staeffa had created a package of 30 projects and became the biggest compared to its similar subsidiary acting in Central Europe.

The projects implemented were mainly in the following groups:

- reconstruction of boiler houses owned by residential cooperatives, changes for new boilers, modern regulation systems;
- conversion of steam-based centralized heating systems to on-site hot water-based heating systems at railway stations owned by Hungarian Railways; and
- implementation of gas engines and wood chip-based heating at a district heating company.

The concept spread rapidly to project financing, obtaining support from contractors with heating regulation and monitoring systems for investments for project implementation as well as for energy savings monitoring and guarantee.

Landis and Staeffa was bought by Siemens in 2004 and merged with the Siemens Building Technology branch. Since that time, the TPF component has decreased.

Honeywell Hungary Ltd.

At the end of the 1990s, Honeywell began implementing TPF with the goal of increasing Honeywell product sales. Under such contracts, it was possible to sell several thousands of products manufactured by Honeywell.

Within a few years, Honeywell Ltd. had created seven major third-party financed projects for energy savings. The main project groups were:

- reconstruction of energy production (possible change of boilers, modernization of heat exchangers and pumping systems, introduction of regulatory systems) in some middle-sized district heating companies and full reconstruction of the secondary systems on the tenant side;
- conversion of steam-based centralized heating systems to on-site hot water-based heating systems at the Záhony railway station (Ukrainian and Hungarian border) owned by Hungarian Railways.

The focus on such projects also migrated to project financing, obtaining possible support, project implementation as main contractor with heating regulation and monitoring systems as well as energy savings monitoring and guarantee.

According to Honeywell Ltd., the activities of the ESCO were stopped in 2001.

KIPSZER Kazántechnológia Ltd.

In the early 1990s, KIPSZER Kazántechnológia Ltd. became involved in TPF. Within a few years, several tenders had been won by KIPSZER Kazántechnológia Ltd. from Hungarian Railways for the refurbishment of railway stations and from various hospitals for:

- conversion of steam-based centralized heating systems to on-site hot water-based heating systems at railway stations owned by Hungarian Railways; and
- reconstruction of hospital heating systems with implementation of gas engines.

KIPSZER Kazántechnológia Ltd. Was later bought by ENER-G Combined Power and converted first as KIPSZER Energy Technology Ltd., later to ENER-G Energy Technology Ltd. at the beginning of this century.

Energy Efficient Technologies Ltd. (EETEK)

The company was established at the beginning of 2001. Its activities were planned not only for Hungary but for Central Europe as well. Finally, only in Hungary and Romania were projects implemented. The company's fields of activity can be divided in three areas:

- Energy-saving projects, which were implemented in chemical factories in Hungary and in Romania.
- Cogeneration projects, which were implemented in one district heating company, two hospitals and in one chemical company. These projects improved the energy efficiency of host organizations, mainly in hospitals with laying down new piping systems and refurbishing heat exchanger stations. Proved—first in Hungary—that it was possible to produce steam from the flue gases of gas engines using heat recovery boilers. Solved the problem of utilization of heat production in summer using absorption devices for cooling.
- Outsourcing of energy supply of industrial companies, upgrading the energy production by implementation of gas engines with 5-10 MW capacity. Capacity, overtaking and decreasing the staff of energy supply and finally giving cheaper energy supply.

EETEK was bought by Energy Star (new Hungarian company with Hungarian owners) in the middle of 2011.

CONTRACTS

Three types of ESCO contracts can be distinguished in Hungary:

1. traditional ESCO contracts;
2. non-traditional ESCO contracts;
3. outsourcing ESCO contracts.

Traditional ESCO Contracts

Traditional ESCO contracts are related to the basics of third-party financed projects in Hungary. The advantage of this approach is that the global technical and financial solution is clear for the hosting company. The disadvantage for the investor is that the safety of the payment is very low and the amount of the work to be performed through monitoring and controlling after implementation is very high. The following elements have been seen as potential sources of conflict with the contracting hosting company: (i) control of a sufficient level of maintenance work; (ii) thorough handling of the whole system by educated labor force; (iii) permanent calculations of savings and results; and (iv) validation with specialists of the hosting companies.

Non-traditional ESCO Contracts

Non-traditional contracts were spread in parallel with combined heat and power production in Hungary, thereby pushing out the traditional TPF component. Combined heat and power production itself was providing for energy efficiency projects. In order to spread combined heat and power production up to a certain capacity (below 6 MW), the government provided support through power prices that ensured the viability of gas engine implementation.

In the case of district heating, most of the energy-saving opportunities were exhausted by the implementation of combined heat and power production. In institutions, energy-saving opportunities were not fully exhausted but thanks to high benefits, contracting partners were satisfied with the results achieved through the implementation of gas engines.

Therefore, the contracts for the implementation of combined heat and power production were different in content from traditional contracts. First of all, there was no technical component to them. Only the capacity of the gas engines was determined and a drawing illustrated the replacement of gas engines. The legal aspect was determined easily and adequately. The lifetime of the contract was equal to the lifetime of the equipment.

The hosting company generally received a renting fee for the territory occupied by the equipment. The produced power was purchased by the investor from the grid owner. The produced heat (in the form of only hot water or hot water and steam) was sold to the hosting company. The natural gas was purchased by the hosting company and partly sold to the investor. The maintenance was the obligation of the investor. Operation and monitoring were ensured by the specialists of the hosting company and financed by the investor. It was possibly the best solution and the cheapest. All the obligations (supply, payment and so on) were fixed on behalf of the investor and on behalf of the hosting company.

The advantage of this approach is that the contract gives the highest safety against non-payment. The cost of natural gas is always higher than the cost of the produced combined heat and power as well as of the rental fee. It is a cheap solution of operation and the best opportunity for monitoring purposes. The investor is not obliged to have staff at the site. A good stimulation system was elaborated. The payback period is relatively short, from seven to ten years. The hosting companies were happy with the renting fee and lower heat prices.

The disadvantage of the approach was that the projects only focused on supply-side energy efficiency and; therefore, demand-side viable energy-saving opportunities were not implemented.

Outsourcing ESCO Contracts

The outsourcing of the energy activity of a facility is understood as being more global than strictly the energy efficiency component. The outsourcing contract is also related to water and sewage management in addition to providing compressed air and even cooling. This form of contract is not spread in Hungary even if there are limited cases known.

Outsourcing provides an incentive to the investor to implement all the viable energy-saving measures. They include implementing combined heat and power production, installing biomass boilers, refurbishing water supply, improving the energy efficiency of compressed air systems, implementing cooling in working places and increasing the energy efficiency of the secondary system within hosting companies. The investor cannot avoid having staff at the site. The solution adopted to implement such contracts is often based on the creation of a new special purpose vehicle. However, the global responsibility for the contract remains with the investor.

The hosting company can be satisfied with the outsourcing approach because it can focus on its main activity while relying on a third party to provide all the necessary services (including energy, compressed air, water, cooling) at a determined price with a strong guarantee. The hosting company is free from problems involving the use of its staff.

MARKETS

In Hungary, the ESCO market was limited until traditional TPF became the only solution. Small-sized projects were not the favorite for investors because overhead costs were not covered by the benefits. The solution for overly large projects was outsourcing.

Over the last 15 years, the dominant sector of ESCO contracts has become district heating. This is because pressure from consumers is now at its highest since the cost of heating district heated apartments is much higher than the cost of heating apartments with natural gas. The energy savings generated from the energy production of district heating was seen as being able to solve this issue through TPF. Unfortunately, the potential of demand-side energy savings was not addressed through TPF due to risk of non-payment by the owners. It is likely that demand-side energy efficiency is going to be addressed only with government support. The progress is very slow due to lack of financing.

In the private sector (non-governmental and non-municipality), TPF is still not widespread. The reasons are not known exactly. One of the reasons might be that, generally, the ratio of energy costs in the competitive sector is low and companies have not identified the opportunity of using energy efficiently. Furthermore, TPF is still not well known and understood enough in that market.

However, it is expected that the dominant ESCO market in Hungary will be the industrial sector. The European Union is engaging member countries to achieve the 20 percent energy-saving goal by 2020. It is also pushing to have 20 percent of energy supply provided through the use of renewable energy and a reduction of 20 percent CO_2 emissions reductions. The Hungarian government committed to achieve its energy savings and emission reduction targets. This will surely have an important influence on the ESCO market in the country as the 20 percent energy-saving commitment cannot be met without the important participation of the industrial sector.

The government recognized the requirement of involving the industrial sector with the help of IPENERG, the society of industrial companies. IPENERG is beginning to create a framework to launch energy-saving activities in the competitive sector. The first meeting with the steering committee took place in 2011. The meeting even included members of the Parliament, the President and some well-known senators. The members of the steering committee are rectors from major universities.

In 2012, a kick-off meeting will be arranged and work will begin. If the stimulation is resolved (reduction of energy tax), the energy savings will accrue and several new and existing TPF companies will be able to reorganize and begin new activities. In such a case, the competitive sector would become a dominant ESCO market within a few years.

The pressure on the district heating sector now comes from the government, which has frozen the price of district heating. It is forbidden for district heating companies to increase their prices. That way, district heating companies are obliged to invest in energy savings or renewable energy to reduce their operating costs. Several district heating companies are already converting their energy production from natural gas to geothermal energy (when possible) and/or to biomass firing. In this relationship, TPF is already active in Hungary. The energy savings that could be achieved by consumers through district heating have not materialized yet due to lack of financing.

FACILITATORS

In Hungary, facilitators do not seem to be needed at this time. Furthermore, there is no ESCO association or energy efficiency agency. On the other hand, there is a government organization called Energy Center, able theoretically to facilitate TPF or control the results of stimulation, if needed. Several energy companies as well as skilled and experienced specialists are ready to continue some version of TPF. The largest TPF companies are still working on increasing their market. The activity will increase in 2012 and beyond if the stimulation is targeted at energy savings. The activity of IPENERG will likely accelerate the volume of TPF in the upcoming years.

India

Pradeep Kumar

ACTIVITIES

The concept of EE project implementation through EPC as practiced by ESCOs has been around for more than 20 years internationally and for a few years in India. Many pilot projects have been tried in the

Indian private and public sectors and efforts are underway to develop a national- or state-level program based on this concept. Presently, the ESCO business model is well established in developed countries but is still struggling to grow in developing countries, such as in India.

In the last six to eight years, ESCO operations have remained at a nascent stage in India and the true impact and advantages of such business model have not yet been realized. This is due to (i) quality limitations on the part of service providers; and (ii) a lack of clear understanding of the business model (especially during the performance period) and of banks'/financial institutions' lending policies.

ESCO operations face barriers such as M&V protocol, financial structuring, repayment mechanisms, ability to borrow funds etc. ESCOs typically do not have large assets to bank upon; therefore, while they have the technical capability to identify and design EE projects, they are often unable to convince project developers, investors and bankers about the certainty of energy savings. Most ESCOs are small companies with limited geographical reach that cannot meet the most common demand related to investing in projects. The end-user often wants the contractual guarantees to be backed by bank guarantees before project implementation. This requires huge amounts of capital to be secured in order to provide collateral to the bank that will extend the guarantee.

Over the last few years, intervention by bilateral and multilateral funding agencies such as USAID, the World Bank, Alliance to Save Energy and the World Resources Institute (WRI) have helped and promoted the ESCO business model through their EE programs. While there are over 110 ESCOs [1] presently in India, the number of projects completed so far is very low.

LEGAL FRAMEWORK

Considering the vast potential of energy efficiency for energy savings and benefits, the Government of India enacted the Energy Conservation Act, 2001 (52 of 2001)[2]. The Act provides for a legal framework, an institutional arrangement and a regulatory mechanism at the central and state levels to embark upon an EE drive in the country. Five major provisions of the Energy Conservation Act relate to designated consumers, standards and labeling of appliances, energy conservation building codes, creation of the Bureau of Energy Efficiency, and establishment of the Energy Conservation Fund. The Energy Conservation Act was amended in 2010 to further strengthen its framework.

The government of India set up the Bureau of Energy Efficiency (BEE) on March 1, 2002 under the provisions of the Energy Conservation Act, 2001. The mission of the BEE is to assist in developing policies and strategies with a thrust on self-regulation and market principles, within the overall framework of the Energy Conservation Act. The primary objective of the BEE is to reduce the energy intensity of the Indian economy. This will be achieved with the active participation of all stakeholders, resulting in accelerated and sustained adoption of energy efficiency in all sectors. The Ministry of Power and the BEE have brought energy conservation and management to the center stage of India's power sector with different schemes and programs. As per the Energy Conservation Act, state-designated agencies (SDAs) at state level are created to implement and support the BEE's various programs.

The following are BEE ongoing programs that support energy efficiency implementation through performance-based contracts:

- standards and labeling;
- agriculture demand-side management;
- municipal demand-side management;
- EE retrofit for existing buildings through ESCO mechanisms;
- star rating of buildings.

The National Mission for Enhanced Energy Efficiency (NMEEE) is one of the eight missions under the National Action Plan on Climate Change. The objective of the mission is to achieve growth with ecological sustainability by devising cost-effective strategies for end-use demand-side management. The Ministry of Power and the BEE have been entrusted with the task of preparing the implementation plan for the NMEEE and upscaling efforts to create and sustain an EE market as a means to unlock investments of around INR 740 billion (USD 15 billion).

The NMEEE is scheduled to undertake the following initiatives relevant to ESCO development as well as the policies and programs for energy efficiency implementation by the BEE:

Perform, Achieve and Trade (PAT) is a market-based mechanism designed to enhance the cost-effectiveness of EE improvements in large energy-intensive industries and facilities through certification of energy savings that can be traded. Targets for EE improvements will be set under Section 14 of the Energy Conservation Act, 2001 in a manner that reflects fuel usage and the economic effort involved. Designated consumers in eight industrial sectors will be subject to mandatory participation in the 1st cycle of the PAT scheme to be implemented from 2011-12 to 2013-14.

An Energy Efficiency Financing Platform (EEFP) to help stimulate the necessary funding for ESCO-based EE delivery mechanisms. The costs will be recovered from the energy savings, which will also reduce the subsidy bill of the state government. The BEE has undertaken the following measures, in addition to those relative to the implementation of demonstration projects in government buildings with a view to stimulating the market:

- putting in place a government-supported standard methodology that covers the entire project chain from audit to performance M&V;
- designing a standard performance contract;
- developing appropriate financial mechanisms to fund projects;
- implementing projects and evaluating their impact;
- building capacity of ESCOs and project owners.

The **Framework for Energy Efficient Economic Development (FEEED)** seeks to develop fiscal instruments to promote energy efficiency, including such innovative fiscal instruments and policy measures as the Partial Risk Guarantee Fund (PRGF), the Venture Capital Fund for Energy Efficiency (VCFEE), public procurement of energy-efficient goods and services, utility-based demand-side management, etc. The PRGF is a risk-sharing mechanism that will provide commercial banks with partial coverage of risk exposure against loans made for energy efficiency projects. This will reduce the risk perception of banks towards lending for new technologies and business models associated with energy efficiency projects. The VCFEE will ease a significant barrier from the viewpoint of risk capital availability to ESCOs and other companies that invest in the supply of energy-efficient goods and services.

MARKETS

Between 2003 and 2016, India's energy demand is expected to climb by 60 percent[3] because of rising incomes, accelerated industrialization, urbanization and population growth[4]. Meeting this greater demand by increasing supply only will lead to adverse environmental, economic and security impacts. The BEE considers the promotion of delivery mechanisms for energy efficiency services as one of ten "thrust areas" in its action plan. The Bureau also recognizes the strong potential of the EPC model in delivering energy savings. India's ESCO industry has already seen growth over the past five years. The WRI estimates a compounded annual growth rate of 95.6 percent from 2003 to 2007[5], with ESCOs saving clients an average of 20 to 25 percent on baseline energy costs.

The Indian energy efficiency market is growing very fast. The major segment for potential EE projects are industries, with a share of 48 percent of energy consumption and a 25 percent potential for energy savings. The commercial sector consumes 20 percent of energy, with a 50 percent potential for energy savings[6]. Furthermore, there is significant room for implementing energy efficiency projects through the ESCO mode, although not many projects have been implemented as yet through this route.

Market Potential of Energy Efficiency Activity in India

Energy efficiency offers promise for a robust market to exist under the PPP model (ESCO-based). According to a study led by the Asian Development Bank, the overall energy efficiency investment market size under the EPC model in India as estimated at INR 140 billion (USD 2.8 billion). Until now, only 5 percent of this market has been tapped through the ESCO mode, mainly in the areas of lighting and industrial applications.

FACILITATORS

Alliance for Energy Efficient Economy (AEEE)

The Alliance for Energy Efficient Economy launched in 2007 is an industry association created to facilitate collaboration among India's energy efficiency industries and service providers, thereby enabling them to come together under a policy research and advocacy platform. The AEEE supports the government in working towards (i) implementing energy conservation and efficiency improvements in the country; (ii) reducing energy intensity; (iii) developing innovative business models including performance contacting solutions; (iv) enhancing energy security through the adoption of clean energy; and (v) mitigating the impact of climate change.

The AEEE identified the need for creating a pool of qualified energy professionals integral to India's energy efficiency business transformation. In response to industry demand, the AEEE organized M&V training programs based upon the IPMVP, developed and conducted by EVO. Up to date, the AEEE has conducted six M&V awareness workshops across the country and four CMVP certification training courses and exams on the IPMVP, in association with EVO and the Association of Energy Engineers. As of now in India, around 250 professionals from various sectors (industries, building, government, NGOs, ESCOs, EE vendors, energy professionals, etc.) have been trained in M&V concepts and, out of those, 65 professionals are CMVPs.

Other Key Market Enablers

- The roll-out of the ISO 50001 standard in India is expected to boost the implementation of EE projects through performance-based contracts. The ISO 50001 standard includes a "set of interrelated or interacting elements to establish an energy policy and energy objectives, processes and procedures to achieve those objectives." The ISO 50001 standard is based on the plan-do-check-act (PDCA) continual improvement framework and incorporates energy management practices into everyday organizational activities.
- The Industrial Development Bank of India and the WRI[7] have launched a financial product aimed at driving investments into energy efficiency projects performed by Indian ESCOs.
- The BEE & the Hong Kong Shanghai Bank (HSBC) India signed a memorandum of understanding (MoU) in 2010 to work closely on the EEFP[8]. The EEFP will provide instruments like bankable detailed project reports and other risk mitigation measures to enhance comfort for lenders towards aggregated energy efficiency projects. The MoU will strengthen the EEFP, which seeks to overcome barriers to EE project financing through risk sharing strategies and capacity building of financial institutions.

GOVERNMENT ACTIONS

Establishment of Energy Efficiency Services Limited

In order to develop a viable ESCO industry, and also to undertake the market-based implementation actions set out under the National Mission, a new commercial organization named Energy Efficiency Services Limited (EESL) was created in 2009. The EESL aims to facilitate market implementation of energy efficiency projects. It will work as a Super ESCO, as a consultancy organization for the CDM and energy efficiency as well as a resource center for capacity building of SDAs, utilities, financial institutions, etc. The EESL will also lead the market-related actions of the BEE and the NMEEE.

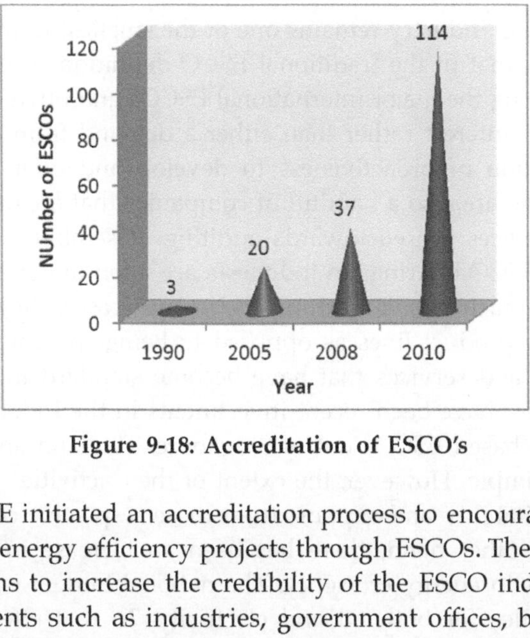

Figure 9-18: Accreditation of ESCO's

The BEE initiated an accreditation process to encourage the adoption of more energy efficiency projects through ESCOs. The accreditation initiative aims to increase the credibility of the ESCO industry among potential clients such as industries, government offices, building owners, municipalities and enterprises as well as among financial institutions that would consider providing capital to ESCOs.

A total of 114[9] ESCOs have been given ratings based on criteria involving success in implementing energy efficiency projects, the ability of technical manpower and financial strength to invest in such projects. Overall, this accreditation process is expected to give ESCOs the incentive to improve so that they can attain higher ratings on the next accreditation round. The ESCO accreditation also serves as a guide for financial institutions when making decisions regarding the creditworthiness of an ESCO.

Investment Information and Credit Rating Agency of India Limited) have carried out this accreditation process.

Indonesia

Christopher Seeley

ACTIVITIES

The ESCO industry remains one of the smallest in the region with few activities that fit the traditional ESCO definition. There is a mix of presence among the major international ESCOs and often their presence is more of an interest rather than either a demand from the market or a representation of proactiveness to develop and offer ESCO-related services. There are also a handful of companies that are offering energy efficiency services geared towards auditing. Resembling a less mature market, the ESCO offerings in Indonesia are often undertaken by direct product manufacturers. Unfortunately, in such cases, they focus on the sales of their product lines as opposed to being neutral or they focus on solutions and services that have become standard in more mature markets. There have been recent investments in the Indonesian market of regionally based ESCOs out of Singapore, Thailand and even Hong Kong for example. However, the extent of their activities are limited to resource availability and support from these respective countries.

On the other hand, there has been an increase in both interest and activities in energy efficiency. The national budget is burdened by heavy subsidies for both oil and electricity. To that effect, the recent announcement that these will be reduced has caused energy efficiency to gain an unprecedented level of attention. In August 2010, the Indonesian Ministry of Energy and Mineral Resources set up the Directorate General for Renewable Energy and Energy Conservation to develop policies and programs that focus on these issues. As a spin-off from the Directorate General for Electricity and Energy Utilization, this new Directorate General is hoped to have more power in taking the lead in laying out policies supporting the development of renewable energy and energy conservation.

The combined energy efficiency market across both the commercial and industrial sectors in Indonesia is estimated to be roughly worth USD 4 billion, which is the largest market for energy efficiency in the whole Southeast Asian region.

In order to successfully implement and sustain projects with performance-based contracts, the project design needs to be simple and allow for flexibility in implementation so as to adjust to changing conditions during the course of a project. It is difficult to promote to financiers projects that are implemented by ESCOs since the Indian ESCO industry is young and lacks the strong credit history preferred by lenders. To date, banks and other financial institutions in India have not become fully convinced that EE project implementation can generate sufficient cash flow to repay the project loans.

Based on the experience gained and the lessons learned from various projects to date, the following are some of the key strategies to overcome such challenges:

- A successful ESCO model requires a strong and well-understood public-private partnership (PPP), which will entail overcoming mutual distrust between these two sectors in India. In addition, there is a general lack of understanding about the full cycle of ESCO projects along with the "roles and responsibilities" of the various parties involved in the project.
- There is a need to adjust the existing contractual frameworks to enable synergies and promotion of EE financing as well as project implementation through the ESCO model.
- Indian ESCOs face severe financial constraints and large international ESCOs are skeptical about investing in Indian public sector projects due to payment security risks and perceived/real risks about maintaining the sanctity of the legal contract. There is a need for secured repayment mechanisms such as trust and retention accounts or escrow accounts.
- Access to financing—most lenders and investors have a limited understanding of the dynamics of ESCO business models and technical issues. Appraisal of ESCO projects requires different techniques and tools from those involved in the appraisal of projects designed to add to productive capacity. Financial institutions should recog-

nize energy efficiency lending as a business strategy and adopt cash flow-based project evaluation for ESCO projects rather than asset/collateral-based lending.

- Long periods for project development have often been cited as one of the biggest barriers that ESCOs face in working with the public sector. The inability of ESCOs to "close the deal" with the government despite negotiations lasting months (or years) means that the total cost of doing the project is much higher than the cost initially estimated by ESCOs.
- Developing and promoting a standardized approach with fixed milestones (project kick-off meeting, site visits by ESCOs, pre-bid meeting, initial proposal, letter of intent to award the project, investment grade audit, final proposal, signing of the contract) can greatly help reduce the project development period.
- EPC entails risks by both parties entering into the contract. A good M&V plan lays out how the baseline will be developed and how the project's performance will be verified to ensure that savings are quantified accurately and within the bounds of uncertainty, as determined by both parties. It also assigns risks to the two parties and provides a framework for how those risks will be managed. The IPMVP can be used as a reference document for such purposes.
- Third-party M&V (for large-scale projects) is growing in popularity worldwide as well as in India. However, when specifying a third-party M&V requirement, it should be made clear that the responsibility of the third party agency is to carry out the M&V plan agreed upon by the two parties and not to question the M&V plan itself.

CONTRACTS

Although both ESCO models are available and used in Indonesia, the predominant one is to provide a performance-based guarantee. However, on several occasions, the contracts that are signed are modified to resemble more of a sales contract rather than the traditional energy performance contract. Typically, they are without any post-implementation plan but do have an extensive monitoring and verification schedule, and/or penalties associated with non-performance over the long term.

LEGAL FRAMEWORK

There is a lack of an existing local legal framework or legal expertise to support clients that are considering undertaking EPC. The clients are left to rely on the explanation from the service provider and the use of their standard contract, which does not allow for a clear and transparent understanding of the finer details pertaining to the contract itself. There are very few sources available in the market that can be used as reference points. This adds an additional risk, not to mention the perceived risk of never having entered into a similar type of contract before as well as the additional costs of engaging legal review from outside Indonesia.

There are currently no specific government-led financing programs or incentive programs to encourage ESCOs or EPC projects (or energy efficiency projects in general for that matter).

It would be ideal for the public sector to take the lead in the use of the EPC and ESCO models in Indonesia and, in this way, demonstrate their value. However, the current procurement framework does not allow for the traditional ESCO model to be used for the retrofit of existing public facilities. There is a requirement for the complete separation of the entity designing a project from the party undertaking its implementation. Hence, two separate bids would need to be released with specific clauses about the inability for one entity to be eligible to undertake both components.

MARKETS

The dominant ESCO market remains in the industrial sector, which reflects a tremendous potential in that regard. The main sectors that have engaged ESCOs are in the textiles, tobacco, food and beverage processing/bottling, and rubber-related product manufacturing segments. The majority of ESCO activities in these respective segments are being undertaken by multinational entities that most likely have had experience with this model.

Projects average an investment of approximately USD 2 to 3 million and they mostly focus on a rather simple scope of work. The latter include equipment replacement or the installation of additional equipment to improve efficiency, such as variable-speed drives, improved process automation devices as well as refrigeration and high-speed freezing systems. Very often, these projects are made up of a single technology application applied to multiples of the same existing equipment. They rarely consist of larger and more complex solutions such as cogeneration, combined cycle on site power plants or new/ innovative sector-specific technology applications.

FACILITATORS

There are a number of facilitators in the Indonesian market addressing various areas relating to energy efficiency activities, including government policy, technical training and development, access to finance and the development of green building codes. However, there is a distinct absence of an association or agency that is specifically focused on, or that represents, the ESCO industry. Additionally, there is no such association or agency which acts to help develop and promote the adoption of EPC as a feasible model for delivering energy efficiency projects in either the industrial or commercial sectors.

While actively promoting increased awareness for improved energy efficiency by the public sector, there is very little information available on ESCOs in Indonesia. Furthermore, there are certainly no dedicated resources that the private sector can approach for support, technical assistance or even general information. There is a common perspective among regional industry practitioners that there are no 'real' ESCOs in Indonesia, although this is perhaps an exaggeration. The Green Building Council of Indonesia (www.gbcindonesia.org) was established in 2009. It has recently launched the Green Building Certification program "Green Ship," which could become a platform for knowledge sharing among ESCOs and other experts involved in energy efficiency services.

GOVERNMENT ACTIONS

The largest and most dominant factor which has resulted in the very slow growth of the ESCO industry in Indonesia is the continued subsidization of electricity costs, applicable to both commercial and industrial usage. This continues to dis-incentivize the uptake of energy efficiency projects and results in these investments being commercially unattractive in most cases. The ESCO projects that have taken place despite this occur when there is already a need for, or planned, capital replacement as well as when a budget has already been set aside for that purpose. Consequently, the marginal cost of investing in more energy-efficient technologies or equipment remains financially viable when including the future capital cost avoidance.

However, this is likely to change for the better. In 2010, the Ministry of Energy and Natural Resources introduced a regulation, which raised industrial electricity prices by 18 percent. This is likely to be introduced into the commercial building sector as well. Nevertheless, there are several barriers to fully enforcing this specific regulation, such as the socio-economic repercussions that are being outlined by these corporate building owners.

The government has embarked on a planned expansion of the current natural gas pipeline, which will pass through the main city area of Jakarta. This pipeline has already spurred the interest of large commercial property owners (hospitals and hotels) in considering the use of technologies such as absorption chillers and even the implementation of small-scale cogeneration systems. These projects represent much larger investments compared to traditional electric chiller systems. Further, the technology application is not familiar to many project owners. As a result, there have already been preliminary discussions about the use of ESCOs to provide a performance guarantee and help mitigate the risks associated with such projects. The government has also supported these activities by offering specific financial assistance for the costs of undertaking the feasibility study and the assessment.

Ireland

Alan Ryan, Cian O'Riordan

ACTIVITIES

Although the Republic of Ireland has a fledgling EPC market, it is at a point from which it may grow and develop rapidly as the public sector turns to ESCOs to guarantee savings and provide alternative financing options for energy efficiency.

During the 1990s, a number of energy supply companies were established to build, own and operate CHP plants at client facilities. These were mainly medium-sized installations supplying Ireland's brewing and dairy processing industries in addition to universities. The model employed was generally electricity supply using an agreed pricing formula, with heat supplied "free."

During the 2000s, a biomass market was established and a number of hardware suppliers used heat supply contracts to help develop the market. This approach continues to develop.

Around the year 2000, Dalkia established operations in Ireland, which were initially focused on traditional facility management services. However, Dalkia quickly moved to promote energy efficiency as a value-added service. This marked the entry of the first international ESCO to the Irish market.

In 2005, the Sustainable Energy Authority of Ireland (SEAI) published an "Assessment of the Potential for ESCOs in Ireland." The latter identified 11 companies that were providing energy supply contracts, generally serving the CHP and biomass markets described above. Only two companies were providing energy performance contracts; i.e., contracts involving a commitment to reduce energy use and with the payment linked to savings.

A recent report by Accenture explored "Barriers to Financing Energy Efficiency in Ireland." Among its key findings, items of relevance to EPC were:

- IMF rules prohibit the public sector from taking on debt;
- energy efficiency is not treated as a priority due to a lack of penalties and adequate incentives;
- a lack of baseline energy data to evaluate potential energy savings from EPC;
- a lack of proven EPC case studies;
- a lack of ESCO services in the market;
- a lack of experience in ESCO procurement, contracting and off-balance sheet solutions; and
- a limited pool of personnel experience in negotiating finance and contract structures.

It is expected that the next five years will see the development of an EPC market, particularly in the public sector, with a number of different models being demonstrated. Many of the barriers to market development will have been addressed and there will be a number of small and large ESCOs actively participating.

As the market is still being developed, a range of energy performance contracts will be tested and the market will decide which models are most suited. These will range from basic energy guarantees for a particular product or service to full energy performance contracts such as variable contract term, guaranteed savings and shared savings.

SEAI will take a lead role in facilitating market development and pilot initiatives. Additionally, it is envisaged that these pilots will involve the development of energy performance contracts, which will eventually evolve into contract templates. A trade organization, Energy Management Professionals of Ireland, has recently been established. To that effect, a number of its members are moving to either become ESCOs or facilitate EPC deployment as independent consultants, including M&V of savings using the IPMVP. Other groups such as the Irish BioEnergy Association and the Western Development Commission are working to develop standard biomass energy supply contracts.

MARKET

Although rising energy prices and greater environmental awareness increased interest in energy efficiency, there was very little appetite for complex energy performance contracts in an economy experiencing rapid progression, new building development, growing government budgets and availability of credit. The economic downturn in 2008 reversed the situation entirely and, combined with a government target for 33 percent energy efficiency improvement in public sector organizations by 2020, rekindled interest in alternative methods to finance energy efficiency, particularly in the public sector.

The aforementioned SEAI study estimated the potential ESCO industry market size at USD 65-135 million per annum. As a result of the above market drivers, it is expected that the EPC market will initially develop to serve the public sector, which spends circa USD 600 million per annum on energy and has a target to achieve 33 percent energy efficiency improvement by 2020. This sector consists of 10,000 buildings, including offices, schools (4,000 spending USD 80 million per annum on energy), healthcare facilities (USD 80 million per annum), water services (USD 65 million per annum), public lighting (USD 35 million per annum) as well as police stations, prisons and barracks.

Market Developments—Public Sector

To date only a small number of energy performance contracts involving guaranteed savings and energy efficiency measures have been established. In 2009, Stewarts Hospital became the first publically promoted public sector energy performance contract (see Case Study for further information).

However, the government has committed to the development of the ESCO market in its Infrastructure and Capital Investment Program (2012-2016). SEAI has established a Public Sector program to assist public bodies in achieving their 33 percent energy efficiency target by 2020. It is also establishing an energy measurement and verification system to report progress by individual organizations. SEAI is actively promoting EPC as a vehicle to help organizations reach their targets. Due to national budgetary constraints, SEAI is also working to unlock TPF options.

An advantage of Ireland's relatively small physical size and population (4.6 million) is that key actors can be pulled together to identify and implement solutions. A number of best practice working groups (WG) have been established by SEAI to develop a coordinated approach to energy efficiency improvements. These are potential ESCO client groups. They include a public lighting WG, a water services WG, an information & communications technology WG and a WG to explore ways of financing retrofit in the public sector.

For instance, a water services WG has been established to address energy efficiency in a sector that spends EUR 50 million (USD 66 million) per annum on electricity. The group has established that, in many facilities, the opportunity exists to achieve 30-50 percent energy savings. A number of the participating local authorities have expressed an interest in bundling 20-40 pumping stations for an EPC-based energy efficiency retrofit trial.

SEAI and the Department of Education and Skills have commenced a pilot project to demonstrate the energy (heat) supply contract and energy performance contract models in schools. It is planned to stimulate the market for deep energy efficiency retrofit projects in cohorts of schools using these innovative contract models.

In a separate development, the Health Service Executive (HSE) is in the process of establishing an energy performance contract for approximately 40 facilities in the south of the country. This will involve engaging an experienced utilities management company to maintain internal environmental conditions and deliver a pre-agreed level of energy savings. The management company will source and install new equipment required to maintain conditions and deliver energy savings. This is a pilot initiative and it is expected that the model will be reused for groups of other HSE facilities in due course.

Market Developments—Residential Sector

In the residential sector, the government has advanced plans to introduce a pay-as-you-save scheme, whereby the loan to fund energy efficiency upgrades in dwellings will be allocated to the utility meter (rather than to the home owner, who may sell the home), with a mechanism to recover this loan over a number of years via utility bill savings.

Market Developments—Commercial and Industrial Sector

The EPC model has been deployed in a small number of facilities. However, it is not well established and is likely to take a lead from pilot projects planned for the public sector.

GOVERNMENT ACTIONS

The 2011 National Energy Efficiency Action Plan identifies planned government actions in the public sector, including a number which will contribute to providing assistance in addressing the barriers to EPC identified above. These include:

- Continuing to develop the Public Sector program, including the WG approach as well as the energy measurement and verification system. The latter item will provide valuable baseline energy data and increase accountability.
- Introducing obligations on all public bodies to develop and implement energy management programs appropriate to their organizations.
- Providing a Better Energy grant program that will deliver retrofit measures to 1,000 public buildings by 2020. Savings will be

achieved, among other things, through utilizing innovative delivery and financing mechanisms.

- Developing an inventory of public sector buildings.
- Refurbishing at least 3 percent of the public buildings each year, requiring each refurbishment to bring building energy performance to within 10 percent of the national building stock.
- Encouraging the development of industry representative groups and work with these to address market barriers.
- Establishing a national advisory service on financing retrofits in the public sector using innovative procurement models, such as energy performance contracts. This will include the development of standard documentation/templates.
- In advance of establishing this advisory service, SEAI will provide expert advice and support to potential demonstration projects for innovative procurement models.
- Investigating options for the provision of private and public finance for energy efficiency projects through innovative project-based financing arrangements.

CASE STUDY

Client:	Stewarts Hospital
ESCO:	Dalkia
Measures:	Fuel conversion from oil to gas. New energy center and district heating system incorporating two 1-ton steam boilers, a 700 kW heating boiler and a 140 kWe CHP. New main distribution board and emergency generator. Controls to optimize the boilers and provide 55 heating zones. Energy monitoring system. Energy-efficient lighting. Second 140 kWe CHP at leisure center.
Contract:	Shared savings contract, covering energy supply and energy efficiency. Financed and owned by the ESCO. The latter invoices on a monthly basis to cover the cost of energy supply, financing, operation, maintenance and energy management. Replacement guarantee on failure. Lower energy costs help offset financing costs. 15 years.
Benefits:	New, reliable plant with no upfront capital cost. Energy savings arising from supply-side and demand-side effi-

	ciencies help offset monthly payment costs. Operational and maintenance risks transferred to contractor. Greenhouse gas reduction.
Investment:	USD 2 million.
Savings:	Circa USD 135,000 per annum.

Israel

Z'ev Gross

ACTIVITIES

MNI's efforts were commenced in 2005 with the help of consultants. A program was mapped where preliminary capacity building efforts would take place, followed by a pilot project in the government sector, to be funded by MNI. It was felt that a pilot "on the ground" would be the best means of introducing the methodology and overcoming the yet ungauged sensitivities to the methodology.

The pilot was carried out in four government facilities which agreed to participate in the project—two police facilities, the offices of the Taxation Authority in one city and one department of a government hospital. The RFP was issued at the beginning of 2006 and successfully carried out during that year by the awardee. The projects were carried out under the guaranteed savings mechanism, with MNI taking upon itself the bulk of the risk—the project costs were covered by government budget (MNI), with the guarantee structured in a way that left the ESCOs with the least exposure. The savings were guaranteed to cover the investment within 2.5 years, a payback period validated by the results.

Much was learned from the pilot, including: (i) the difficulty of government entities to grasp the concept of a procurement process which is, ostensibly, not defined by the tender, but by the response to the tender; (ii) the difficulties involved in creating an audit process within the RFP methodology (extensive, seemingly unnecessary costs for the ESCO—a large risk factor); (iii) implementation issues involving interference with ongoing operations; (iv) M&V issues; and (v) supervision issues.

The stage was set for creating the "economic environment" that would enable the growth of the energy services sector in the field of energy efficiency. The initial work was aimed at further capacity building efforts, with a number of workshops carried out, targeting ESCOs and the financial community. In parallel, MNI created a registration process for ESCOs that was intended, for the most part, to (i) "assure" some sort of financial strength for the ESCOs; and (ii) require ESCOs to adhere to a minimal set of "good practice" rules (which were adopted, actually, from a similar document developed by the city of New Orleans in the US).

However, the major breakthrough came in 2008, when two important government resolutions were adopted in the area of energy efficiency. First it was resolved that the 2020 reduction target for energy efficiency would be 20 percent or 16 TWh per annum. The aggregate reduction (assuming attainment of the goal on a straight-line basis) was to exceed 100 TWh up until and including 2020. This was, clearly, a serious challenge. The resolution also created an inter-ministerial committee to determine the actual program for attaining this goal.

The resultant program was adopted by the government in a resolution dated September 18, 2008. The gist of the resolution was a detailed demand for energy efficiency in the government sector, a call for a similar, but not detailed, program in the municipal sector (a government resolution does not bind the municipalities), as well as general language aimed at creating regulatory and financial "tools" for the support of energy efficiency projects.

Although not said in so many words, the upshot of this resolution was to put all resource acquisition activities in the field of energy efficiency in the hands of the private sector, availing it of all the rewards for its activities while also placing the risk of success at its doorstep. Provisions of the resolution essentially freed the government of the requirement that it have budgets to finance energy efficiency infrastructure investments as a prerequisite for tendering. Government entities were allowed to go into "off-balance-sheet" arrangements. Supporting provisions of the resolution called for dedicated budget line items for energy procurement, for systematic budget reductions in those line items, for incentivization of employees of government units attaining savings and for maintained budget levels for units in the course of M&V stages under the shared savings mechanism (in order to assure funds for payment to the ESCOs).

In addition, MNI included performance contracting transactions in its "micro-projects" program, making the implementation of the performance contracting methodology, in and of itself, a sufficient prerequisite to qualify for funding.

MNI also sought other tools to create the "environment" necessary for the performance contracting methodology. They were:

- Initial steps were taken to determine the feasibility of a loan fund or a guarantee fund. MNI and Econoler interviewed members of the financial community in order to determine their interest as well as their comments. An initial document was drawn up, although it was not taken up by the Ministry (probably in light of failure by another ministry to create a similar type of fund). Attempts to differentiate between the two instances were unsuccessful and the matter did not move forward.
- A concerted effort was put into creating taxation incentives for ESCOs. The issues involved included recognition of investments as business expenses (rather than capital investments—allowing immediate write-off), harmonization of legal and accounting definitions of equipment ownership (which would assure the ESCO's right to claim depreciation on equipment not yet transferred under shared savings agreements) and accelerated depreciation. Certain provisions were found to already appear in the tax code (immediate write-off of yet unclaimed depreciation in situations of equipment replaced during "depreciation life" and abandoned by the entity). Certain requests were taken under advisement to be dealt with on a transaction-by-transaction basis. Only accelerated depreciation was accepted outright, which (depending on the view of the Taxation Authority with regard to the accounting "nature" of the shared savings agreement) would be an advantage in the hands of the ESCO or the client.
- A fund similar to the System Benefit Charge (or charges similarly titled or administered) was proposed. This fund, inter alia, was to have funded various programs in various sectors and driven the creation of a more professional ESCO sector. The fund would still only support private sector efforts where it was felt funding could overcome material barriers. The fund initially proposed by MNI's energy efficiency division called for a surcharge on the electricity bill (with a similar charge on other fuels), paid by the consumer.

MNI's management did not agree (there had been, during that time, complaints about stiff surcharges on water and MNI was sensitive to the issue), and instead, proposed a payment into the fund from the utility. The failure to come to terms with the Israeli regulator on the issue of recognition of this payment as a utility expense (to be factored into electricity rates), as well as the issue of involving the utility in energy efficiency without decoupling provisions, led this effort to fail.

In parallel to these "regulatory" efforts, there was an attempt to introduce performance contracting into the market via projects. As described above, the first endeavor was in four government projects and was carried out by MNI using the guaranteed savings mechanism (whereby the project was paid for out of the MNI budget and a surety issued for the savings by the ESCO). After the success of this effort, MNI incorporated performance contracting into its "micro-projects" program. Under this program, energy consumers could submit projects to be performed under three headings, and receive funding, if found eligible. The three headings were (i) electricity projects (aimed at efficiency in electrical systems); (ii) performance contracting (any performance contracting mechanism for any type of energy); and (iii) renewable energy (specific renewable energy technologies with certain caveats). The combination of percentage of project cost and overall cap on the grant led to projects of a general scope of about USD 100,000.

The micro-projects program was quite a success—it averaged about 35 projects a year, consistently exploiting the available budget fully. Performance contracting represented about 35 percent of the projects. One company actually accumulated a portfolio large enough to offer it as an aggregated package to a financial entity for refinance.

The government entities also began showing an interest. Towards the end of 2008, MNI was approached by the Ministry of Health (MoH) to assist it in creating a project for all government general hospitals (a total of 11) at once. MNI, commenced developing this project with MoH. The project went through many phases of development, including its expansion to include an additional eight hospitals belonging to Israel's largest health fund. Issues of budget prevented the project from being developed under the guaranteed savings mechanism, and issues of employee rights prevented the project from being developed under the "chauffage" mechanism. As a result, it was decided to develop the project under the shared savings mechanism.

MoH issued a pre-qualification document relating to a project in all 19 hospitals in September of 2011. The solicitation was international and it was expected and hoped that some of the experienced, international ESCOs would participate.

In parallel, as was hoped, the Accountant General of the Ministry of Finance (AG), the official responsible for the management of government assets, saw the potential in the performance contracting methodology, and established an inter-ministerial tender committee to implement this methodology on all relevant structures under the jurisdiction of the Government "Housing" Administration (administers properties occupied by government offices).

The first tender was issued in 2010 for a project in a group of six courthouses, an educational campus and the offices of the Ministry of Finance in Jerusalem. Due to the inexperience of the Tender Committee, the project was not developed sufficiently well. As a result, only one ESCO responded to the tender. The Tender Committee, however, decided to continue its efforts, and, after a series of attempts to try and better understand the market, is developing a roster of structures for projects, which, it is expected, will be substantially more successful.

Another government office which decided to implement the shared savings mechanism is the Ministry of Education in a project for its offices in Jerusalem. The tender for the project was to be issued by the end of 2011.

There are not much data on what is happening in the private sector. There are a number of projects reported using the shared savings mechanism and some using a "chauffage" type contract for the purchase of energy or hot water (recently solicited by a large hotel chain). The consensus, however, is that the sector has not "taken off." There is now an attempt to unite the ESCOs under a professional organization, which will attempt, at the outset, to create greater demand for energy services.

A number of issues plague the energy services sector. The use of the shared savings mechanism has created unwarranted expectations. Clients today see ESCOs as banks and view the transactions with them as strictly financial. Efforts are now being made to re-educate the public as to the nature of the energy efficiency transaction, the nature of the ESCO and the importance of the bilateral relationship between the client and the ESCO, without which the transaction will fail.

A second issue concerns finance. The local banks do not understand the concept of the "cash flow from savings." Coupled with the relatively small size of Israeli projects (which, therefore, have relatively high transaction costs), they are not willing to give non-recourse (or, for that matter, even partial recourse) project financing. The result is not so much the cost of the finance, but the sureties demanded by the banking system in order to grant loans. Attempts were also made to approach institutional investors. Here, the issue was the relative size of the projects as well as the very short timetables associated with them.

A third issue is that of measurement and verification (M&V) of savings. This matter is a complete mystery to both the average client as well as the financier—and to a great extent also to the average Israeli ESCO. The M&V plans submitted have been sketchy at best. This has led to mistrust on the part of the clients, as well as skepticism on the part of the financial community.

GOVERNMENT ACTIONS

In November 2010, the government had a "change of heart" and adopted a resolution aimed at cutting back GHG emissions. At the core of the program was the program issued by MNI for energy efficiency, even though the differences between the economies of GHG reductions and energy efficiency in a capacity starved environment are great and clear. The government also budgeted approximately ILS 300 million (USD 80 million) for the years 2011-2012, and a total of ILS 2.2 billion (USD 0.6 million) until 2020. Approximately ILS 160 million (USD 42.7 million) is budgeted for projects in non-residential sectors.

The major change of heart is the extent of government intervention in the market. Inasmuch as the projects will be more heavily supported by the government, serious ESCOs should require less bank finance to carry out their projects. In addition, clients will not have to invest as much, possibly making them more amenable to such projects. The requirement for validation imposed by the government support programs should pave the way for more intensive M&V training—which, in turn, should give ESCOs the tools to deal with a growing energy services sector. It should be noted, however, that the non-residential project programs require CDM-type measurement processes and methodologies, which do not necessarily comply with other more common methodologies in the field of M&V (IPMVP).

Notwithstanding the differences, it is hoped that the requirement for M&V will become more inculcated into the energy efficiency project environment, leading to more accurate project design and implementation.

The Israeli ESCO saga is essentially sourced in a technical absurdity found in Israeli law. Under Israeli law, for a piece of legislation to bind the government as an operating entity, the legislation must clearly state that it applies also to government. This provision was lacking in the Energy Resources Law 5750-1989, the law which governs matters of energy efficiency in Israel. The resultant situation was that regulations aimed at attaining energy efficiency applied to every consumer EXCEPT government consumers. An attempt in 2005 by an Israeli non-governmental organization (NGO) to challenge this situation in the High Court of Justice failed.

Thankfully, this absurdity was rectified in 2011 by an amendment to the Energy Resources Law. However, the Ministry of National Infrastructures (MNI) found itself, even before this court challenge, unable to persuade government entities to move forward on energy efficiency projects. This led MNI to turn to performance contracting as a possible solution and to extend its energy efficiency goals to include the creation of a robust energy services sector.

Italy

Ettore Piantoni

ACTIVITIES

Italian energy services date back to 1990 when Law no. 10, 1991 introduced energy services contracts (Contratto Servizio Energia) mainly applied by primary fuel suppliers in the residential and tertiary sectors (public administration buildings). Their strategic intent was to move up in the value chain by securing the sale of primary heat energy by including in the selling price operations and maintenance (O&M) costs of energy equipment. In case of investments in new equipment, the contract duration was pluriannual with a yearly reimbursement of the cost of capital.

Since then, market needs and new legislative provisions designed to serve a liberalized energy market have caused the market to evolve. The opening of the market to private companies in 1999 triggered the entry of more specialized players who started to operate either on the supply or on the demand side of the energy market.

ESCOs began to develop as a product or service extension of the core competencies of companies already active in the market, such as utilities, installers, engineering equipment suppliers, facility management, and consultancy companies.

Applicable Legislation

In 2004, the innovative mechanism of tradable certificates, "Energy Efficiency Certificates or White Certificates" (Decree of July 20, 2004 issued by the Ministry of Productive Activities jointly with the Ministry of the Environment and Land Protection), was enacted as an incentive for energy efficiency measures. These certificates are known as Titoli di Efficienza Energetica or TEEs. Electrical energy and gas distributors have the yearly obligation to deliver to the Autorità per l'Energia Elettrica ed il Gas (AEEG) a certain number of TEEs (proportional to their sales volume). The TEEs are issued by the Gestore Mercati Energetici (GME) for certified energy savings of primary energy.

There are four types of TEEs, which have a nominal value of one ton of oil equivalent (toe):

1. Type I, certifying the achievement of primary energy savings through projects reducing final electricity consumption;
2. Type II, certifying the achievement of primary energy savings through projects reducing natural gas consumption;
3. Type III, certifying the achievement of primary energy savings through projects other than electricity and natural gas (coal, fuel oil);
4. Type IV, certifying the achievement of primary energy savings through projects in the transportation sector.

Electricity and natural gas distributors may achieve their mandatory targets, set yearly by the AEEG (based upon the previous year sales volume), both by implementing energy efficiency projects (and obtaining TEEs) or by purchasing TEEs from the following third parties:

- companies controlled by electricity or gas distributors;
- ESCOs
- end users (with an appointed energy manager in accordance with Law 10/1991).

The GME issues the TEEs after AEEG approval, it organizes and manages the trading platform of TEEs and, jointly with the AEEG, it formulates the operation rules of the energy efficiency certificates market.

In the energy efficiency certificates market:

- distributors may purchase or sell certificates, while fulfilling their obligation;
- ESCOs or end users may sell the certificates that they have obtained through their own projects, as they do not have to fulfill any obligation and may thus make a profit by selling their certificates on the market.

The AEEG website (www.autorita.energia.it) and the GME website (www.mercatoelettrico.org) provide reports on the TEE mechanism and market.

More recently, Decree no. 115/2008, adopting European Directive 2006/32/CE about energy efficiency, sets the ground for the development of the current ESCO market.

Figure 9-19 reports the main legislative sets of laws applicable to ESCOs. The laws include European directives that were expected to be issued in late 2011 or the beginning of 2012.

Regulations have a great impact on the profitability of energy efficiency services and therefore on the ESCO business. For comprehensiveness, a detailed analysis is needed periodically as values and rules may change over time.

The fiscal regulation applicable to ESCOs refers to a value added tax (VAT) and an "accise" (inland excise tax) imposed on the use of energy products.

While the VAT is "ad valorem" (proportional to economic value), the inland excise tax is applied to the end user's consumption and is specific per unit of measure (EUR/kWh for electricity, EUR/Sm3 for natural gas, etc.). The inland excise tax applies to the final users of electricity, natural gas, fuels and biofuels in relation to a specific use

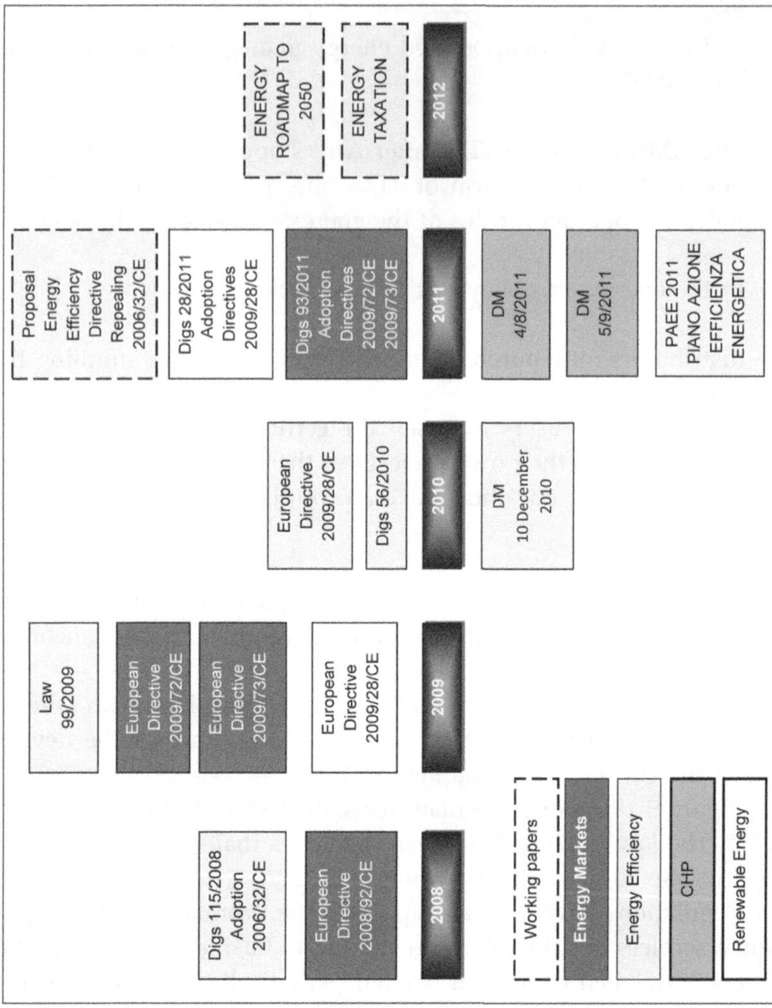

Figure 9-19: 2008-2011 Legislative References

(natural gas used in cogeneration, electricity production, heat production in industry, heat production for buildings, district heating).

Major barriers to a more solid and sustainable ESCO industry are:

- lack of information and understanding of EPC at broad level (customers, consultants, financial institutions, procurement);
- lack of commitment to energy policy at broad level;
- bad will from failures of unqualified ESCOs;
- uncertainty of legislative, fiscal and regulatory framework.

CONTRACTS

In Italy, energy service providers can offer a wide variety of contractual agreements. The energy service provider has to achieve improved energy efficiency.

The general work flow of an ESCO is a typical stage-gate process, as reported in Figure 9-20.

The ESCO, certified under national standard UNI CEI 11352, has to deliver an EPC project with guaranteed efficiency improvements and must meet other agreed upon performance criteria. Customer payments to the ESCO are proportional to the actual energy efficiency improvements delivered by the ESCO.

Tables 9-11 and 9-12 summarize the minimum requirements of an energy efficiency service and the risk assessment of an EPC project.

These tables should help customers identify the EPC model in comparison to other energy service agreements.

In Italy, the most commonly used contract is the energy service contract (or heat service contract "Servizio Energia" also known as "chauffage"). In this agreement, the customer has to pay a service fee (irrespective of energy consumption) and a supply charge of EUR/MWh related to energy consumption. The service fee may include provisions for the investment recovery of the new equipment (lease, TPF, own investment from the supplier) for which ownership is transferred to the customer at the end of the contract.

While the amount of energy consumed by the customer is typically linked to quantitative parameters (degree-days for heating), most of the time, there is no commitment to energy efficiency improvements versus the baseline.

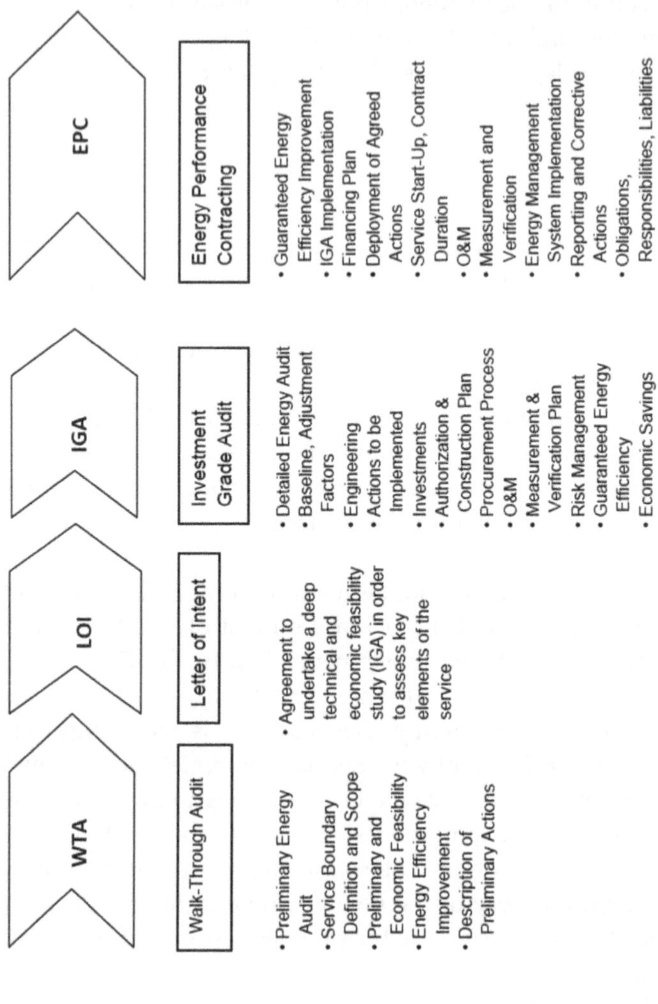

Figure 9-20: ESCO Workflow

Table 9-11: Energy Efficiency Services Check List

Process	Outcome	Actions	
Energy Audit	Baseline	Data gathering	✓
		Measurement campaign	✓
		Scope of services	✓
		Baseline and adjustment factors	✓
		Energy efficiency measures	✓
Energy Efficiency Measure Implementation	Guaranteed energy efficiency improvement	Equipment technical specifications	✓
		Operation and maintenance	✓
		Continuous improvement	✓
		Guarantees, liabilities, obligations	✓
Measurement and Verification	Definition of new energy efficiency level	Measurement, verification and reporting plan	✓
		Meter accuracy and calibration	✓
		Effect of non-compliance with the contract	✓

Table 9-12: Risk Assessment

Risk Retention	Customer	ESCO
1. OPERATIONAL		
Non-compliance with energy efficiency improvement commitment	☐	☐
Improper design (regulatory, safety and environmental maintenance)	☐	☐
Overrun O&M costs (ordinary – extraordinary maintenance)	☐	☐
Improper equipment or plant performance versus design conditions	☐	☐
Reliability – working time versus committed	☐	☐
Non-compliance with applicable legislation, standards (safety, environment)	☐	☐
Difficulties in obtaining requisite licenses or permits	☐	☐
Responsibility for measurement and verification	☐	☐
Customer consumption variances	☐	☐
2. FINANCIAL		
Energy prices, index – volatility	☐	☐
Investment cost overrun	☐	☐
O&M cost overrun	☐	☐
Delay in service start-up	☐	☐
Change of law	☐	☐
Customer payments	☐	☐
Effects of non-compliance with contractual obligations – liabilities	☐	☐

In the energy service contract plus model (heat service contract plus, "Servizio Energia Plus"), an upgrade of the contract discussed above, the supplier has to implement energy efficiency measures to get a minimum of ten percent savings of primary energy (winter heating) versus the theoretical value of the energy efficiency certificate of the building. Both the energy efficiency certificate and energy efficiency improvements are calculated with reference to standardized methodologies and do not refer to meter measurements.

EPC agreements are scarcely offered in the market and it will take time to have them widely spread across industry and the tertiary sector.

MARKETS

The market with the most potential for energy services is the "utility market" since all final users have the same need for reducing energy costs while meeting other agreed performance criteria (comfort).

Figure 9-21 analyzes market participation. In addition, there are several special purpose vehicles (SPVs) that were created for a specific project.

The market is being served by many players (about 150-200), from which a few are subsidiaries or departments of large international companies and the remaining are small and medium enterprises, mainly local, with a very limited number of projects.

Only a limited number of companies have the technical and financial capabilities to provide and sustain a long-term EPC contract. It is therefore difficult to estimate the size of the market. There have been different estimations made from business information companies, associations and government institutions (Enea). Unfortunately, they do not use the same definitions and the market estimation is not always comparable. In a recent publication from a market research company, the market value is estimated at about EUR 450 million/yr, (USD 592 million/yr) which includes revenues of companies whose activities derive from energy efficiency improvements. The market is growing at 10 percent per annum, sustained by public lighting projects, while it remains stagnant in the industry sector.

Including other energy-saving activities not based on EPC contracts, the total market value is about EUR 1.8 billion/yr (USD 2.4 billion/yr). Other sources estimate this market in the range of EUR 5-6

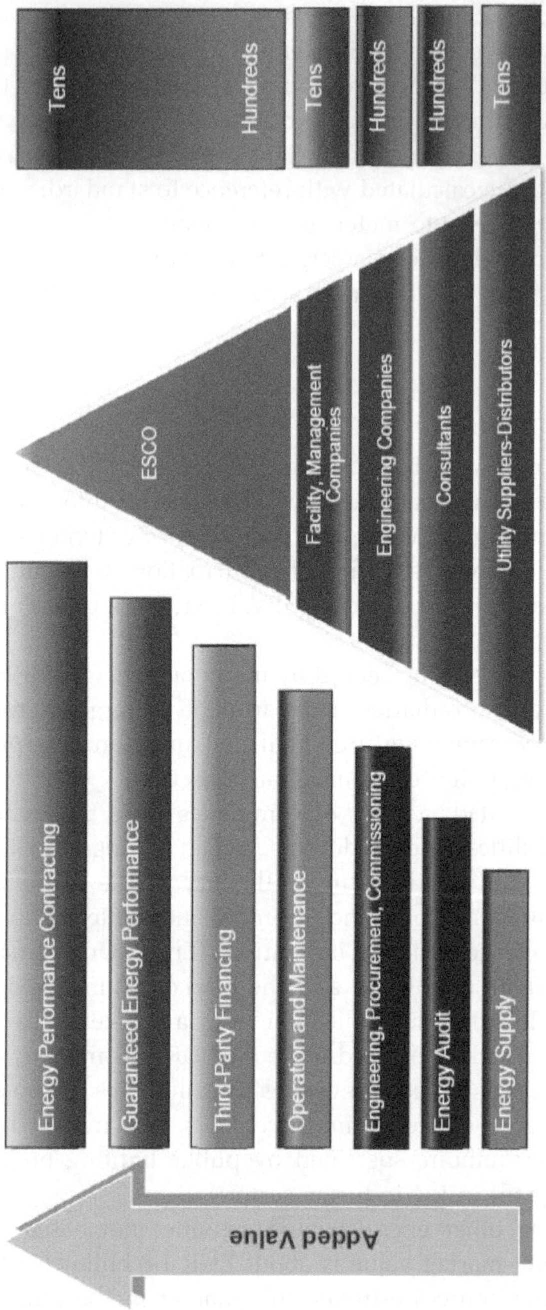

Figure 9-21: Market Participation

billion/yr (USD 7-8 billion/yr) (PAEE report, 2011). Hence, the ESCO market is still underdeveloped. However, there is a huge potential for growth by replacing existing service contracts with contracts that do not have any energy efficiency guarantee and adding the new market potential. This has been identified in the government document "Piano Azione Efficienza Energetica" (PAEE) issued in June 2011.

FACILITATORS

Since the market is served by different players from utility suppliers and technology providers to ESCOs, there are numerous associations involved in the ESCO business. The following associations can be found within CONFINDUSTRIA (Federation of Industrial Enterprises):

- AGESI, Associazione Imprese Facility Management ed Energia;
- ANIMA—ITALCOGEN (Associazione Costruttori e Distributori Impianti Cogenerazione, cogeneration plant installers and distributors);
- ANIE (Associazione Imprese Electrotecniche, electrotechnical companies);
- ASSOELETTRICA (association of electrical energy producers);
- ASSOGAS (association of gas producers and related services);
- FEDERUTILITY (utilities' producer association).

COGENA (Associazione Italiana per la Promozione della Cogenerazione) is part of CONFCOMMERCIO (federation of commercial enterprises).

There are also independent associations, such as AssoEsco and FederEsco.

This scattered situation does not support unique initiatives at legislative level to develop a qualified market.

GOVERNMENT ACTIONS

The major actions that will support market growth and unlock investments are:

- a stable legislative, fiscal and regulatory scenario;
- an effective communication program to create knowledge about energy management standards and the ESCO business model to enhance long-term competitiveness.

These actions should allow financial institutions to shorten the TPF evaluation time and develop a financial structure (tenor, grace period, repayment schedule) around EPC.

Jamaica

Eaton Haughton

ACTIVITIES

Energy saving companies in Jamaica date back 30 years when Eaton Haughton founded Econergy Engineering Services in Ocho Rios, St Ann, in 1981. The main aim of Econergy was to provide maintenance services primarily to the island's hotel industry but with an eye on providing energy conservation services. By natural progression, the two services were eventually separated with the energy conservation department being registered as Caribbean ESCo in 2004. Econergy Engineering and Caribbean ESCo now co-exist as separate entities.

Over the years, Econergy Engineering has spread its wings across the Caribbean region. At the inception, in 1981, Econergy caught the eyes of hoteliers who were concerned with their high energy bills and Club Caribbean, a 180-room resort, became the first property to request an energy audit from the new company. The subsequent recommendations were partially implemented but despite obvious savings available, the owners at the time seemed to lack the will to fully implement the proposals.

For the next two decades, little happened. This could have been caused by the lack of preparedness of Jamaicans to embrace the concept of an energy saving company. Another factor could have been the lack of financial incentives and public education provided by the government on the importance of energy-saving practices, despite the country's oil bill being the main consumer of scarce foreign exchange.

There is also a school of thought that suggests that Jamaicans, especially in the commercial and industrial sectors and the government, believe that there should be a separation between consultant and contractor in the energy services process. In the true sense, however, it is better for the ESCO to deal with the entire process rather than separating consultant and contractor.

Despite the apparent lack of interest, Haughton, who could aptly be described as the pioneer of the ESCO concept in Jamaica and the Caribbean region, maintained his presence, keeping his company afloat through its maintenance services portfolio. In addition to these lifesaving activities, he kept his hopes and business alive by carrying out individual energy efficiency improvements and renewable and alternative energy projects in various industrial, commercial, hotel and government sectors. These included energy audits, major solar water heating projects for health facilities and hotels, air conditioner heat exchangers, solar PV street lighting, hotel chilled water cooling systems, 3 MW hotel cogeneration systems, etc. Effective efforts for survival also included direct contracts and subcontracting work with CARICOM Secretariat, OLADE and international consulting establishments such as, DSI (US), Econoler Inc. (Canada), ALLPLAN Gmb (Austria) and Egis Bceom International (France). These contracts engaged his work in all of the English-speaking territories across the Caribbean region over a period of two decades.

In 1996, it appeared as if the government was finally prepared to establish legislation to deal with energy saving with the draft of an energy efficiency building code and, at about the same period, a cogeneration policy. However, 15 years later, it is still a draft.

The drought was basically broken when, in 2001, the state-owned Petroleum Corporation of Jamaica (PCJ), under the stewardship of the late Dr. Raymond Wright, commissioned Econergy Engineering Services to conduct an energy audit and implement recommendations. This was the first Jamaican ESCO experience and it became a success story.

Econergy obtained a loan of USD 219,000 from E&Co, LAC (a facility that provides financial support to special energy-related projects in developing countries worldwide) to implement the recommendations of the PCJ energy audit. The loan was to be repaid over 42 months. One of the main objectives of E&Co was to highlight the significant impact ESCOs could have on the Jamaican economy and that of other developing countries. The project included optimization of operations and maintenance (O&M), installation of an energy management system, lighting retrofits, variable-frequency drives for fans and pumps, high-efficiency chillers, and a 1,800 ton hours, partial storage thermal energy storage system.

The project was implemented in its entirety and PCJ, satisfied with the level of savings it was getting, was able to repay the loan over an 18-month period.

The University of the West Indies (UWI), Mona campus, in 2006 became the second Jamaican ESCO experience, somewhat on a partial basis. Initially seeking external funding, the institution later reverted to self-funding selected energy projects included in the recommendation. Eleven projects were chosen, of which four have been implemented. The campus has recorded substantial savings as a result. In fact, despite significant growth in the student population over the last five years, the institution has managed to keep the 2011 energy consumption at about the 2006 level. The projects involved the implementation of an energy management system, an air conditioning refrigerant change from Freon (R 22) to hydrocarbon (R 22a), power factor correction as well as lighting retrofit by replacing T 12 fluorescent tubes and magnetic ballasts with T-8 tubes and electronic ballasts. The benefits of the refrigerant change on electricity demand are shown in Figure 9-22.

Both scenarios underscore the importance of ESCOs to the bottom line of any energy-consuming operation. Despite this, the last 30 years in Jamaica have remained virtually stagnant in terms of ESCO progress.

At a recent presentation in Kingston, Haughton highlighted several hurdles to energy project implementation. These include lack of consistent energy policies and strategies, absence of energy expertise, potentially high debt burden that makes guarantees difficult, low business acceptance of renewable energy technologies due to cost structures favoring fossil fuel and reluctance of banks to fund energy efficiency projects.

Other obstacles underlined included the lack of awareness of the value of energy efficiency measures and the unavailability of more than one ESCO in the market place.

However, there could be changes ahead. The government, through funding from the IADB and the EU, has embarked on a new program of project financing and funds are now in place for companies

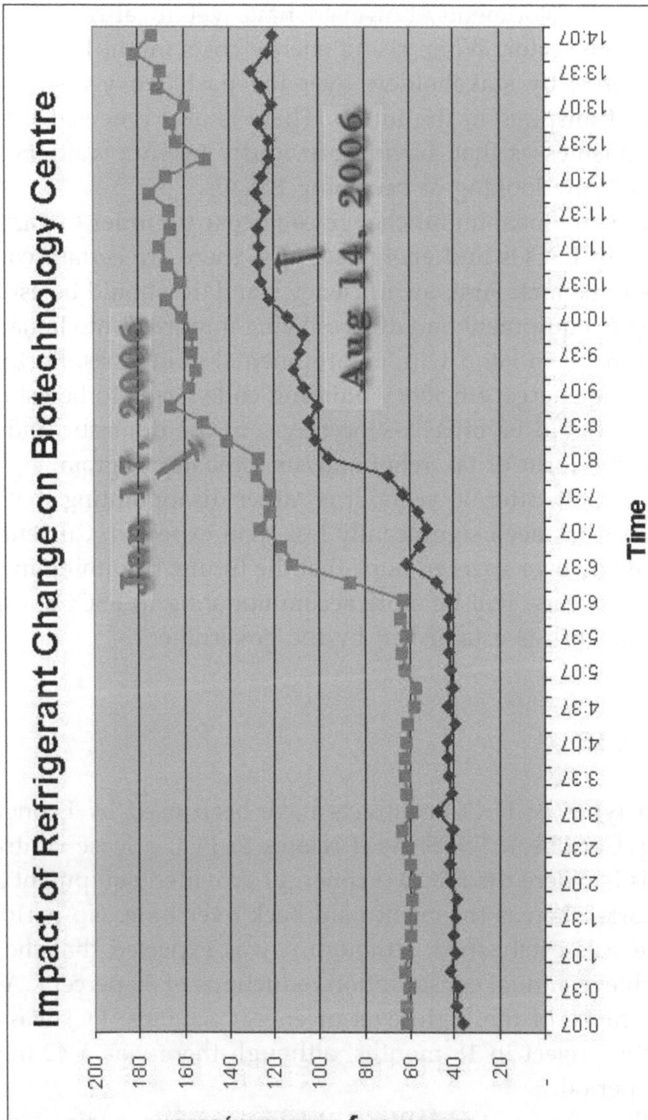

Figure 9-22: Impact of Refrigerant Change on Biotechnology Centre

to finance their energy audits and implement necessary recommendations. Commercial banks have also been taking a more serious look at providing financial support through specific lending programs.

These developments, however, have yet to attract the interest of the private sector. While rising energy costs might have forced renewed interest by stakeholders over the past two years, it is to be seen what transpires in the future. There is also renewed interest by technical businesses that, having previously taken a wait-and-see approach, are now looking at becoming ESCOs.

Before any meaningful change can occur (in order to garner public interest in ESCOs and energy saving), there are certain conditions that have to be met. First, an efficiency standard should be established pertaining to equipment and devices being imported into Jamaica. This should go hand in hand with environmental guidelines. Furthermore, a stand-alone energy efficiency building code needs to be established. There also should be effective incentives to the domestic and private sectors in the form of tax relief and/or a rebate program.

As a result, after 30 years, it is rather disappointing that growth in the sector has been significantly less than expected. Current indications have however sparked hope that the future, beginning in the next two to three years, will be more accommodating to ESCOs, especially if the framework is established by the government.

CONTRACTS

Two types of ESCO contracts have been used by Econergy Engineering/Caribbean ESCo. As it relates to PCJ, a lease contract was used. This is where the ESCO (Econergy) provided equipment and did all the work whereas the client paid back over time. No performance guarantee was established, although it was expected that the project would achieve annual consumption reductions of 30 percent. As stated earlier, because of the high level of energy savings, PCJ was able to pay for the project in 18 months, although there was a 42-month repayment period.

In the other instance, the UWI was offered a direct purchase contract. The UWI was able to fund the implementation of the project on its own. As stated previously, in 2011, the university recorded energy peak demand and consumption levels at about the 2006 rates, a major accomplishment that could only have been achieved through the implementation of the energy efficiency improvements.

LEGAL FRAMEWORK

No legal framework has been established on a national basis, despite a draft in 1996. The only legal framework therefore, relates to contracts that might develop between the ESCO and the facility owner. The current situation sees local entities relying on the use of US templates to conduct business. The situation is one that needs urgent attention by the government.

MARKETS

ESCOs should be able to offer energy (BAS) management systems, high EER AC and chillers; solar PV and thermal (water heating); high-efficiency refrigeration systems; lighting retrofit for industrial and government sectors as well as cogeneration systems in hotels and food processing industries.

FACILITATORS

Potential facilitators are present but have not been impacting the sector in any significant way. Among them are the Jamaica Solar Energy Association, which has been dormant for the past three years despite having personnel in positions of chairman and secretary; the Jamaica Association of Energy Engineers, which was recently developed in association with the University of Technology (UTech) in Jamaica (the association membership at this time is limited in field experience); and the Jamaica Institute of Engineers, which has shown no strong interest in the field. It is to be noted that there is no local energy efficiency agency.

As it relates to IADB assistance, this is extended mainly to government agencies, although recently small and medium enterprises have been able to benefit from an audit program which should springboard to project implementation. The same applies to EU funding.

GOVERNMENT ACTIONS

Government non-action rather than action could be blamed for the relative stagnation in the industry. There is no energy efficiency standard in place and the Jamaica Bureau of Standards (JBS) has failed to wholly deliver on the energy efficiency building code.

A cogeneration policy which had been established was later cast aside. In 1995, the Government of Jamaica approved and published an Energy Sector Policy and Strategy paper, which was to have been revised in 2005. The paper was essentially a description of the existing situation and also focused on the potentials from indigenous energy supplies and energy efficiency measures. One of the main pillars of the policy was the use of cogeneration, which would have been pursued. This has never been implemented. It is believed that the privatization of the state-owned Jamaica Public Service company (JPS) made the policy dormant. There has also been indecisiveness on tax incentives for energy conservation and alternative energy systems and equipment.

There has been no real support for the ESCO sector, but this could be due to ignorance of the concept and benefits.

Japan

Chiharu Murakoshi, Hidetoshi Nakagami, Ji Xian

ACTIVITIES

It has been 15 years since the ESCO industry in Japan began (1996). As a first step in implementing the ESCO development phase, the ESCO market grew year by year supported by the government. However, after the 2008 financial crisis, the economic recession seriously hit the development of the market. Most Asian countries tried to reconstruct the ESCO market in the shortest period of time. By contrast, Japan is still slipping into stagnation as this is written. After the March 11, 2011 earthquake and the nuclear power plant accident, a gradual decline of the power-reserve margin and a reduction in power consumption have now become a significant problem in East Japan. Because of this situation, the government ordered large facilities served by Tokyo Electric Power Company and Tohoku Electric Power Company to reduce peak weekday electricity consumption, and undertook a "Large Energy Saving Campaign" urging people to save power. Therefore, due to the limiting conditions for operation of nuclear power plants, Japan is under pressure to fundamentally reconsider and review its energy policy and targets for reducing GHG emissions. Against this background, although energy saving becomes relatively more important, for now, the energy efficiency business is still limited in its growth.

The ESCO market was seriously hit by the 2008 financial crisis and, to date, there has not been any major sign of recovery. And after this financial crisis, there has been a significant drop in capital spending and delays in public market development. All these have led to an economic recession affecting the ESCO industry's market which is down all over the country.

Since the March 11, 2010 earthquake and the nuclear power plant accident, the Japanese government has been preoccupied with responses to sensible solutions. More than a year and a half has passed since the disaster, and to date there have not been any drastic and concrete policies put forth by the government. Therefore, in order to address the energy challenges that we are facing now, a new national energy plan and policy should be set up for the future.

To expand the size of the ESCO industry new policy proposals are indispensable, such as the Supplier Obligation and Emission Trade System designed to induce investments in energy saving. Increases in investments in energy efficiency and related policies to reduce default risks are also required. Finally, central government facilities should begin to get into ESCO projects to expand the public market.

Japan is under pressure to fundamentally reconsider and review its energy policy and targets for reducing GHG emissions. Moreover, as the necessity for measures against global warming becomes strong and regulations are strengthened, the ESCO industry is expected to expand.

CONTRACTS

During the initial phase, most ESCOs used the guaranteed savings approach but the shared savings approach started spreading from 2000 and accounted for 68 percent in 2002. The industrial sector's shared savings contracts accounted for 68 percent of the total ESCO contract order value in 2002. However, this proportion fluctuated and fell down to 47 percent in 2004 with a dramatic increase to 71 percent in 2005 before eventually dropping steadily from 2006, resulting in a 23 percent share of industrial shared savings contracts by 2008. Sharp shrinkage of equipment investment in each company after the 2008 financial crisis was considered as the main reason.

On the contrary, the commercial sector almost always used the guaranteed savings approach in the beginning (1998 to 1999). It accounted for about half of all ESCO contracts in 2000, 37 percent in 2001 and continued to shrink to 8 percent in 2007 before it picked up to 35 percent in 2008. Expansion of the ESCO market led to a higher growth rate for the shared savings approach, by comparison, shrinking of the market resulted in a higher growth rate for the guaranteed savings approach.

Overall, the industrial sector was slightly larger in project scale than was the commercial sector. The scale of each project in the industrial sector showed a wave-like curve, with USD 2.4 million in 2003 and USD 2.3 million in 2007, while it went down to USD 0.8 million in 2008. By contrast, the commercial sector has remained relatively stable over the past seven years, with USD 0.4 million in 2002 and USD 1.8 million in 2007, while it dropped by half in 2008 (USD 0.9 million).

MARKETS

In 2001, the Japan Association of Energy Service Companies (JAESCO) began an ESCO market survey of its members. The authors of this study were in charge of both the proposal and implementation plan for this survey. To date, the data collected is the only existing database pertaining to the ESCO market in Japan.

Figure 9-23 shows the trend of the market size over the past ten years. ESCO projects are broken down into two major categories: projects with performance contracting (PC) and energy efficiency projects. Projects accompanied by PC and ESPs are exclusively classified as "ESCO Market," and the rest are designated as "Ordinary Energy-Saving Retrofitting Projects." The ESCO market occupied almost the whole country; however, energy efficiency retrofit projects were not as widespread. At present, the database for energy

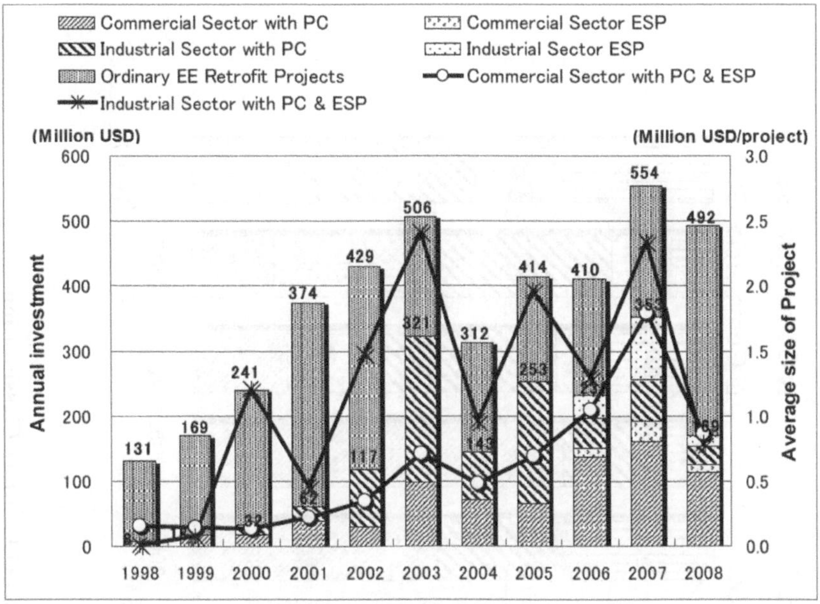

Figure 9-23: Scale of ESCO Market in Japan

efficiency retrofit projects is not yet available in Japan. Therefore, collecting such information at the national level is projected to be very costly. As shown in Figure 9-23, the ESCO market grew by leaps and bounds in 2001, and reached nearly USD 100 million. In 2002, it recorded a new high of USD 117 million, and the industry sector showed a gradual growth.

Based on the above facts, the following analysis is separated into two parts: "Initial Phase" and "Later Phase."

Initial Phase (Before 2001)

By sector, the scale of the ESCO industry expanded and almost doubled from a level of USD 8 million in 1998 to USD 62 million in 2001. At that time, the commercial sector had begun to grow at a rapid pace, which accounted for 63 percent of the whole ESCO industry market.

During the initial phase, most ESCOs used the guaranteed savings approach. The shared savings approach started spreading in 2000 and accounted for 48 percent in 2001, as shown in Figure 9-24.

Meanwhile, ESCO promotion programs became more active in-

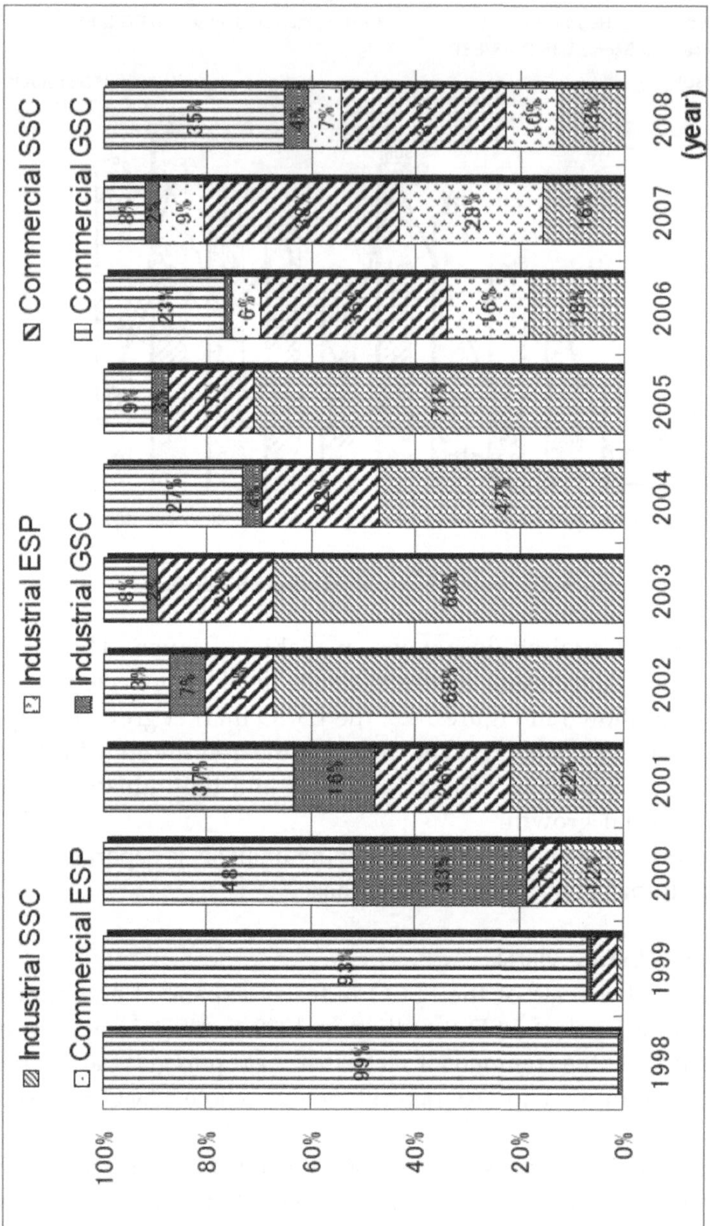

Source: Japan Association of Energy Service Companies

Figure 9-24: Ratio of Order Value by ESCO Contract Classification

cluding capacity building, feasibility studies, demonstration projects, standard contracts, M&V guidelines as well as explanatory meetings. With respect to policy measures, the ESCO industry was positioned as one of the pillars against global warming and JAESCO was established. Awareness of ESCOs was improving and general enlightenment activities were just getting started.

Later Phase (Since 2002)

The market of the industrial sector showed a rapid growth in 2002. Its size reached the equivalent of USD 117 million, nearly doubling that of the previous year. In particular, the industrial sector expanded 3.7 times and accounted for 75 percent of the whole ESCO contract order value. In 2003, this tendency became even more remarkable. The scale of the ESCO market grew 2.5 times over 2002 and reached USD 321 million. Both the commercial and industrial sectors showed a high growth. The industrial sector reached its highest level with USD 233 million. The market showed favorable growth until 2003. In 2004, it fell back to USD 143 million, decreasing by 49 percent compared to the previous fiscal year. A depression in the industrial sector accounted for the contraction (35 percent of the 2003 level). Although not as severe as in the industrial sector, the commercial sector was down 22 percent from the preceding year. Meanwhile, crude oil prices went up from USD 30.8/bbl in January to USD 41.9/bbl in November 2004. Diffusion of petroleum cogeneration systems was dealt a serious blow, which led to a further shrinking of the industrial sector market. Crude oil prices have kept on rising and reached USD 54.8/bbl in December 2005. Higher prices resulted in more ESCO activity and reached USD 253 million in that year. The industrial sector reached USD 187 million, 92 percent of the 2003 figure.

In 2007, the ESCO market achieved a new record again with USD 353 million and the commercial sector alone reached USD 192 million, contributing to more than half of the whole ESCO contract order value. However, the ESCO market was once again hit by the 2008 financial crisis. The performance in 2008 was only half that of the previous year (USD 169 million). In that year, the industrial sector merely reached 28 percent over 2007 (USD 46 million).

Jordan

Amjad Saleh Alkam

ACTIVITIES

The Ministry of Energy and Mineral Resources was established in Jordan in 1984 to manage energy policies, planning and energy efficiency issues. To enhance the awareness of consumers about energy efficiency matters, the Ministry established the energy and electricity information and advisory centers (EEIACs) in 1986. Branches for these centers in other cities like Irbid and Zarqa were also launched in the following years. Moreover, the Higher Council for Science and Technology established the National Energy Research Center in 1998. In addition to its research activities, the center got involved in energy services benefiting public and private sector entities.

The first private ESCO in Jordan was Energy Management Services (EMS) company, which was established in 1991 by two Jordanian entrepreneurs: Khaled Bushnaq and Samir Murad who recognized the need for an ESCO to provide the Jordanian market with skills for enhancing the efficiency of energy use. A full technology transfer was conducted by a Canadian company called "ADS" (now Econoler International). EMS did not limit its services to Jordan but also expanded to cover the gulf region.

In recent years, more ESCOs have been established in Jordan as a reaction to the huge increase in energy prices and the government plans to reduce subsidies gradually, which will increase demand for ESCO services.

The following is a list of ESCOs working in the Jordanian market:

- Energy Management Services Int. (EMS)—www.ems-int.com;
- Green Tech for Sustainable Environment—www.greentech.ae;
- Eco Engineering and Energy Solutions (EcoSol)—www.ecosol-int.com;
- Terra Vertis, El Tayeh, Kilani and Partners Co.—www.terravertis.com;
- Nur Energy and Environmental Services—www.nursolarsys.com;
- ETAmax Energy and Environmental Solutions—www.eta-max.com.

These ESCOs provide several services related to energy efficiency like energy audits, renewable energy, water management, environmental consultancy, energy value engineering and green building solutions. Some of these companies provide performance contracting to their clients through loans from commercial banks, leasing companies or through their own internal funds.

CONTRACTS

The most common type of contract used is one which the ESCO finances the whole project against a share of the actual savings achieved. This includes the cost of the study (energy audit). This type of contract was necessary at the beginning to convince clients to proceed with the energy management program by making it more attractive to them as they did not have to pay capital cash or operating costs (no capex and no opex). However, they paid a share of the actual savings achieved. The client was requested to sign an EPC agreement with the ESCO in which the ESCO guaranteed that it would pay its share if the promised savings were achieved. Typical shares of savings to the ESCO were in the range of 50-70 percent of monthly savings and the contract period was usually four to six years. The ESCO took care of the financing, which was usually provided by local commercial banks at normal lending rates. This type of energy performance contract is expensive and very risky to the ESCOs, yet it is very profitable if the project performs as planned.

Currently, ESCOs are trying to find more secure investments that are still attractive to their potential clients, such as asking the client to pay the cost of the study.

LEGAL FRAMEWORK

No real legal framework is currently in place to cover ESCOs other than normal business practices, such as registration with the Ministry of Industry and Trade and municipality licensing. However, there have been some recent attempts towards more organized and professional energy efficiency service businesses including:

1. The new energy law which includes clearer targets and requirements to support and regulate renewable energy and energy efficiency in Jordan. This law also includes establishing a separate entity called Jordan Renewable Energy and Energy Efficiency Fund (JREEEF) to support and promote renewable energy and energy efficiency projects. Currently, this law is in parliament for discussion and approval and expected to be finalized soon.

2. The National Energy Research Program under the Higher Council for Science and Technology is playing an important role in the energy sector of Jordan. It provides several services including research and development, testing and measurements, specifications, training in addition to participating in several energy efficiency and renewable energy projects in Jordan.

3. An Initiative aimed at improving Jordan's energy, water and environmental productivity was launched in 2009. This initiative is called "EDAMA," an Arabic word for "sustainability." The initiative was supported by the USAID Jordan Economic Development Program, in cooperation with Jordanian private sector businesses. EDAMA Association was established in January 2010. It is expected that the association will play an important role in advocating renewable energy and energy efficiency issues and creating more business for ESCOs.

4. The Jordan Energy Chapter (JEC) was established in March 2010 in cooperation with the Association of Energy Engineers (AEE). The chapter is currently working under the umbrella of EDAMA Association and hosts 142 members (trained and certified energy and carbon reduction managers). A new course "Certified Renewable Energy Professional-REP" is being conducted.

5. The Jordanian Engineers Association (JEA) through its Energy Committee and Engineers Training Center (ETC) also contributes

to the development of Jordanian engineers' qualifications and plays an important role in increasing awareness about energy efficiency issues as well as building awareness of the need for energy-related regulations among decision makers and the engineering society in Jordan.

6. The Jordanian Society for Renewable Energy (ESTEDAMA) was established in 2008 and is also contributing to national efforts towards better organization of the energy services sector in Jordan. ESTEDAMA has organized several events and activities for its members and also for the general public.

MARKETS

Figure (9-25) illustrates the breakdown of energy consumption in Jordan and it is obvious that the potential markets for ESCOs (industry and services) represent 29 percent of total country net energy consumption. The residential sector can also be added as a market but it requires special types of ESCOs and services. A more accurate indication about the ESCO market in Jordan and its breakdown can be found in Figure (9-26), which shows the pattern of electricity consumption.

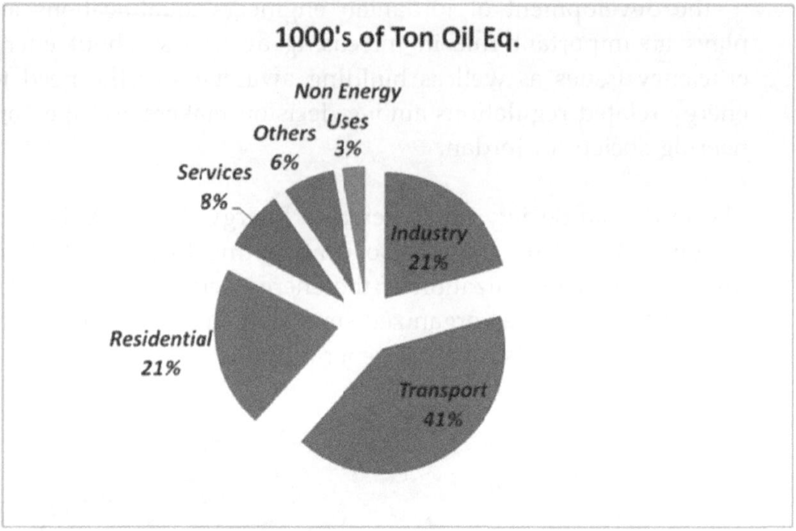

Figure 9-25: Jordan Energy Balance (year 2010)

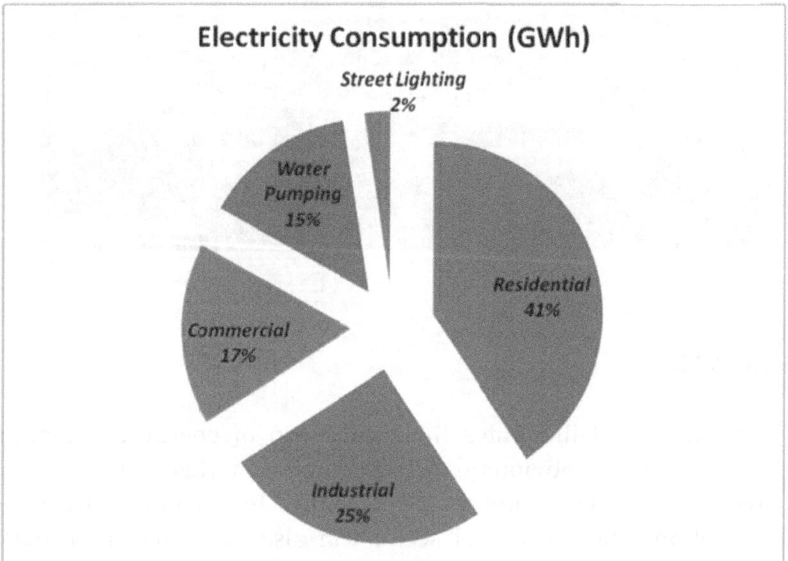

Figure 9-26: Jordan Electricity Consumption Breakdown (year 2010)

Historically, the industrial and commercial sectors were the main markets for ESCOs in Jordan. Recently, the Ministry of Water and Irrigation invited ESCOs to bid for implementing an energy-saving program for one of its main pumping stations as a pilot project for implementing the same concept (performance contracting) in other pumping stations as well.

Quick estimates of the energy-saving market size in Jordan indicate that the market has an annual market size in the range of USD 200,000,000 for the commercial, industrial and pumping sectors only. Half of this market value can be covered by Jordanian ESCO targets and eligibility conditions based on experience.

CASE STUDIES

Although the ESCO business in Jordan is still new and growing, several successes were achieved by existing ESCOs, as can be seen in Table 9-13, provided by EMS. It should be mentioned that more in-depth information and analyses are needed to analyze the accurate profitability of ESCO projects. However, the provided information shows that performance-based energy-saving projects have a very attractive profit margin if properly implemented and maintained.

Following is also a summary of results of a project implemented by another ESCO in Jordan (EcoSol).

Project Type: University Facilities

Description: The university consists mainly of seven buildings, five female dormitories and cultural foundation forums (ARENA), the total built area of the university is 72,868 m^2.

Contract Type: performance contracting.

The audit showed a potential 24 percent reduction in the energy and water bill. The following figure depicts the energy-saving measures implemented at the university.

Table 9-13: Results of a Sample of Performance Based Energy Saving Project in Jordan

Project	*Average Savings/Month*	*Total Contract Savings*	*Investment (Equipment & MP)*	*ESCO Average Income/Month*	*ESCO P/B Period (Years)*
Large Hospital	12,159	729,556	49,781	4,935	0.8
Printing Facility	1,678	100,678	25,164	1,226	1.7
Tobacco Industry	2,695	161,695	24,414	1,887	1.1
Aluminum Industry	7,605	456,271	39,831	4,194	0.8
Tissue Processing	1,647	98,814	26,253	1,153	1.9
Ready Wear Manufacturing	3,020	181,186	41,472	2,226	1.6
Paper Mill	10,554	633,220	96,106	6,860	1.2
Cable Industry	3,160	189,576	51,723	2,212	1.9
Paint Manufacturing	6,648	398,898	33,260	4,654	0.6
Government Building/Offices	1,568	112,881	32,175	1,144	2.3
Average/Project	**5,073**	**306,278**	**42,018**	**3,049**	**1.1**

All figures are in USD and provided for indicative purposes only.
Source: Energy Management Services Int. (EMS) project sheets

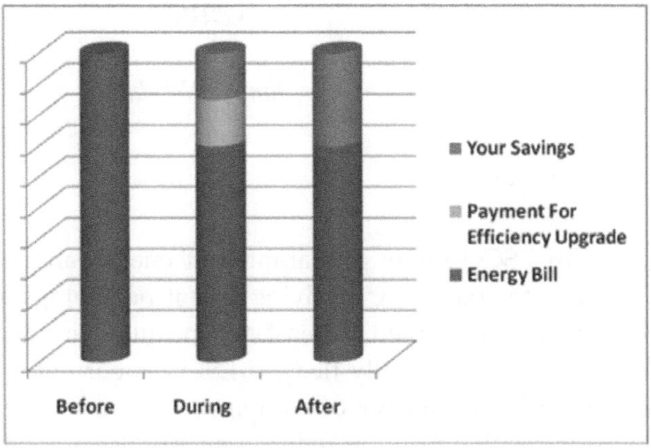

Figure 9-27: Energy-Saving Measures Implemented

Table 9-14: Examples of energy conservation measures identified with an encouraging pay-back period (less than two years).

Energy Consuming System	*Energy Conservation Measure*	*Payback Period in Months*
Lighting System	Replacement and upgrading of the low efficiency internal and external lighting fixtures by high efficiency lighting fixtures.	9-14 Months
	Control the operation of lighting fixtures using special controllers.	6-10 Months
Pumping System	Control the operation of the main feed-water pump according to the demand.	18 Months
	Control the operation of the ARENA's swimming pool filtration system based on the International Standards.	15 Months
	Modification and control of the operation of water heating pumps for the swimming pool, Jacuzzis, and domestic hot water.	14-16 Months
Thermal System	Control the operation of the heating boilers according to the demand and installing a new controller to utilize the solar system as a backup for the boilers.	6-10 Months

Kenya

James Wakaba

ACTIVITIES

An ESCO offers a guarantee of energy and/or monetary savings in return for payment, implying that if no savings are realized then the ESCO will not be paid, at least not in full. Going by this definition, it is true to say that no fully fledged ESCO is currently operating in Kenya, despite a large number of companies and individual consultants offering services that result in reduced energy use among consumers. Indeed, some of the users that contract the consultants directly (as opposed to through a government or donor-aided program) insist on clauses in the contract that peg payment of the consultant to the identification of a specified quantity or percentage of energy savings for a given production or service level. However, payments take place regardless of whether the identified energy-saving measures are implemented, largely because the users are not willing to commit themselves to investing in the measures.

Most of the recent initiatives aimed at reducing energy use have been funded by the Government of Kenya and multilateral donors, such as the World Bank, the United Nations (UN) and the Global Environment Facility. They have been implemented by the Ministry of Energy and/or the Kenya Association of Manufacturers (KAM). Under these programs, the implementing agency then contracted with consultants to carry out energy audits, assist in implementation and/or monitor the results. The contract was, therefore, between the implementing agency and the consultant, which does not result in ESCO-type contracts as typically there is no requirement for the consultant to demonstrate actual energy savings before they are paid. Whereas the government or donor will require that the implementing agency demonstrate some results. These will be typically in the form of bankable projects identified through the audit process, and do not necessarily involve actual savings arising from implemented projects. This again is due to the fact that the user is not obliged to implement the measures identified given that there is expenditure involved.

The other types of ESCOs, which supply energy for a fee, do not exist in Kenya and are unlikely to do so in the foreseeable future. Such an ESCO requires significant capitalization which, as we will see later, is a key hindrance to ESCO development in the country.

Interest in reducing energy use in Kenya dates back to the oil crisis of the early 1970s, when oil prices skyrocketed as a result of political crises in the Middle East. The government responded by developing awareness campaigns encouraging people to use less fuel. In the 1980s, the government introduced a 'power alcohol' program to blend ethanol from the sugar industry with petroleum fuels. While not an energy efficiency measure, this helped reduce dependency on fossil fuel and raise awareness on the impact energy use was having on the economy.

From 1987 onwards, donor interest in energy efficiency started with the World Bank's ESMAP program that helped the Ministry of Energy carry out energy audits in industry to identify energy-saving opportunities. The program was extended through KAM in the 1990s to include more industries. Another program called the GEF-KAM Industrial Energy Efficiency Project was implemented from the year 2001 at KAM. Unlike earlier programs, this one specifically encouraged private consultants (as opposed to employees of government and KAM) to become energy efficiency professionals through the following activities:

- Training—consultants were trained in energy auditing, finance and specific energy technologies. A special training program involved preparing consultants for the Energy Manager Certification from the Association of Energy Engineers.
- Energy audits—professionals were subcontracted to carry out energy audits in industry.
- ESCO development: consultants were engaged in the process of ESCO development through discussion forums that were part of developing a business plan for an ESCO.

The program precipitated the establishment of an ESCO by its former employees and provided some seed funding. However, while the company so established remains quite active in energy efficiency, it has not evolved into a full ESCO.

While no company carries out the full range of activities that define an ESCO, there are quite a number of companies engaging in ESCO-like activities. These include, in alphabetical order:

- Energy Track Associates;
- EMS Consultants Ltd.;

- Integrated Energy Services Ltd.;
- Greenworld Energy Ltd.;
- Recon Associates;
- Sustainable Energy Initiatives Ltd.; and
- Synchroconsult.

The first four were set up or are run by former staff at the GEF-KAM program, while the rest are owned by engineering consultants, who have diversified into energy management mostly as a result of training in energy efficiency carried out by the same program. The list is by no means exhaustive.

The bulk of the work done by these companies emanates from government or donor programs at the Ministry of Energy or at KAM. These programs are heavily subsidized by the government, or donor programs, to the energy user, with the user typically paying less than 20 percent of the cost. The downside is the generic nature of the contracts entered into with the consultant, which are not specific to the individual user and not pegged to energy savings identified or implemented. However, an increasingly observed trend is where these consultants are entering into bilateral contracts with energy users. This is largely because the latter want to be in control of the outcome and are likely to enter into more stringent contracts with the consultants. As mentioned, the contracts observed recently have clauses pegging payment to quantities or percentages of energy savings identified. It is just a matter of time before the contracts are pegged to energy actually saved, which will put the contracts firmly in the ESCO category.

Barriers

- *Donor funding*: Whereas donor and government funding has had an overall positive effect on energy efficiency in the country, enabling hundreds of energy users to understand what opportunities there are for energy savings, it does not necessarily augur well for the development of ESCOs. The contracts drawn with the consultants are not exacting, are not pegged to actual savings and are generic as opposed to company specific. Due to the minimal investment in the audit by the user, there is less incentive to implement saving measures. There is also less aggressive marketing by the consultants among end-users, which would result in more awareness and more result-oriented activities.

- *Contract context*: The contractual context for the country is quite weak, with drawn contracts either unenforceable or, in the event of a breach, likely to result in protracted legal processes, which the fledging ESCOs are ill-equipped to go through. There is little expertise in contract law, especially performance contracts, in a rapidly changing energy price/production volume environment where verifying results is hardly an exact science.
- *Capitalization*: An ESCO needs to be adequately capitalized to make an investment on behalf of an energy user and be able to take in the occasional downside when expected savings are not realized. The consultancies carrying out energy efficiency activities in Kenya do not have such capitalization, and in situations where the returns are not very clear, few are willing to put equity into such ventures. Getting debt to finance ESCO activities is almost out of the question given the perceived (and real) risks and the conservative nature of financial institutions in the country.

Favorable Factors

- *High energy prices*: Like other countries in East Africa, Kenya is experiencing high electricity prices caused by a large dependency on thermal power (up to 60 percent during drought), high oil prices, a weakened shilling that has lost more than 20 percent of its value to the dollar this year and high inflation, which is factored into the electricity bill. Fuel prices are also very high making a good case for implementing energy-saving measures. ESCOs are in a good position to pitch for their services to be accepted by industry.
- *Energy scarcity*: In addition, there are indications that demand is outstripping supply. This year, Kenya has already experienced power rationing and power failures have become common. To conserve energy, the government is encouraging implementation of measures through instruments like energy audits and an energy efficiency accord, which encourages industry to voluntarily adopt energy efficiency measures in return for some incentives.
- *East African countries integration*: More and more business is being done across the five East African countries of Burundi, Kenya, Rwanda, Tanzania and Uganda, which have signed a cooperation framework. Many consultants are also scaling up their operations to cover the region, giving them the exposure and the economies of scale to be able to undertake ESCO-type investments.

- *New energy efficiency regulations*: These regulations have been developed by the Energy Regulatory Commission. They require large energy users to, among other things, have an energy audit carried out at least once every three years. Furthermore, they require implementation of not less than 50 percent of identified feasible energy-saving measures within the same period. This will need quite some efforts from large users and may well require the services of ESCOs to achieve full implementation.
- There are opportunities in the energy supply market that could trigger energy supply ESCOs. Many agro-based industries generate agro-waste that could be used to produce electricity or heat, but are themselves reluctant to venture into that because it is outside their core business. These include the sugar industries, fruit, vegetable and flower processors as well as beef and dairy processors.

Lebanon

Pierre El Khoury

ACTIVITIES

The long-term objective of the Lebanese Center for Energy Conservation (LCEC) is to create a market for ESCOs, whereby any beneficiary can directly contact a specialized ESCO to conduct an energy audit, implement energy conservation measures and monitor energy-saving programs according to a standardized energy performance contract. Currently, LCEC is helping in funding a study on energy audits and thus is linking both beneficiaries and energy audit firms. LCEC also targets the creation of a special fund used for the implementation of energy conservation measures resulting from the study.

In 2007, LCEC, through funding by the United Nations Development Programme (UNDP), launched a national campaign to promote the concept and application of energy audits. The campaign had a clear target (engineers, architects, contractors, interior designers, industrialists and commercial businessmen) as well as a clear strategy for application (TV news reports, radio interviews, press interviews and targeted seminars).

LEGAL FRAMEWORK

In the case of Lebanon, the Lebanese Draft Law for Energy Conservation set the threshold for mandatory participation in energy audits at 400 toe. The present report seeks to justify this choice based on results from previously conducted energy audit campaigns and experience of countries with identical company size. The analysis of the results also shows that these facilities are not concentrated in one sector and that the energy consumption levels cannot be used to characterize different economic sectors.

FACILITATORS

In Lebanon, there is no ESCO association. In fact, the only ESCO momentum has been created through the "Energy Audit Program" (2005-2009) launched by LCEC, a young energy center created by UNDP and the Ministry of Energy and Water (MEW). LCEC has been in operation since 2002, more active since 2005, on a project-by-project basis as well as financed by the GEF and the MEW in addition to other bilateral donors, under the management of UNDP.

In 2010, LCEC was officially registered as an independent national agency affiliated with the Lebanese MEW. LCEC is the national energy agency in charge of EE and RE matters in Lebanon. LCEC has succeeded in establishing itself as a focal point for energy conservation issues within the Lebanese MEW.

As a matter of fact, LCEC has supervised more than 120 audits for major sites like the Beirut International Airport, Casino du Liban and Hôtel Dieu de France. LCEC established successful partnerships with the MEW, Electricité du Liban (EDL), local power companies, the Industrial Research Institute (IRI), the Council for Development and Reconstruction (CDR) and the Order of Engineers and Architects in Beirut (OEA).

Representing Lebanon, LCEC is a founding member in the Regional Center for Energy Conservation and Renewable Energy (RCREEE). Under the MEW, LCEC provides free, practical advice to businesses and public sector organizations to help reduce energy use. The institution's goal is to reduce GHG emissions in Lebanon by improving demand-side EE.

GOVERNMENT ACTIONS

The National Energy Efficiency Action Plan (NEEAP) is "an official document that outlines energy conservation issues to be taken in the country in order to reach national energy targets."

The NEEAP of Lebanon was developed by LCEC in October and approved by the Lebanese MEW on December 21, 2010 and by the Council of Ministers on November 10, 2011.

It includes 14 initiatives and was developed in accordance with the different points mentioned in the declaration of the Lebanese government of 2009 relating to EE and RE, namely the strategic target of 12 percent for renewable energy by 2020.

Initiative 13 of Lebanon's 14 initiatives in the NEEAP is dedicated to the development of an ESCO business. Initiative 13 is entitled "Paving the Way for Energy Audit and ESCO Business" with the objective defined as: "This initiative aims to support the development of the Energy Service Companies (ESCOs) working in the energy audit business and provide them with financial, fiscal, and technical incentives to remove barriers and promote energy audit activities."

On the other hand, the National Energy Efficiency and Renewable Energy Account (NEEREA) is an innovative mechanism launched by the Government of Lebanon at the end of 2010, with the technical support of LCEC and BDL. It is a flexible mechanism to fund EE and RE projects of any type, all over Lebanon. This action benefits from a EUR 12.2 million (USD 16 million) grant by the European Union.

The slogan of NEEREA is: "Finance your energy efficiency, renewable energy, or green building project through Lebanese banks with 0 percent interest rate and a repayment period of 14 years."

This attractive financing method makes NEEREA successful to nearly all stakeholders (bankers, developers, promoters, architects, investors, communities). NEEREA covers all types of buildings: residential households, industry lots, service buildings (offices, hospitals, etc.). Eligible buildings can exist or be under project/construction.

Figure 9-28 is an example of the potential savings for 58 facilities considering various technologies.

Figure 9-29 shows the overall procedure for the NEEREA initiative.

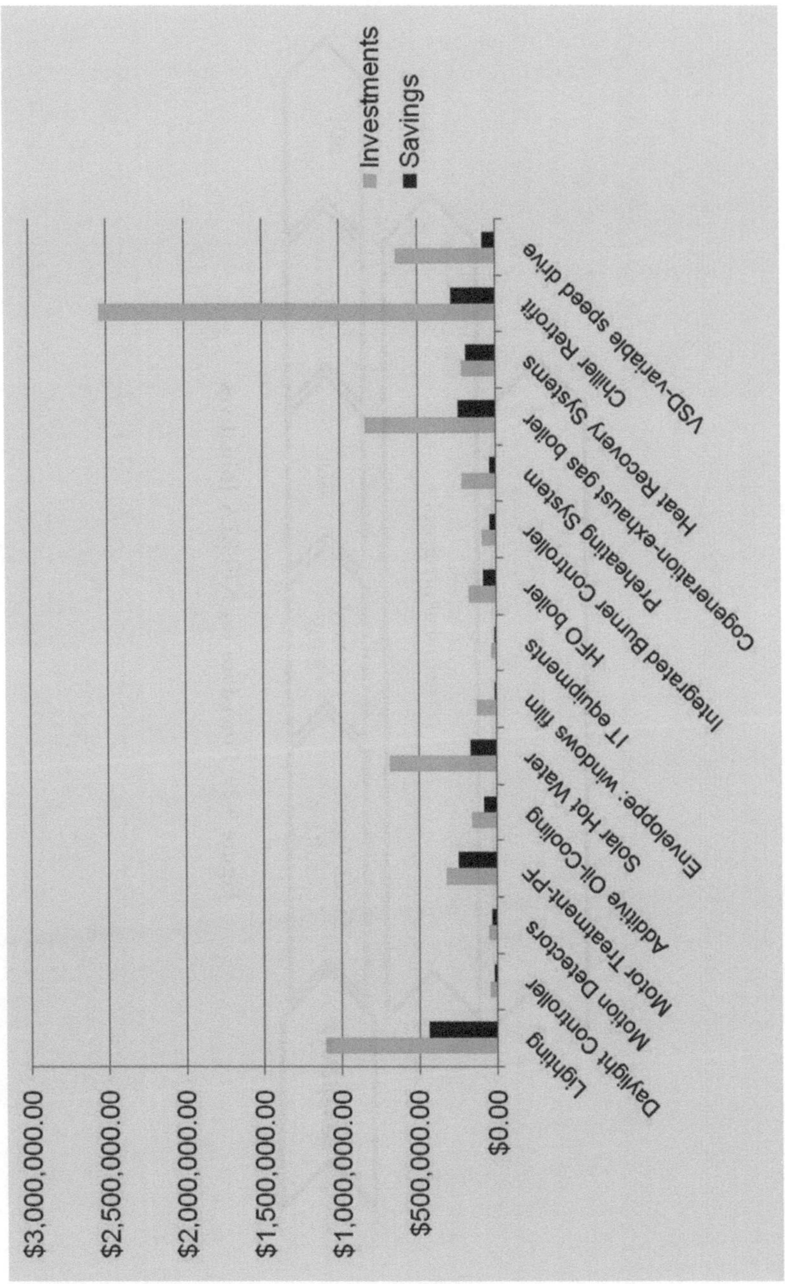

Figure 9-28: Potential Savings by Measure

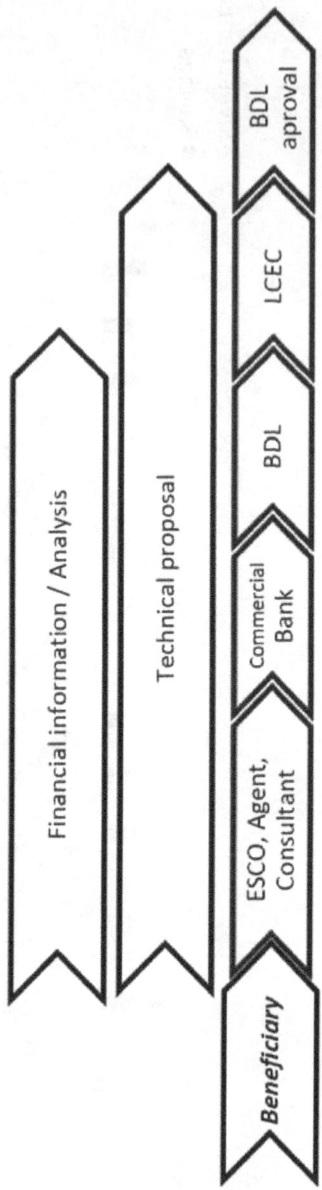

Figure 9-29: Procedure for NEEREA Initiatives

Mexico

Javier Ortega
Pierre Langlois

ACTIVITIES

The ESCO concept emerged in Mexico in the 1990s. At that time, FIDE (Trust Fund for Electric Energy Savings), through the support of a loan from the IADB, supported projects aimed at promoting electricity savings. A strategy was then developed to support and promote the development of a private sector offer from consulting companies, manufacturers and distributors in order to address the different market barriers related to the implementation of energy efficiency projects. Additionally, FIDE engaged the interest of financing organizations in order to secure such projects and in parallel established a link with the newly developed ESCOs.

This early effort led to different alliances among equipment suppliers, international ESCOs and other stakeholders. These alliances resulted in (i) the development of the first model contracts; (ii) the design of financing options; and (iii) the preparation and implementation of demonstration projects. These efforts helped demonstrate the viability of the concept.

As the newly formed ESCOs developed their capacities in project evaluation, development and implementation, they worked with an association called "Camara Nacional de Empresas Consultoras" (CNEC). The latter was to certify the consulting firms and increase their exposure and credibility in the market. Unfortunately, access to financing limited the growth of the EPC market at that time.

Even though the economic trends of the last few years favored energy efficiency projects, very limited investments have been made in this sector. ESCOs are still generally thinly capitalized and unable to finance or obtain financing for project implementation. These ESCOs are therefore incapable of providing energy savings guarantees that are considered of value to their customers or lenders.

At the present time, it is possible to identify about 15 ESCOs in operation. A few US ESCOs have been exploring joint venture arrangements with their Mexican counterparts to bring additional expertise and credibility to the market. Unfortunately, most of these proposed partnerships have failed to materialize to date.

CONTRACTS

Mexican ESCOs offer identification of energy efficiency opportunities, development, financing, installation and O&M services at end-user facilities on a performance basis. The end-user is offered no-cost financing through EPC. Most offer a shared savings structure; however, a couple of well-established ESCOs with good track records are able to get end-users' commitments to pursue the guaranteed savings structure.

MARKETS

Most Mexican ESCOs have focused on specific sectors where their expertise has a significant effect on energy and/or water consumption reduction. For example, some ESCOs have successfully developed projects for the commercial sector, especially in hotels and hospitals. Others have focused their activities on specific technologies such as heat recovery systems, sea water for cooling systems, lighting, peak generation and power factor.

The size of projects in Mexico range from USD 100,000 to USD 5 million. This presents a problem since these projects are too big for micro finance programs and too small for typical transaction costs.

The most typical financing sources for performance-based projects in Mexico are (i) the end-users' own money; (ii) the participation of private trust funds, development banks or commercial banks; and (iii) the ESCOs' own equity. Accessing project financing by project developers and ESCOs has been particularly difficult.

FACILITATORS

FIDE is a private trust fund based in Mexico. Its mission is to foster savings and the efficient use of electric energy in order to contribute to economic and social development. It also works towards the preservation of the environment in the industrial, commercial, municipal and residential sectors. Among its activities, FIDE provides technical support and financing to end-users wishing to develop energy efficiency initiatives. This includes energy audits, assessments, acquisitions and installations of energy-efficient equipment. FIDE has been supportive of the development of EPC in Mexico for many years.

The National Commission for Energy Efficiency (CONUEE) is a decentralized administrative agency under the Secretary of Energy, with technical and operative autonomy. It aims to promote energy efficiency and establish itself as a technical body in terms of sustainable use of energy. CONUEE was created under Law for Sustainable Use of Energy enacted on November 28, 2008. The law states that all human and material resources of the National Commission for Energy Savings (CONAE) are to be allocated to this new commission. Among CONUEE's responsibilities has been the support the development of EPC in Mexico. For instance, it has helped facilitate various projects through business matchmaking (linking facility managers to potential ESCOs, project developers, financing sources, technology companies, NGOs and others). In addition, facilitated the development of efficient financial vehicles which allow ESCOs, end-users and project developers to access financing and reduce transaction costs.

In 2011, in order to promote energy-saving projects, Mexican consulting companies and electric equipment manufacturers set up the Consultant Companies Association of Power Services in Mexico (AMESCO) with the main purpose of working as a team sharing the knowledge required to promote the development of ESCO projects.

Morocco

Abdelmourhit Lahbabi

ACTIVITIES

Morocco is largely dependent for its energy supply on fossil fuel imports, which represent more than 95 percent of its global energy demand.

The national energy balance is characterized by the dominance of petroleum products, accounting for 60 percent of domestic consumption. Strong economic growth in the 1970s benefited mainly petroleum products which saw their share increase from 70 percent in 1970 to 83 percent in 1985. However, the share of petroleum fuels, largely produced by local refining of imported crude oil, has experienced a continuous fall.

Growth in primary energy demand since 1980 has mainly benefited coal, which has registered an average annual growth rate of 10 percent. Thanks to competitive prices, fuel switch conversions made on coal power and cement plants and the development of new coal-based power generation capacities (Central Jorf Lasfar) have allowed coal to gradually increase its share of the energy balance to 26 percent in 2008, from a mere 8 percent in 1980.

In 2010, the Moroccan energy consumption per capita has been about 0.49 toe. This is low not only with respect to the world average (1.9 toe/capita in 2010), but also compared to the average of the African continent (0.67 toe/capita in 2008). This low level of energy consumption is partly explained by the heavy use in rural areas of traditional fuels: firewood and charcoal. It is estimated that wood fuel represents nearly 30 percent of the national energy balance.

Electricity is the driving force of the energy sector in Morocco. Electricity demand has experienced a sustained annual growth rate of 7 percent[1] against an average of 5 percent for primary energy. In fact, the generation of electricity, which has shaped the national energy mix over the past decade (push coal at the expense of fuel oil) will continue to orient the major changes in the sector: (i) driving the growth in the demand for primary energy; (ii) strengthening the share of natural gas in the mix; (iii) developing renewable power (wind and solar); and (vi) strengthening the socio-economic development of rural areas.

With most of its energy supply imported, Morocco is highly vulnerable to rising international energy prices. Indeed, Morocco's energy bill has increased fourfold since the early 2000s to MAD 68 billion (USD 8 billion) in 2008 under the combined effect of rising oil prices and a strong energy demand increase registered in the last decade. Moreover, the subsidy for petroleum products that has reached a record level of MAD 20 billion (USD 2.4 billion) is a heavy burden on public finances.

Facing these difficulties, the Government of Morocco adopted an ambitious and comprehensive energy strategy to secure the country's energy supply and to promote energy efficiency policies and large-scale renewable energy development programs.

The main objectives of the Moroccan energy strategy are as follows:

- Reduce energy consumption in buildings, industry and transportation by 12 percent by 2020 and 15 percent by 2030.
- Increase the share of the installed capacity of renewable energy-based power generation to 42 percent (14% solar, 14% wind and 14% hydro) by 2020.

The only past initiative worth mentioning about ESCOs' establishment in Morocco is a joint venture between ADS Groupe Conseil and local company, Innov Projet, back in 1993. ADS Maroc[2] was then created through a partial Canadian International Development Agency (CIDA) financing for the development of TPF of EE projects and ESCO business in Morocco. ADS Maroc aimed to implement EE projects resulting from energy audits carried out as part of Projet GEM[3]. GEM began as a major energy demand management project that was financed by USAID and managed by a private American company as subcontractor. The project targeted energy savings in the industrial sector through energy audits, training and awareness technical workshops as well as implementation of energy-saving projects.

Despite the important number of good cost-effective projects and the promotion efforts deployed to attract financial institutions to join the ESCO tour de table, ADS Maroc could not secure the financing resources needed to carry out its ESCO business plan.

Commercial efforts were then oriented toward public buildings (hospitals, schools and administrative buildings). Energy audits were carried out by ADS Maroc for some 20 public buildings and simulations were carried out on the national energy savings potential in public buildings based on the results of the implemented energy audits.

With the comprehensive energy reform implemented by the Government of Morocco covering the institutional, legislative and financial aspects of the energy sector, the current situation in Morocco is very favorable for the development of ESCO activities. The enacting of a new EE law setting mandatory energy audits for important energy consumers, the set-up of financial incentives for EE activities and the creation of an EE fund will certainly help develop ESCO activities in Morocco.

To date, private service companies have not developed full ESCO-type services in Morocco. The following table presents the main limited service initiatives that have been developed so far in Morocco.

ESCO financing is one of the main barriers to the development of performance-based energy services in Morocco. The risks associated with client payment based on energy savings are perceived as very high by commercial banks. In this regard, alternative financing is needed for the development of ESCO activities.

It is worth mentioning that a public Energy Investment Company [4](SIE) was created in 2010 within the energy sector reform. SIE's main mandate is to manage allocated funds for financial support to public and private operators in the field of renewable energy and energy efficiency. SIE is planning the set-up of a new EE dedicated fund for cost-effective investments in EE projects and participation as

Table 9-15: Current Energy Services

Company	*Type of Service Offered*	*Type of Contract*	*Remark*
Rio*	Investment, maintenance and follow-up for EE projects	Fee based on the energy savings achieved	Rio started its activities on efficient lighting projects for private companies. The model could not be implemented for public buildings. Rio stopped its ESCO offers following contractual problems on savings evaluation with a major client.
Cofely** Affiliate of GDF-Suez	Service for utilities management	Fixed-fee contract based on the service provided	Developed its activities with the utilities' management for Renault's Tanger Med plant. Does not yet have energy performance contracts but expects to develop full ESCO services.
Utility companies including ONE***	Investment, maintenance and management of public lighting	Long-term contracts with local communes based on investment in public lighting and fixed-fee payments	Limited service to public lighting.

*www.no.ma
**www.cofely-gdfsuez.com
***www.one.org.ma

shareholder in ESCOs.

The provisions of the new EE law and the set-up of the EE fund should help develop ESCO activities in Morocco.

LEGAL FRAMEWORK

EE Law no. 47.09 has just been enacted in Morocco [5]. One of the main provisions of the law is mandatory energy audits for important energy consumers with annual reporting of energy consumption and energy savings achieved. The law also requires the certification of the ESCOs that will be allowed to carry out energy audits and energy impact assessments. The new law defines the requirements for the certification of these companies.

As for ESCO services for public institutions, there are no legal provisions for TPF schemes, performance-based contracting and service payments. This constitutes the main barrier to the development of ESCO activities for public institutions.

MARKETS

Despite the various EE activities that have been carried out in Morocco during the last decade, project implementation rates have been very low and thus the full potential of EE in Morocco in terms of energy savings is still untapped. With the new favorable institutional, legal and financial framework, EE offers cost-effective investment opportunities. The latter should help develop and sustain ESCO-type activities in the following sectors:

- manufacturing industries;
- energy-efficient public buildings;
- energy-efficient housing; and
- public street lighting.

OVERALL ESCO MARKET ASSESSMENT

The following ESCO development SWOT chart summarizes the Strengths, Opportunities, Weaknesses and Threats of ESCO business development in Morocco.

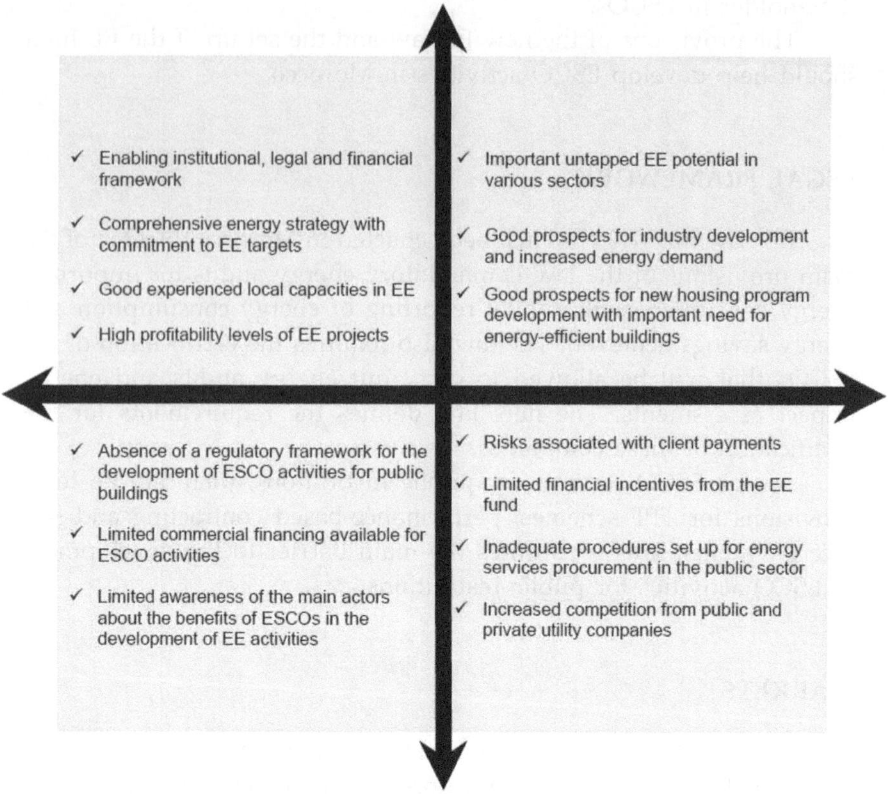

Figure 9-30: ESCO Business Development SWOT

FACILITATORS

International experience in ESCO activity development has shown that the commitment of public authorities is crucial for sustaining the ESCO market. In this regard, renewable energy and EE agency ADEREE as well as SIE could play an important enabling role in ESCO development in Morocco by setting the regulatory framework for the implementation of public building EE programs through ESCO activities.

Netherlands

Albert Hulshoff

ACTIVITIES

ESCOs are springing up across the Netherlands. The City of Rotterdam is the first major property owner in the Netherlands to enhance the sustainability of its real property through the use of ESCOs.

Although the Netherlands has seen increasing interest in ESCO services, the development of its ESCO market is still lagging behind countries like Germany, France, Belgium and England. There are several reasons for this. One is the market players' lack of familiarity with ESCO services. Another is the lack of legal requirements for the energy performance of existing buildings. Moreover, the construction and management sectors in the Netherlands are traditionally not used to providing performance-based services.

Despite these market barriers, a number of international ESCOs have been operating in the Netherlands for several years. Major international companies have entered into energy performance contracts for improving the sustainability of individual buildings (offices, hospitals and swimming pools). Besides these major ESCOs, a number of Dutch service providers, such as installation companies and energy utilities, have signed performance contracts for the supply and management of a single energy system; i.e., a cold/heat storage system or a gas-fired heating boiler.

Recently, the Netherlands has observed an increasing interest in guaranteed energy-saving services by ESCOs. This shift is partly due to the first major ESCO tender that took place at cluster level. The City of Rotterdam is the first major property owner in the Netherlands to enhance the sustainability of its building stock through the use of ESCOs. The first contract, signed in the spring of 2011, was focused on improving the energy efficiency of nine municipal swimming pools. This tender has garnered significant attention in the trade press. A number of Dutch municipalities are now considering following in the footsteps of Rotterdam.

In the slipstream of this first major performance contract for the Netherlands, banks and other financiers are showing a growing interest in financing ESCO projects.

Some commercial property owners are also becoming aware of the opportunities that ESCOs can provide in enhancing the sustainability of their building stocks. This already has led to signing by some commercial owners of the first energy performance contracts for improving the sustainability of office buildings. This growing interest of the commercial sector is due in part to (i) the proven rent differences between energy-efficient and energy-inefficient office buildings in the Netherlands (Kok, Jennen, 2011); (ii) increasing pressure from pension funds on real property funds to enhance the sustainability of their real property portfolios (Global Real Estate Sustainability Benchmark, 2011); and (iii) the minimum requirement of energy label C that the Dutch government places on the renting of office buildings.

Still, ESCOs are far from active in the Dutch housing sector. A small number of lease contracts, however, have been concluded for the supply and management of heating boilers in individual homes. These contracts generally do not contain any energy performance agreements.

CONTRACTS

Most performance contracts concluded in the Netherlands are energy performance contracts including a bonus/penalty scheme. This contract model is similar to the EPC model used in many countries.

Aside from EPC agreements, several leading market players have recently started using the maintenance and energy performance contract (M&EPC) model. This type of contract includes performance targets that must be met by the ESCO in terms of the energy systems and the management and maintenance of the entire building. Models of a number of performance contracts of this type are available in Dutch at www.agentschapnl.nl/esco. The M&EPC model used by the City of Rotterdam, including a process description, will be available in mid-2012. In energy performance contracts or M&EPCs, responsibility for energy procurement and associated price risks, in general, does not rest with an ESCO.

The Government Buildings Agency (Rijksgebouwendienst) uses a performance-based design, build, finance, maintain and operate (DBFMO) contract for the construction and renovation of large office buildings. An interesting example of such a project is the renovation of the Ministry of Finance in The Hague. The term of this performance contract is 25 years.

The energy supply contract, where a procurement contractor purchases energy on behalf of a building owner or tenant and creates price advantages for customers including through bulk purchasing and peak shaving, is often used in the Netherlands as well. In this type of contract energy saving is of secondary importance.

MARKETS

The Netherlands is rife with opportunities for sound ESCO business cases for both commercial and public property. This is evident from the 2011 market study on market opportunities and the energy savings potential for ESCOs conducted by the research firm BuildDesk, for the NL Agency. In its report, BuildDesk also concludes in more general terms that the use of ESCOs can make an important contribution to improving the energy performance of existing non-residential buildings in the Netherlands. Moreover, BuildDesk finds that the development of the ESCO market can be significantly expedited if the Dutch government and municipalities act as launching customer without having to deploy additional policy instruments such as grants.

The following is the estimated market turnover in ESCO contracts in the Netherlands per year (order size estimate by market experts, BuildDesk 2011 market study "CO_2 savings potential for ESCOs in non-residential construction"):

- 2010: EUR 4 million (USD 5 million);
- 2011: EUR 18 million (USD 24 million);
- 2012: EUR 69 million (USD 91 million).

The information provided in Table 9-16 prompts the following observations.

Commercial Property Sector

Research by Kok and Jennen in 2011 shows that the presence of a public transport hub has a positive effect on the rent. Thus, it can be concluded from these market research data that a relatively higher

Table 9-16: Estimated Market Turnover in 2012, per Sector

Overview of Market Potential in 2012							
Indicator	*Education*	*Healthcare*	*Shops*	*Offices*	*Industrial Buildings*	*Other*	*Total*
Building types	Primary schools Secondary education Higher professional education	Care Cure	Small shop Supermarket Department store	Small, medium, large	Industrial buildings, non-residential buildings	Hotel, café, restaurant, sports, leisure, culture	
Floor space (mil m^2)	29.1	29.2	28	46.5	31.4	18.7	182.9
CO_2 reduction (Mton CO_2/year)	0.18	0.32	0.11	0.35	0.06	0.06	1.07
Potential market turnover in 2012 (EUR mil/year)	11	9	6	36	4	3	69

rent can be asked for office buildings located near a public transport hub and that this is more likely to result in sound ESCO business cases for sustainability at these locations.

Schools

Current financing arrangements for primary schools (where municipalities are financially responsible for the shell and school boards for the inside of the school building) can stand in the way of reaching a good ESCO agreement for this specific building segment.

LEGAL FRAMEWORK

If the contracting authority is a government body, the tendering ESCO must take the European tendering rules into account during the tendering phase. ESCO contract prices are usually so high that government bodies are obliged to put public contracts out to tender at European level. The main tendering rules are laid down in EU Directives No. 2004/17/EC and No. 2004/18/EC. The tendering rules apply to public procurement contracts, supplies and services whose value exceeds a certain threshold amount. Delivery of energy systems and building improvements, including support services, fall under the item "Contracts."

In addition, the Dutch government applies a number of sustainable procurement criteria to the purchase, rent and facility management of office buildings. These criteria can be found at www.rijksoverheid.nl/duurzaam-inkopen (in Dutch).

Due to the Dutch environmental law (Activiteitenbesluit), entrepreneurs are obliged to take all energy efficiency measures with a payback period of five years or less. This requirement applies to companies with a minimum annual energy use of 50,000 kWh or 25,000 m^3 (natural gas).

During the implementation phase, the ESCO service provider must comply with various mandatory inspections and legal requirements. The guideline that the NL Agency has put together (Leidraad Prestatiecontracten Beheer en Onderhoud Gebouwen) provides an overview of these legal obligations. The document is available in Dutch and can be downloaded from www.agentschapnl.nl/esco.

FACILITATORS

The NL Agency, a (central) government organization, which implements laws and regulations and incentive programs, including in the area of energy savings in the built environment, has been commissioned to stimulate the development of the Dutch ESCO market. The Agency does so by bringing together market players, providing informational films, articles and seminars, commissioning market studies and providing support to leading market players. The NL Agency then posts all the information derived from these activities on its website (www.agentschapnl.nl/esco).

According to the BuildDesk study noted above, lack of familiarity with ESCO services is one of the reasons that the development of the ESCO market in the Netherlands has been lagging behind other countries. To eliminate this market barrier and provide market players with greater insight into ESCO services, the NL Agency and various market players are currently conducting a study on the feasibility of developing a Dutch platform for ESCO services. This platform aims at providing property owners with practical information, ensuring support in tendering processes and making standard tender documents available to the market.

GOVERNMENT ACTIONS

Compared to (some) other countries in Western Europe, the Netherlands does not have an energy-saving requirement for existing buildings that would effectually stimulate a market for ESCOs. Nor does the government act as a powerful launching customer for ESCOs (as the German government does), although a noticeable shift is taking place in this regard.

Following the success of the City of Rotterdam (see following case study), several municipalities are presently exploring the possibility of enhancing the sustainability of their municipal property portfolio through the use of ESCOs. This is expected to result in a number of ESCO tenders by municipalities in 2012.

CASE STUDY

The results of the first ESCO tender in the framework of Rotterdam's Green Buildings were presented in the spring of 2011. The purpose of the tender was to improve the energy efficiency of nine municipal pools.

The results were as follows:

- guaranteed energy cost savings of 34 percent;
- 15 percent savings in maintenance costs;
- structural improvement of the indoor climate of the swimming pools.

These measures are laid down in a 10-year M&EPC and will be implemented without costing the City of Rotterdam, the owner of the pools, any additional money. It is precisely this combination of significant energy reduction and indoor climate improvement that is seen as a breakthrough in the Netherlands for the sustainability of buildings through the use of an ESCO. This contract has been awarded to Strukton, an ESCO, which will make the swimming pools significantly more energy efficient and manage them during the term of the performance contract.

New Zealand

Mike Underhill
Terry Stewart

ACTIVITIES

New Zealand lies in the southwest pacific approximately 1,500 km east of Australia. Consisting of two main islands stretching approximately 1,800 km north to south, it is home to some 4.4 million people spread over an area similar in size to the United Kingdom.

The non-residential building stock reflects the population size and geographic spread where 70 percent of New Zealand's is primarily located in 16 cities spread across the country. The largest city, Auckland, with nearly 1.4 million people far outweighs other cities whose populations are less than 400,000 (NZ Dept. of Statistics, 2011). A recent study by the Building Research Association New Zealand (BRANZ) suggests that there are 50,539 non-residential buildings with only 564 larger than 9,000 m2 (BRANZ, 2010), which in itself suggests that the size of energy efficiency projects are likely to be smaller than they are in major international cities like New York or London.

For the calendar year 2010, total consumer demand for energy, excluding transport energy supply, was 326.5 gross PJ. From this, 233.56 gross PJ, or 71.5 percent, was consumed by the industrial and commercial sectors, with 184.24 gross PJ and 49.32 gross PJ, respectively. Within these two sectors, electricity is the largest energy source at 88.29 gross PJ, renewables at 55.22 PJ, natural gas at 46.46 PJ, coal at 22.72 PJ and oil at 20.86 PJ (Ministry of Economic Development, 2011).

The annual electricity consumption amounts to 9,980 kWh per person and the Ministry of Economic Development rates New Zealand as 9th highest, but behind countries such as Australia (11,582 kWh) and the United States (13,731 kWh)—Ministry of Economic Development (MED), 2011.

On the supply side, 74 percent of electricity generation came from renewable resources, and according to analysis of data from the International Energy Agency conducted by MED, this places New Zealand third behind Iceland and Norway in terms of the largest percentage of total primary energy supply from renewables (Ministry of Economic Development, 2011).

Electricity prices for commercial customers are estimated at just over 15.3c/kWh and 10.4c/kWh for industrial consumers (Ministry of Economic Development, 2011), which according to the New Zealand Energy File produced by MED, suggests that our electricity costs are lower than in many other Organization for Economic Co-operation and Development (OECD) countries.

These factors, while not game-breakers, do put economic constraints around the commercial viability of an ESCO industry, particularly the establishment of a nationwide offering. However, as is the case in many countries, the focus on energy efficiency is growing rapidly.

While many participants in the energy efficiency sector have an understanding of what an ESCO means, there is no clear and consistent interpretation. Some take a purist view citing characteristics such as "must be energy services," "must offer performance contracts," "must offer finance," "must guarantee savings," "savings must pay for the work," while other market participants adopt a more liberal translation which can be as broad as "any company providing energy services." One area of common understanding seems to be that a company exclusively selling energy-efficient products does not meet the minimum threshold for inclusion. In the absence of an ESCO industry body to act as a reference point, the term ESCO may well remain an ambiguous term.

While a number of organizations undertake projects that meet the purist view of an ESCO, there are no known organizations that solely operate in this manner. Rather, some energy service providers will offer an energy performance contract during the selling process when the project economics make sense and the customer is receptive. By this very nature, this suggests that market demand is insufficient to sustain a business relying on EPC alone. This is reflected in comments from a long-term market participant who observed that companies do not come to you with the idea: you have to go out and sell it.

During the late 1990s, the Energy Efficiency Conservation Authority (EECA) promoted EPC and the ESCO concept to industry via a series of workshops conducted by Kiona International around the country. While market participants who attended felt they were well patronized and the message was well received, they also felt there was no immediate uptake of EPC at project level. Despite reports of a lack of immediate traction, a quick industry scan suggests progress has occurred and perhaps some momentum is building at project level in two key areas: commercial buildings and the industrial arena.

In the commercial building sector, feedback from market participants suggests that around 20-25 energy performance contracts have been put in place over the last 12-15 years. These projects have typically focused on upgrading individual equipment in commercial office buildings, universities and hospitals, such as HVAC, lighting upgrades, chiller replacements and building control systems. The size of these energy performance contracts typically range from around NZD 200,000 (USD 165,000) to NZD 2 million (USD 1.6 million) each, are focused on guaranteed savings and have a life span of up to six years. One internationally branded organization has implemented over a dozen with a total capital value of around NZD 10 million (USD 8 million).

In the industrial area and at the larger end of the scale, the industrial sector has examples of projects; e.g., co-generation facilities and heat generation plants, which are completed under a design, build, operate (DBO) or a build, own, operate (BOO) model. In the DBO model, a provider designs then builds the equipment on a customer's site and then operates equipment under a performance contract typically for 20 years. Under the BOO model, a provider builds the equipment on a customer's site at its cost, retains ownership, and then under a performance contract, operates the equipment to deliver the agreed energy. One award-winning example is a co-generation facility developed and built for a local hospital by a subsidiary of an electricity generator. The project is briefly described on their website as "... plant built at no cost to the hospital operates in a 20 year agreement to supply the ... Hospital with base load electricity (3.6 MW net), heat (4.2 MW) and standby power (2.8MW)." (Energy for Industry, 2010).

Another development is the market intervention by government in electricity efficiency which is encouraging energy service providers and some product suppliers to adopt models more closely aligned with performance contracting. For example, EECA operates a program for the commercial building sector. This program provides grant funding to break down barriers preventing electrical energy efficiency measures from being implemented. Delivered through 16 designated private sector providers, the program requires a three-year guarantee of the electrical energy savings to EECA from either the end customer or the provider. Currently focused on efficient lighting upgrades, HVAC system enhancement, refrigeration, continuous commissioning of building management systems, and monitoring and targeting systems, the program is delivering measured savings of over 50 GWh per annum through more than 400 buildings. These providers are a mix of local subsidiaries of large multinational companies and a number of locally owned and operated private companies.

Financing mechanisms remain somewhat conservative with most organizations in the private sector financing projects from their internal capital pool. While emission reductions and energy efficiency are a growing part of many organizations' sustainability programs, the primary driver for capital allocation to energy efficiency projects is driven by the financial returns with the primary measure being simple payback. Struggle for capital within organizations remains competitive and as energy efficiency projects are not always termed "strategic," an energy efficiency project will often need a shorter payback period to win funding. In the private sector, payback periods of three years or less are a common requirement putting additional pressure on the economic viability of an energy performance contract.

Government organizations designated Crown Entities are able to access another EECA program entitled "Crown Loans Scheme." This scheme allows crown entities to access low interest loans for energy efficiency projects with a payback period of five years or better.

Factors that prevent the adoption of an energy performance contract model are cited by market participants as a general lack of understanding of EPC, difficulties establishing and agreeing on energy baselines and the resultant savings (particularly avoided energy costs), reluctance by operational people to outsource key services, reluctance by organizations to use (and/or the ability to find) TPF and the organization's ability to build a robust business case for management. Additionally, because energy performance contracts rightfully focus on guaranteed outcomes, this can also mean that project outcomes receive higher degrees of scrutiny from the organization. This means the operational manager not only has to manage, but also to demonstrate and explain the financial outcomes to other functional heads. This can be particularly problematic when the technical nature of M&V methodologies do not necessarily conform to an industry standard. This is compounded when they are not readily understood by other functional managers, have moving baselines that may at times record savings greater than the energy bill, or are open to some degree of interpretation. These barriers often lead people to undertake energy efficiency measures as and when the maintenance and/or capital replacement budget allows.

MARKETS

While the market opportunity for ESCOs has not been formally evaluated, a number of studies have been undertaken to identify and quantify the size and scope of energy efficiency opportunities. For example, a report commissioned by the Electricity Commission calculates the economic potential for electrical energy efficiency across the industrial, commercial and residential sectors at 6,437 GWh per annum (KEMA, 2007), demonstrating that the market for energy efficiency initiatives should be attractive to energy consumers and energy service providers.

GOVERNMENT ACTIONS

Government policy indirectly supports ESCO development in that it provides significant direction, encouragement and support for energy efficiency, which in turn should translate into more favorable conditions for ESCOs to operate. For example, the current government policy on energy efficiency is laid out in the New Zealand Energy Strategy 2011-2021 (NZEES). A companion strategy, the New Zealand Energy Efficiency and Conservation Strategy 2011-2016, focuses specifically on the promotion of energy efficiency, energy conservation and renewable energy. Furthermore, it articulates policy such as the following: "The Government will continue to support energy efficiency initiatives for businesses with measures such as energy audits, support for energy efficient purchasing, grant and subsidy programs, and building sector capacity and capability in energy management" (Ministry of Economic Development, 2011). While many market participants yearn for a mandatory energy rating scheme for commercial buildings, the government position remains one that "...encourages development and use of voluntary industry standards to rate building energy performance" (Ministry of Economic Development, 2011).

Additionally, the government provides significant support through funding to EECA, which is staffed with around 100 personnel and has a broad mandate toward energy efficiency. As part of executing these responsibilities, EECA links with numerous other industry organizations including the Energy Management Association of New Zealand (EMANZ), the New Zealand Green Building Council and the Institute of Refrigeration, Heating and Air Conditioning Engineers (IRHACE). While these organizations do not specifically focus on developing an ESCO sector, they do support the development of a healthy and vibrant energy efficiency industry.

The profile and awareness of the need to focus on energy efficiency continue to grow. The industry continues to develop through both public and private sector funding and innovation, while information sharing among market participants allows for deeper savings to be achieved. The lack of traction on developing a specific and sizeable ESCO industry may be a reflection that the current conditions are not commercially conducive to the widespread delivery of energy efficiency measures via this mechanism at this point in time. Whether the ESCO environment can move from a project-by-project scenario to a whole-building focus by fully fledged ESCO companies, only time will tell. However, the current momentum around energy efficiency initiatives will provide a very strong base for its emergence if it delivers value to end-users and energy service providers.

Norway

Sami Siltainsuu
Filip Medhammar

ACTIVITIES

Norway is different from other Nordic countries in many ways. First of all, Norway has been self-sufficient in the field of energy production for decades. The country has vast oil resources in the North Sea, while high mountains provide more hydropower than the country needs. As a matter of fact, Norway is able to export oil and power to its neighboring countries. It has 20 percent of the hydropower resources, 40 percent of the gas resources and 60 percent of the oil resources in Europe. Self-sufficiency has decreased pressure to raise utility prices. However, the price of electricity is expected to increase and move closer to the average price in Europe, as the transmission capacity to the European market is expected to go up in the coming years.

Another differentiating factor is that Norway is not part of the EU. However, Norway is closely related to the EU and its legislation through its membership in the European Economic Area (EEA).

Taking these facts into account, it is easy to understand that the ESCO business moved forward at a somewhat slower pace compared to the rest of the Nordic countries. The first project was implemented in 2004 in the public sector. From 2004 to 2011, the total market was estimated at NOK 136 million (USD 23 million) with NOK 64 million (USD 11 million) being generated in 2010 and 2011 alone—proof of the recent rapid growth of the ESCO market. Today, there are roughly 10-15 projects either completed or being implemented.

CONTRACTS

As is the case with other Nordic countries, the guaranteed savings model is the most common in Norway. The shared savings model is almost non-existent. Not only is it due to the fact that it is riskier for vendors, but customers also understand that due to the greater capital expense and risks involved, they get less with the same investment (compared to the guaranteed savings model). It is customary that the customer finances projects from its own budget or finds a financial institution to finance the project.

LEGAL FRAMEWORK

Common, standardized and general terms of contracts regulate agreements, projects and services countrywide. Guaranteed savings and performance assurance are beyond the scope of general terms of contracts and they are not applicable as such. The length of the assurance period, comprehensiveness of guaranteed savings and the definition of the baseline have to be agreed upon separately. In addition, special circumstances (renovation or sale of a building) are dealt with on a case-by-case basis. However, a standardized contract structure for ESCO projects is being developed and is expected to be finished in 2012.

The law of public procurement regulates purchasing in the public sector. It has the same characteristics as the law of public procurement in the EU. The main purpose of the law is to give equal opportunity to every vendor in the market to bid against RFPs.

The law allows municipalities to use quality criteria with different weighting percentages (instead of price criteria) in order to identify the best possible supplier. As in Finland, in reality, indisputable quality criteria are very difficult to find. In worst cases, public biddings become contests that are won by the vendor who bids for the lowest price and offers the largest savings. Unfortunately, the ability of a vendor to deliver a quality project and long-term performance assurance becomes a secondary consideration.

The Norwegian market has not yet faced some of the problems

and disturbances found in other Nordic countries. So far, there have not been many law suits filed by discontented vendors who have lost bids. If the atmosphere in the market remains healthy, rapid growth is expected to continue. It is worth noting that the size of Norway's ESCO market has exceeded that of the Finnish market, despite the fact that the ESCO business model started later in Norway. That is partly due to the fact that customers feel secure to move forward with ESCO initiatives (without the threat of law suits looming, as is the case in Finland).

MARKETS

ESCO business is practiced in both private and public sectors. At this point in market development, it is difficult to state which market is bigger. It was not possible to find a market study on the private sector but the size of the public sector ESCO market is estimated at approximately NOK 4,700 8,100 million (USD 807-1,390 million). The public sector can be divided in the three major clusters shown in Table 9-17.

Table 9-17: Major Clusters in the Public Sector

	*Gross Market Potential (NOK)**	*USD*
Municipalities	2,400-4,500 million	412-772 million
Counties	700-1,100 million	120-189 million
Hospitals	1,600-2,500 million	275-429 million
Total	4,700-8,100 million	807-1,390 million

Savings have been estimated at 18-25 percent with a payback period ranging from eight to ten years. To date, the total value of completed and ongoing ESCO projects is NOK 136 million (USD 23 million), which amounts to 1.7-3 percent of the public sector gross market potential. If competition and market growth remain healthy, there is significant work to be done.

Compared to other Nordic countries, a savings potential of 18-25 percent might seem a bit high although it can be explained from a historic perspective: in past decades, energy prices in Norway have been relatively low and there has been little focus on energy efficiency in buildings.

FACILITATORS

As in Sweden and Finland, Norway has an energy agency facilitated by the government. The agency, called Enova SF, became operational in early 2002 under the Royal Norwegian Ministry of Petroleum and Energy. As Enova states on its website, "Enova's main mission is to contribute to environmentally sound and rational use and production of energy, relying on financial instruments and incentives to stimulate market actors and mechanisms to achieve national energy policy goals."

Enova focuses on both supply- and demand-side issues in order to achieve objectives given by the Norwegian Parliament in 2000. These objectives are to:

- limit energy use considerably more than if developments were allowed to continue unchecked;
- increase annual use of water-based central heating by using new renewable energy sources, heat pumps and waste heat;
- increase wind power capacity and environmentally friendly land-based use of natural gas.

GOVERNMENT ACTIONS

As a member of the EEA, Norway is obliged to adopt most EU directives and legislation. In practice, more than 99.5 percent of the EU directives have been implemented in Norway's national legislation. EU directive 2002/91/EC: Directive on the Energy Performance of Buildings defined national energy requirements for new buildings and buildings under renovation. This directive was implemented in Norway as Energiloven (Energy Act) in January 2010. The directive stated that energy certificates for new and existing buildings would become mandatory beginning in 2012. For public buildings and buildings in public use, the energy certificates must be made visible.

There are discussions in Norway about implementation of the EU's 2006/32/EC: Directive on Energy End-Use Efficiency and Energy Services in the legislation as well as in order to improve the market for energy efficiency in the end-user market.

Enova has regular subsidy programs in place for both the renovation of existing buildings and new construction projects. Subsidies are aimed at energy conservation and heating source conversion projects. Heat source is switched for example from electricity to district heating or from oil to bio fuel.

Norway has set a goal to be carbon neutral by 2030. In order to reach that goal, the country needs to reduce carbon emissions by 15-17 million tons by 2020. It is likely that government support to energy efficiency projects will continue.

Philippines

Ron Alan Go-aco

ACTIVITIES

Resulting from the energy crisis in the early 1990s and through government promotional activities and programs with international funding and grants, the ESCO concept was brought up for public and private sectors to consider. An ESCO was defined in the Philippines as an aggregator of energy services that includes developing and arranging financing of energy efficiency projects. Many trading firms have called themselves ESCOs merely because they supplied energy-efficient technologies. Some engineering design firms also have considered themselves ESCOs because they designed energy-efficient systems. Regrettably, many of these types of firms did not have the attributes of packaging financed EE projects.

Electric utilities were among the best positioned firms in the private sector to become ESCOs. MERALCO, the largest private electric utility in the Philippines, considered putting up an ESCO as an offshoot to its DSM activities in mid-1990. MERALCO Energy, Inc. (MEI) was officially established in June 2000 and is considered to be the first utility ESCO in the country. CESTCO, another electric utility, evolved from its parent company (CEPALCO). Other ESCOs from the private sector came from global manufacturing companies such as Danfoss, Honeywell, Johnson Controls and Trane. These manufacturing ESCOs started to promote and implement hybrid energy performance contracts much earlier than did the utility ESCOs. Towards the late 1990s, Danfoss was among the first known manufacturers actively promoting financed EE projects utilizing variable-speed drives (VSDs) for motors. The primary downside to these manufacturing ESCOs in the Philippines has been their limitation on brand (they only focus on the products they manufacture).

By mid-2000, some of the manufacturing ESCOs had re-structured their operations and decided that they would no longer directly finance EE projects. They realized that financing was not their core business. As such, these manufacturing ESCOs sought partnership with other trading companies, utility ESCOs and engineering ESCOs such as Tropical Engineering, Inc. (TEI).

In May 2005, an association of ESCOs named ESCOPhil was established and registered under the Securities and Exchange Commission (SEC). The association was meant to organize firms engaged in energy services. Specifically, the purposes of the creation of ESCOPhil were to[1]:

- provide a forum for the effective exchange of information about industry trends and practices including introduction and propagation of new technologies for the industry;
- promote energy efficiency and demand reduction technologies, thereby creating tangible economic values;
- develop strategic advocacy positions with government agencies;
- initiate policies geared towards increasing business opportunities for its members; and
- educate and accredit other firms and organizations as new members.

Unfortunately, ESCOPhil has been inactive since 2006 without clear leadership, direction and cooperation among its members and officers.

Coinciding with the National Energy Efficiency and Conservation Program (NEECP) of the Philippine Department of Energy (DOE), a list of four public energy audit service providers and 16 ESCOs were identified.

The Philippine DOE issued a circular in September 2008 officially calling for the assessment and accreditation of ESCOs. The meaning of "ESCO" is defined better, in contrast to other types of contractors. Guidelines have been set on accreditation of ESCOs. To date, there are only eight ESCOs accredited by the DOE[2].

Between 2008 and 2009, UNDP, through its Philippine Efficient Lighting Market Transformation Project (PELMATP) with the Philippine DOE, developed a model ESCO demonstration project on efficient lighting systems. The mandate of the project included collaboration with local financial institutions (LFIs) and efficient lighting manufacturers. The project successfully retrofitted the lighting systems of two manufacturing plants of an industrial firm.

By 2009, the Philippine government had attempted to create a public "Super ESCO," operating as a sub-component of the Philippine Energy Efficiency Plan (PEEP) through a loan from the Asian Development Bank (ADB). The Super ESCO, a subsidiary of the Philippine National Oil Company (PNOC), was intended to finance EE projects for government accounts. At the same time, it was meant to technically and financially support the operations of other ESCOs looking to develop the private sector. Similar concepts were successfully implemented in China and India. Unfortunately, after the national elections in 2010, a new administration had a different position on the Super ESCO concept, which led to the program being shelved.

The International Finance Corporation (IFC), a member of the World Bank group, continued to develop the ESCO market in the private sector. In early 2011, IFC held a capacity building workshop for ESCOs as well as for service and technology providers. IFC strategically partnered with LFIs such as the Bank of the Philippine Islands (BPI) and Banco de Oro (BDO) through its risk-sharing agreements in order to develop EE in the public sector.

The ESCO market in the Philippines is quite promising considering that the country has the highest electricity tariff in Asia. The ESCO market, however, is not yet mature. First, the market itself is difficult. Potential clients' interest in energy efficiency is wavering and is most intense only when there is an energy crisis. During stable energy supply periods, the commercial and industrial sectors in general do not put high regard on energy costs mainly because of a mindset that the costs can be passed on to consumers. Such mindset is a weakness common to a country where the market is local and is unlike in exporting countries where competition is stiff and all costs of production are strategically monitored and minimized. The market is therefore reactive rather than being proactive to energy efficiency. It takes excellent marketing skills for ESCOs to propel the EE business in various sectors.

Secondly, many local ESCOs have been used to promote the shared savings EPC model, but they have limited financial resources. Compared to mature ESCO markets from other countries, such as in North America and Europe, LFIs have high interest rates on loans. Moreover, loans are subject to asset-based collateral, with which ESCOs have difficulty complying. Hence, for long-term and multiple projects, ESCOs in the Philippines will have difficulty financing projects by themselves. Philippine ESCOs need to develop the marketing skills needed to promote a guaranteed EPC model to meet this challenge.

Thirdly, there is a lack of understanding and acceptance of internationally accepted M&V protocols in the Philippines. This often leads to parties (ESCO and client) mutually agreeing on simplified terms and eventually arguing on energy savings calculations. Consequently, the projects are either put on hold or terminated once an M&V disagreement arises, leaving an impression that the ESCO business is not an attractive undertaking. Promoting the IPMVP and the CMVP program would address this barrier.

There are many other essential aspects that need to be addressed by the government and stakeholders in order for the Philippine ESCO market to progress. To name a few, tax/duty exemptions or subsidies on EE technologies and other financial incentives on EE projects can greatly bolster the growth of ESCOs. An effective legal, technical and financial framework in the country is much needed.

CONTRACTS

The main type of ESCO contracts introduced in the Philippines in the early 1990s, also generally known as energy performance-based contracts, were the "shared savings" and "guaranteed savings" EPC models. Under the shared savings model, the ESCO borrows from a financing institution such as a bank in order to finance and implement a turnkey EE project for a client. The client therefore will not provide initial cash outlay to the project but pays the ESCO an agreed periodic amount based on actual energy savings derived from the project over a contract period. The model is best illustrated in Figure 9-31.

Often, this is the model that ESCOs adopt in the Philippines with the rationale that it clearly differentiate ESCO services from that of their contractor and supplier competitors. The shared savings payments can be a fixed percentage of the actual periodic energy savings. In some cases, the shared savings payments are annually declining percentages of actual energy savings. The contract period can be as long as ten years. Despite the implied guaranteed savings and obvious financial risk that the ESCOs are exposed to under this model, some clients still seek for guaranteed minimum actual energy savings. ESCOs are penalized if the minimum guaranteed savings are not met. ESCOs adopting this model have a limited number of projects because

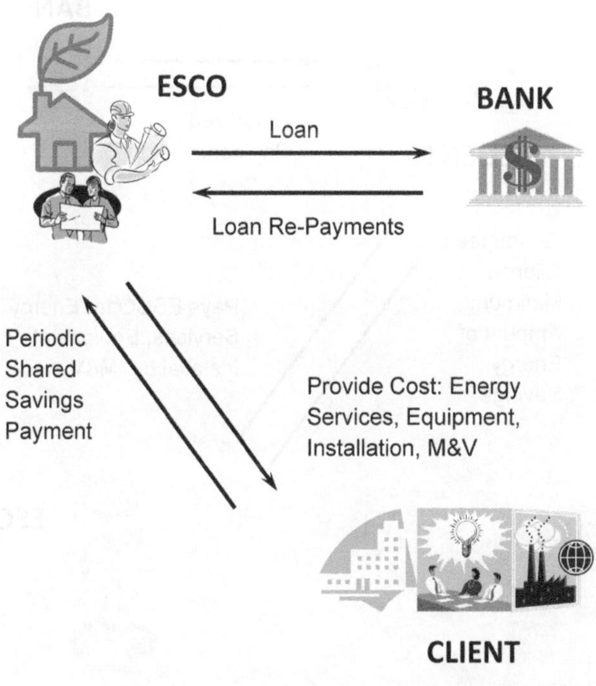

Figure 9-31: Shared Savings Model

they are not able to provide more collateral for additional loans from banks, hence they rely on internal funds.

Another Philippine EPC model is "guaranteed savings," which requires the ESCO to arrange financing between the client and a financing institution. The bank extends a loan to the client which will be utilized for payment to the ESCO's services. The client pays the bank an amortized amount. The ESCO guarantees the energy savings will be greater than the scheduled amortized payment by the client. This model is better illustrated below.

LEGAL FRAMEWORK

The country's current legal framework on EPC is presently insufficient. Government accounts potentially represent a large ESCO market share. However, financing through EPC for public accounts

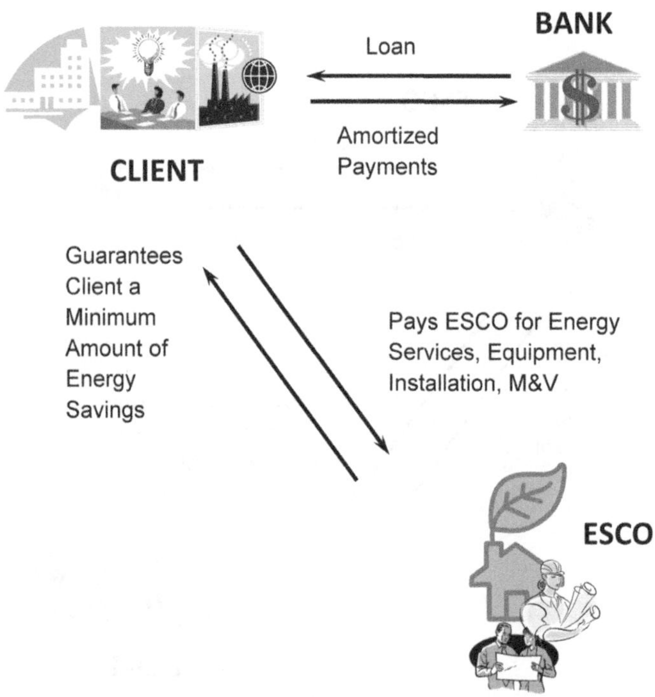

Figure 9-32: Guaranteed Savings Model

is currently impossible. Under the country's Commission on Audit (COA), the government's standard accounting procedure is not flexible to EPC terms. All public accounts are confined to fixed amortized payments on loans, hence, variable periodic payments as in the case of EPC agreements are not allowed.

Compared to international or even Asian region standards and labeling, the minimum energy efficiency standards imposed on certain appliances, such as air conditioners, refrigerators and CFLs, are not quite competitive and with limited coverage.

In contrast to other countries where ESCO markets are mature, there is a lack of incentives to promote EE, such as rebates, subsidies and tax exemptions, to name a few.

MARKETS

The ESCO industry in the Philippines is generally classified as two major markets; i.e., the commercial and industrial sectors. The institutional sector (hospitals, schools, universities and the like) normally falls under the category of the commercial sector. The residential sector is a difficult market to penetrate mainly because of the numerous accounts that require substantial amounts of time and marketing. The industrial sector seems to be more enticing to ESCOs as the longer operating hours of most industrial firms make payback periods shorter.

In order to attract foreign investments in the industrial sector, the government introduced incentives in special economic zones, which included lower electricity rates. This made ESCO business opportunities less attractive in the industrial sector. The industrial sector's sensitivity to opportunity costs arising from possible interruptions inherent to ESCO project implementation make it more challenging for ESCOs to market EE projects.

Presently, the energy consumption from the commercial sector is evidently increasing much faster than it is in the industrial sector. MERALCO's data in 2010 indicated that energy consumption from its commercial customers accounted for 39 percent of its total supply against 29 percent from its industrial customers[3]. The commercial sector continued to experience a constant increase in electricity rates and, therefore, is becoming a dominant ESCO market in the country.

EE projects in our tropical region are focused more on air conditioning comfort and energy solutions followed by efficient lighting systems. The major energy-intensive industries at present include business process outsourcing, retail trade, wire harness and semiconductor manufacturing.

FACILITATORS

Besides the Philippine DOE, the presently active facilitators of EE and energy conservation in the country include the Energy Efficiency Practitioners Association of the Philippines (ENPAP), the ASEAN Energy Manager Accreditation Scheme (AEMAS) and the European Chamber of Commerce in the Philippines (ECCP).

ENPAP (i) regularly conducts training, seminars and energy audits; (ii) provides energy management consultancy services; (iii) prepares feasibility studies; and (iv) undertakes local and international projects related to energy efficiency. AEMAS, the world's first regional certification system, conducts training for Certified Energy Managers (CEMs). In collaboration with IFC, ECCP has recently been actively promoting the concepts of EE and EPC through training courses and forums.

The Department of Science and Technology, particularly through the Integrated Program on Cleaner Technologies, is providing a range of business development support services to SMEs. The Philippine Business for the Environment (PBE), an NGO, assists companies and organizations in identifying their carbon footprints using the GHG Protocol Corporate Accounting and Reporting Standard developed by the World Resources Institute (WRI) and the World Business Council for Sustainable Development.

The Development Bank of the Philippines and the Land Bank of the Philippines (LBP) have the scope for becoming involved in climate change-related initiatives. Both provide loans to local government units (LGUs), cooperatives, private corporations and small-scale producers. The Development Bank of the Philippines (DBP) through its Industrial Pollution Control Loan Project II, a policy-based lending program, supports SME investments in efficient production and environmentally sound technologies. The Credit Line for Energy Efficiency and Climate Protection (CLEECP), a Land Bank product supported by KfW, is one example providing sub-loans to reduce electricity consumption and GHG emissions. Additional financial resources can also be accessed through project-based carbon trading under the CDM, which is a market-based mechanism included in the Kyoto Protocol. The Department of Science and Technology through its Technology Application and Promotion Institute (TAPI) offers the Venture Financing for Environmentally Sound Technologies Program.

Multilateral development banks, such as the Asian Development Bank and the World Bank, provide additional finance for countries and their respective government agencies to address climate change mitigation and adaptation. The Climate Investment Fund (CIF) is an example of a financing resource that was approved in July 2008 with over USD 6 billion in pledges. The Clean Technology Fund seeks to scale up financing to contribute to demonstration, deployment and transfer of low-carbon technologies with a significant potential for long-term GHG emissions savings. The Strategic Climate Fund provides financing through several pilot programs for new development approaches or to scale up activities aimed at a specific climate change challenge or sector response through targeted programs. The Carbon Partnership Facility (CPF) is designed to target investment programs that have the potential to contribute significantly to a transformation of emission-intensive sectors in client countries including the Philippines. The Global Environment Facility has been the main source of grants and concessional funding for adaptation projects and is of relevance for the Philippines. The Special Climate Change Fund under the United Nations Framework Convention on Climate Change (UNFCCC) was established to support activities in adaptation, technology transfer, energy, transport, industry, agriculture, forestry as well as waste management and economic diversification.

GOVERNMENT ACTIONS

Between mid-1980 up to 1992, the Ministry of Energy (now DOE) successfully implemented the Rational Use of Energy (RUE) project. The project introduced the use of an "energy bus," a vehicle that was customized with energy audit equipment and a mini-library for on-site EE activities. It also pioneered the Junior Exchange Program wherein experts from other Asian countries, such as Thailand, were sent to the Philippines. Local experts from the Philippines were consequently sent to work at the Energy Conservation Centre of Thailand. In 1990, potential ESCOs were convened for the first time, through the RUE project, to impart knowledge on the guidelines for establishing ESCOs. About 50 EE practitioners participated in discussions on the role of ESCOs in generating energy savings for companies.

Emanating from the acute power shortages in the early 1990s, the concept of DSM was developed. It was triggered by the promulgation of the Energy Act of 1992 that mandated the government to ensure sufficient, reliable and continuous supply of energy at affordable costs as well as to promote the judicious conservation, renewal and efficient utilization of energy. The DOE directed the Energy Regulatory Board (ERB) and the electric utilities to develop the DSM Regulatory Framework and DSM programs in 1995. The DSM framework, however, did not prosper well partly because there were no clear provisions of "DSM cost recovery mechanisms" for utilities.

An RP[4]-Hawaii project on EE technology and policy transfer was also implemented in 1998 that aimed at information exchange between the Philippines and the US, which bolstered the development of the Philippine ESCO industry. The US experience, wherein state and federal governments supported energy services and EPC, served as a model for the Philippines to adopt. Unfortunately, until now, no similar legislation has provided incentives specific to EE projects in the Philippines. The lack of information on EE and EPC led the DOE to explore possible retrofitting of a shopping mall as a demonstration project, in collaboration with MERALCO and Honeywell under a USAID grant. There was a lack of understanding about internationally accepted M&V protocols, as a result, the demonstration project did not come about. Towards the late 1990s, utilities began to have excess capacities as a result of allowing the entry of several independent power producers—the government's short-term solution to the energy crisis. This has partly led further to diminished energy conservation consciousness among stakeholders in the country.

The ASEAN Promotion of Energy Efficiency and Conservation in South East Asia (PROMEEC) project is a continuing activity in which the Philippines participated. Under the Energy Efficiency and Conservation program, the promotion of the ESCO business is one of its strategies. Specifically, it intends to achieve the following:

- development of M&V protocol for ASEAN

- development of EPC legal framework and standard form contract;
- development of project management and institutional guidelines;
- development of energy-saving potential indexes (benchmarking);
- e-commerce development for energy services.

The government needs to implement procurement policies that distinctly favor energy-efficient technologies. In addition, the current accounting standard procedures for government accounts should be enhanced whereby EPC models are accommodated. That is to say, repayments through actual and/or variable energy savings equivalent are accepted on EE loans or investments. While the Renewable Energy law (R.A. 9513) was finalized in May 2009, the Energy Efficiency and Conservation law is still underway, which is essential to the growth of the ESCO industry.

Poland

Janusz Mazur
Joanna Toborek-Mazur

ACTIVITIES

The history of the EPC industry in Poland started with the beginning of the democratic transformation that unfolded in the 1990s. Unfortunately, 20 years of experience has still not significantly influenced the popularity of this formula. To date, the number of projects implemented is still very low and does not seem to show a distinct growth trend.

Over that same period, there have been several companies among the pioneers of the implementation of projects under the EPC formula. All such companies have carried out projects focused on the improvement of energy efficiency in housing constructions, military facilities and hospitals (heat source replacement, improvement of wall insulation and replacement or modernization of central heating systems). Probably the largest ESCO project in Poland was also performed at that time through the modernization of about 40,000 street lighting points in Cracow. The consortium that took part in the project was comprised of ES System, Elektrim and Cracow Energy Utility.

In spite of a number of successful projects in the first period of their implementation, a few poorly prepared and unsuccessful initiatives saw their end in a court of law. This fact had a negative influence on the popularity of the formula.

The beginning of the 21st century brought new transformations of the ESCO market in Poland. Companies, agencies of large foreign entities appeared and initiated the implementation of proven foreign patterns, such as the EPC model (shared savings or guaranteed savings agreements). The participation of such entities contributed to making the EPC approach credible. As a result, the approach started to be used by other companies. This especially concerned the street lighting industry, which guaranteed more advantageous payback periods than did thermo-modernization initiatives.

During this period, with the support of the World Bank under a GEF grant, POE ESCO implemented one of the largest thermo-modernization projects. The projects were developed under the EPC approach and involved comprehensive actions in more than 30 facilities (mainly educational buildings) in the Małopolskie Voivodship. In spite of considerable success, the reluctance of the market to use the EPC approach was difficult to overcome. Among the reasons for such a state of affairs were a complicated legal system (unadapted fiscal regulations, balance sheet regulations, energy-related regulations, public procurement law) and the "spoiling" of the market by a wide range of subsidies/grants, in which ESCOs could not participate.

The ESCO industry hoped that the new parliamentary act enacted in spring of 2011 on energy efficiency would create a market favoring ESCO undertakings. However, the simulations and observations so far have shown that it is not causing significant market growth.

CONTRACTS

In Poland, there is no organization or institution that collects official statistics on the ESCO contracts concluded. The major source of information in that respect consists of an analysis of the public procurement procedures in place and interviews with active companies.

This gathered information indicates only agreements in the simplest TPF formulas. Energy performance contracts will be executed in the Polish market and rigorous guarantees for energy savings will need to be provided by the ESCO.

Based on the author's observations, a simpler TPF formula is most frequently applied to street lighting projects. From the very beginning, contract provisions were often phrased so as to enable participation of companies from outside the ESCO industry. After the conclusion of the works, these companies could sell their liabilities from the communes to banks or financial institutions. In such public procurement proceedings, the contractor does not provide guarantees for savings, although the system maintenance services have to be accepted for several years. The contracting authority itself, or its consultants, expect the estimated savings to be achieved as a result of this modernization.

Contracts under this formula were also concluded for projects involving the installation of heat exchanger stations in small district heating companies, the construction of boiler rooms or housing installations and the modernization of furnaces in the industry. Furthermore, they were used in investments for passive power compensation and small CHP projects with a size up to several MW.

An extremely interesting project was prepared and implemented in 2010 in Radzionków commune that provided for a comprehensive modernization of five educational buildings (in the scope of thermal and electrical energy savings). The PPP formula was used with EPC and the Public Procurement Law.

Energy performance contracts are currently far less frequently concluded than are TPF contracts. This often results from a lack of reliable information about energy consumption before project implementation, difficulties in estimating savings or lack of knowledge about standards. Perhaps the start of a Polish version of the EVO website as well as the translation by the Polish Foundation for Energy Efficiency (FEWE) of the IPMVP manual will contribute to changes in this regard.

Recently, in a number of business activity sectors in Poland, a "success fee" formula has become popular. The contractor's remuneration in such contracts is usually specified as a percentage of the effects achieved or fully depends on their achievement. It is, therefore, a modification to energy performance contracts and shared savings agreements.

LEGAL FRAMEWORK

The legal situation in Poland has neither enabled nor facilitated conducting ESCO activities. Dr. Jan Rączka, President of the National Fund for Environmental Protection and Water Management, in his report titled "Towards the Modern Energy Policy, Electrical Energy," wrote the following:

> The structure of ESCO—in spite of its well-known advantages and high efficiency—has not caught on in Poland. Firstly, for over two decades, a wide range of thermo-modernization grants for public facilities has been available. Secondly, preparing a contract for obtaining an energy saving effect requires a strict cooperation between ESCO and a facility owner before the contract conclusion. The provisions of the Public Procurement Law impose a number of constraints on both parties.

The studies administered by the Norwegian Bellona Foundation highlighted the following barriers:

- lack of legal provisions promoting ESCOs;
- poor recognition of ESCO services, excluding the possibility for ESCO participation in state-funded projects;
- lack of understanding of the EPC concept on the part of potential customers; and
- competition regarding grants.

Legal issues mainly relate to clients from the public finance sector (budgetary units). It is worth noting that, in this regard, they are obliged to apply the Public Procurement Law and the Act on public finance. The former is limited to specific conditions for the conclusion of contracts for a period extending four years. The provisions of the second act are limited to the budget year.

Another constraint is the Ordinance of the Minister of Finance, which does not allow energy performance contracts in the public sector to be considered as off balance sheet. Therefore, ESCO services must be considered as an encumbrance when calculating permissible debt and will have to compete with other debts to be incurred by any public sector entity. Unfortunately, this situation is the result of a lack of understanding of the EPC model since the latter may be neutral at worst for the budgets of public finance sector units as their return is paid from the energy savings. The situation in this regard is supposed to change starting in the year 2013, which will neutralize repayment of debt expenses to a certain extent. This should give a major boost to the development of the EPC concept in Poland.

Other barriers that are faced in the market include:

- The lack of possibility for co-financing projects under the EPC model as part of EU initiatives or national ecology funds.
- The EPC concept is not accounted for in the energy law. On the contrary, the system of price tariffs and rates (both in terms of thermal and electrical energy and gas) prevents the implementation of EPC in a way to ensure a return on investment over the periods expected by ESCOs.
- ESCOs encounter doubts in the use of off balance sheet financing in the private sector, based on bad understanding of the accounting regulations. The first problem emerges upon selection of the type of accountancy to be used: should an ESCO that finances investments at the customer's place apply bank accountancy or that of a company conducting construction activities?
- The presence of long-term obligations and receivables on the ESCO balance sheet. Chartered accountants are currently discussing whether they should be valued based on their nominal value or their adjusted price of purchase (discount value). A number of expert auditors think that trade receivables cannot be discounted while obligations with respect to banks can create an unfavorable accounting situation for ESCOs.
- The lack of sound understanding and experience about the legal control over the property of improvements installed by the ESCO.
- The difficulty for ESCOs to write off debts by selling their receivables (factoring). The basic offer put forth by banks and independent financial institutions deals with the purchase of receivables from the public finance sector for a maximum period of up to one year. Such a situation gives an opportunity for development only to companies in good financial standing.

A great source of hope for the ESCO sector is the Act on energy efficiency, which entered into force in August 2011. It imposes obligations on units of the public finance sector within the scope of improving energy efficiency. One of the alternatives points towards the possibility to conclude an EPC standard contract with an ESCO (although the legislation does mean ESCOs, the name itself is not used in the act even once). Unfortunately, it is the only possibility and it is rather improbable that it will be frequently chosen from much simpler contracts as far as implementation is concerned. The act also provides for the obligation to obtain white certificates for companies selling energy (heat, electricity, gas) to end-users. Such certificates may be obtained in exchange for the performance of undertakings within the scope of energy efficiency. Certificates may be obtained for the performance of undertakings to improve the energy efficiency of device auxiliaries, distribution systems (20%) and end-user facilities (80%). In order to be granted a certificate, apart from reporting relevant undertakings, one will have to win a tender (auction) organized by the President of the Energy Regulatory Office. Upon failure to comply with such requirement, the company will be obliged to buy the certificates from entities that have already obtained them (certificates will be subject to public trading) or pay a substitute fee. It is assumed that a new market for the projects implemented by ESCOs could be created that way. The reality of the next few years will show if the act proves to be an impetus to the activation of ESCOs and the development of the market.

MARKETS

At the beginning of the current year, the Norwegian Bellona Foundation conducted a research among active companies that dealt with obstacles and recommendations for the development of ESCO services in Poland. To that effect, among the six companies covered under the study, the main activities of each company revolved around contracts with respect to:

- guaranteed energy consumption reductions;
- guaranteed reductions in energy payments;
- energy management.

There is a very low level of competition observed in tenders for thermo-modernization and heat management. On the other hand, there is fierce competition for tenders relative to projects aimed at reduced electrical energy consumption. Most frequently, tenders target the modernization of street lighting. Their value is usually lower and few entities participate in the bidding process. The financial crisis of the last few years has caused there to be a dwindling number of such projects being implemented. The number of projects executed while ensuring financing or the purchase of receivables have been estimated at roughly ten in a single year. The above information constitutes the basis for estimating the size of the active ESCO market, which represents at least USD 10 million.

This small market, despite its huge potential, does not make for very enthusiastic business owners. In this context, large companies operating in the building sector have little incentive to enter the EPC market.

FACILITATORS

It should be noted that one of the reasons why the ESCO formula has failed to catch on in Poland lies in the small number of facilitators that are ready to help potential ESCO customers set up projects based on that model. Despite successful projects, neither consultants nor NGOs have attempted to popularize the concept on a wide scale. Nonetheless, the last few years have seen transformations beneficial to the development of the EPC market. An increasingly larger number of consulting companies are now providing advice on the preparation of ESCO projects.

Additionally, more and more publications on ESCO activities are made available. Discussions concerning barriers to the development of this sector as well as its future in Poland are very often held at high-ranking conferences.

The project of the Ordinance of the Minister of Economy on the training and examination of applicants to become energy efficiency auditors, which is to be issued in connection with the Act on energy efficiency, covers a six-hour lecture on the ESCO model. The EBRD has expressed initial interest in developing the latter in Poland. The EBRD, PNECA (KAPE) and NFEPWM (NFOŚiGW) are contemplating the creation of a specialized fund designed to buy out ESCO debts or to guarantee financing, in collaboration with the Bellona Foundation.

The European Committee of Manufacturers of Domestic Equipment (CECED) has greatly contributed to the improvement of ESCO operating conditions in Poland. It partook in the public consultation process relative to the Act on energy efficiency as well as collaborated with members of Parliament and the authors of the act. The committee also played an active role in the activities of the unlicensed ESCO association and has been assessing the conditions necessary for the successful development of the sector in Poland.

The activities of the Polish Foundation for Energy Efficiency (FEWE) are no less significant for that matter. The representatives of this organization have conducted numerous research studies and worked on the preparation of many EE programs. The translation and publication of the 2010 version of the IPMVP will work to the advantage of the ESCO sector.

GOVERNMENT ACTIONS

In order to increase the popularity and scope of the EPC concept in Poland, only a few changes within the law are necessary. This would encourage both business owners and governing units of the public finance sector to use it on a wider scale.

As an example, the second NEEAP, published in 2011, indicates that ESCOs will be beneficiaries of the white certificates system owing to the possibility to aggregate energy savings and proceed to take part in tenders on behalf of other entities, as provided for in the Act on energy efficiency. Furthermore, public sector units, which are obliged to apply measures provided for in the Act on energy efficiency, will be able to enter into agreements with entities such as ESCOs for the implementation and financing of energy efficiency projects. To that end, the Ministry of Economy will post the list of ESCOs active in the market along with the contract template for the provision of energy efficiency services. Unfortunately, the ESCO sector is not as optimistic in this regard.

Portugal

Luis Silva

ACTIVITIES

EPC started in Portugal in the early 1990s through a company called Econoler Portugal, a subsidiary of Electricity of Portugal, Petrogal and other minority shareholders. The focus at the time was mainly on cogeneration projects. The company was the only one active in that sector in the 1990s. It stopped its operations and EPC somewhat disappeared as an active market for the next decade.

Currently, almost 100 companies are registered, showing the interest of the sector in executing energy performance contracts with the Portuguese state. From those companies, the majority are represented by energy consulting and engineering design companies. However, it is also possible to identify many energy utilities and multinational companies from the technology and maintenance sectors.

In order to access the new public sector market, ESCOs have to enter a public registration phase. In order to be shortlisted for calls for tenders to set up energy performance contracts with public bodies, the ESCO must comply with a series of criteria defined by the Directorate General for Energy and Geology (DGEG). These criteria are related to the minimum technical, financial and staff qualifications required to insure the success of the procedure. In parallel, a model contract is being drafted to facilitate future contracting procedures.

Although figures[1] on the Portuguese ESCO market are not very clear, preliminary information from major players allow estimating that the overall turnover is between EUR 10 and EUR 30 million (USD 13.1 and USD 39.4 million). This is mainly due to energy performance contracts executed in the industry and building sectors for outdoor lighting as well as cogeneration and micro renewable power plants.

Currently, the major threat to the operationalization of this program for public administration buildings is related to the current difficult financial situation. This in turn introduces significant barriers to getting credit from the banking sector (severely limited) and to ensuring banking loans to finance the implementation of energy efficiency measures as a result of energy performance contracts.

Nevertheless, the Energy Efficiency in the Public Administration Program (ECO.AP) will guarantee support for the creation and development of a real ESCO market in Portugal. The ensuing ESCO market is expected to grow consistently over the next few years and contribute actively to increasing energy efficiency in the Portuguese economy.

LEGAL FRAMEWORK

Decree-Law No. 29/2011 of February 28, 2011 was published in accordance with the National Energy Strategy and the NEEAP as well as with the objective of accelerating the implementation of energy efficiency measures in public buildings. This new piece of legislation established the legal framework for energy performance contracts to be signed by public administration bodies and ESCOs as a result of Council of Ministers Resolution No. 2/2011, which instituted ECO.AP.

Subsequently, in accordance with item 10 of Article 3 of Decree-Law No. 29/2011, an online registration system was developed and made available to all ESCOs interested in entering into energy performance contracts with the Portuguese public administration. The procedure was intended to assure, at an early stage, that companies fulfilled basic minimum requirements to guarantee the effectiveness of the program.

MARKETS

In the near future, it is expected that the implementation of ECO. AP will contribute to the creation of a much more dynamic ESCO market in Portugal, able to meet the target of a 30 percent reduction in public sector energy consumption. In parallel, it is expected that this specific program for the public sector can also contribute to increasing the number of energy performance contracts in the private sector, thereby allowing for the growth of the Portuguese ESCO market.

FACILITATORS

The Portuguese Association of Energy Service Companies (APESE) has been formed as a result of the new market dynamics as well as to give voice to, and represent the interests of, the companies currently working in Portugal's market. The association is led by some of the most active Portuguese ESCOs that have been working under the EPC approach over the last few years.

GOVERNMENT ACTIONS

The Agencia Para A Energia (Portuguese Energy Agency) was created in September 2000 through the transformation of the Center for Energy Conservation (CEC), created in 1984. In December 2001, following Council of Ministers Resolution No. 154/2001 of October 19, the mission, scope and functions of the Energy Agency were adjusted and its name changed to ADENE.

ADENE will certainly be instrumental in supporting the development and growth of the EPC market through its role in the implementation of the new legal frameworks related to this sector. ADENE will also play a key role in providing the necessary capacity building for the different stakeholders.

Romania

Florin Pop
Tudor-Alexandru Socea

ACTIVITIES

The Romanian ESCO market is still quite new but growing at a constant rate mainly due to the government and legislative energy efficiency actions over past years. However, Romania remains one of the most energy-intensive countries within the European Union.

ESCOs only started in the mid-2000s even though the market was almost nonexistent. At the time, only a few organizations had shown interest in the concept and started to promote it.

At the moment, there are around 14 players in the Romanian ESCO market, most of them being small-sized companies and only a few that are larger, such as EnergoBit ESCO and Dalkia Romania. While most ESCOs provide audit and consultancy services, they are limited when it comes to financing large projects. The market track record shows that only a small number of projects estimated above EUR 1 million (USD 1.3 million) have ever been financed, especially because of the Romanian banking system's lack of expertise and knowledge regarding the energy efficiency sector. Nonetheless, experienced finance providers in energy efficiency projects have come to offer support for the Romanian ESCO market. As a result, in 2011, the EBRD signed a EUR 10 million (USD 13.2 million) loan with EnergoBit ESCO in order to finance energy efficiency ESCO projects to be implemented in the Romanian market. The ESCO received a high capital of trust from the EBRD after successfully implementing a series of very complex turnkey projects in the transmission and distribution (T&D) sector as well as for large-scale wind farm power plants.

Another player, Dalkia (formerly Veolia Energy), is engaged with public clients in public district heating systems through a wide range of integrated services that improve and optimize clients' equipment. Dalkia has three subsidiaries: Dalkia Termo Prahova, Dalkia Alba-Iulia and Dalkia Tulcea. Dalkia recently signed a partnership that involves cogeneration and refurbishing of existing installations within the heat production and distribution network for the city of Ploiesti, serving 165,000 people. This project transformed Dalkia into one of the strongest ESCO players in the field of public thermal heating in Romania.

The ESCO business development process is slow and painstaking due to a series of constraints:

- Mistrust of customers caused by the lack of awareness of the ESCO concept and the energy efficiency business culture. In order to stimulate the ESCO market in the long run, the Romanian ESCO market still needs a period of "education."

- Public procurement rules—although many improvements have been made over the past years, for local ESCOs, the procurement process is still complex, ambiguous and time consuming.

CONTRACTS

One of the most common types of contracts used by Romanian ESCOs is the BOOT contract. The BOOT contract has proved its feasibility both in implementing and operating cogeneration-based ESCO projects. EPC and the guaranteed savings approach are still popular, but have been developed only in the industry sector for small to medium-sized projects.

LEGAL FRAMEWORK

PPPs reflect one of the most important legal developments in the European Union. PPPs have been described as an essential legal instrument for the delivery of public services and as the most innovative interface between the public and private sectors. Such relations will provide for infrastructure projects, as well as for many other schemes in areas covering transport, public health, education, public safety, waste management and water distribution. Therefore, this law seems to favor the development of ESCO projects, both for the public and private sectors and it will likely have a direct influence on the development of the EPC market in Romania.

MARKETS

The ESCO market is mostly based at this time on small to medium-sized projects implemented in the industrial and public lighting system sectors, with a few large dimension district heating projects using cogeneration technology.

It is expected that the growth of the Romanian ESCO market will come from projects targeting the public sector due to the partnerships established between public and private entities. This has been the result of strong interest shown by the Romanian government in stimulating investments in public infrastructures and energy efficiency.

FACILITATORS

The Romanian Energy Efficiency Fund is an energy efficiency investment provider that develops energy efficiency projects promoting rational energy use. The Fund assists industrial companies and other energy consumers in adopting and using modern technologies for the efficient use of energy. The Romanian Energy Efficiency Fund seeks to promote projects through the successful implementation of GEF energy efficiency projects while increasing the banking sector's interest in supporting energy efficiency investments in Romania.

GOVERNMENT ACTIONS

The regulatory organization in the energy efficiency sector, the Romanian Agency for Energy Conservation (ARCE), subordinated to the Ministry of Economy, was dissolved following the merger through absorption by the Romanian Regulatory Authority for Energy (ANRE). Its activities will be reorganized within the Regulatory Department for Energy Efficiency of ANRE. Ongoing projects of the ARCE continue within the new organization and it is expected that some kind of support to the development of the EPC market will be offered.

International agencies (EBRD, USAID, UNDP/GEF) have been active in the development of energy efficiency financing. Once a strong finance promoter for energy efficiency projects in the early 2000s, the World Bank itself is not currently active in the Romanian ESCO market. Other financing providers have appeared, such as the Romanian Swiss Fund or the Romanian Norwegian Fund, which sustain municipalities in securing funds for various energy efficiency and environmental projects estimated to have long-term positive impacts.

Russia

Valentin Andrianov
Remir Mukumov
Alexei Zakharov

ACTIVITIES

EPC, as a business model, was almost nonexistent on the Russian terrain up to 2009. Industrial companies either managed energy-saving projects on their own, reframed them to outsource them to external engineering companies, or used leasing schemes. In the public sector, procurement law only permitted them to buy lowest priced equipment while the budget code did not permit public customers to execute contracts if the term went beyond the three year budgeting cycle. It is to be noted, that this could have been avoided if separate dedicated funds had been allocated in the budget for long-term target regional or municipal programs. Or a special decision of the local administration had been taken with dedicated funds attached to it.

At the end of 2010, the Russian government approved the Energy Saving and Energy Efficiency Increase program. The program will be in operation until 2020. The law introduces, among additional requirements, the installation of meters for all participants in the energy market as well as requiring regional and municipal authorities to create energy efficiency programs. Finally, the law introduced energy services agreements as a special type of contract and established requirements within these agreements.

CONTRACTS

Given the lack of comprehensive metering at sub-plant level on the industrial side, most EPC projects are based on deemed savings to establish the baseline level of energy consumption.

The situation is different in the Russian public sector. In the public sector, energy baselines are for now impossible to establish in quite a number of cases because regions have not been able to comply with meter installation deadlines. However, as meters are being rolled out at a rapid pace, it is reasonable to expect that the market will see many more EPC tenders in a couple-of-years' time across all types of facilities nationwide.

It is mandatory since 2011 that all public sector orders (both federal and municipal) of the value above USD 3,000 be announced and tendered through a national public procurement portal (www.zakupki.gov.ru). This portal is a great resource to monitor the public sector market as both past and ongoing EPC procurement notices and results can be searched online.

Municipal utility companies are subject to the kind of tariff regulation that permits only a one-year tariff planning horizon, which creates additional uncertainties and risks for energy efficiency projects. However, the new water supply law to come in force in 2013 envisions return on asset-based tariff schemes and other long-term tariff planning options. A heat generation and distribution law, with similar tariff options built into it, is currently being deliberated in the federal government.

LEGAL FRAMEWORK

After adoption of the federal Law on Energy Efficiency, the energy efficiency market in Russia got a very powerful impulse for its development. The law stimulated demand for energy audits and created the possibilities for new players to enter the market. The market of energy auditing and EPC is self-regulated; i.e., all players must be members of a self-regulated organization (SRO). The Ministry of Energy registers all SROs and their members.

The EE Law for the first time introduced the idea of an energy performance contract called an energy service contract, which has been defined very broadly as a contract whereby the contractor takes on obligations to implement EE measures for the benefit of customers. The law further provides that these contracts may be executed in the residential and public sectors as well as in the utility sector. Energy and water utilities may take on an ESCO role. The industrial sector is not specifically touched upon in the EE Law as regards performance contracting as private enterprises have always been entitled to enter into any types of contracts under the Russian civil (commercial) law.

The EE Law also mandates that energy audits be conducted before the end of 2012 and every five years afterwards for the following types of entities:

- all public entities or publicly owned organizations;
- all regulated enterprises (utilities, railways and other monopolies);
- all producers or carriers of water, gas, electricity, coal, oil and petrochemicals;
- all entities which spend more than ~USD 330,000 annually on purchasing gas, oil, diesel and related fuel, thermal or electric energy.

Entities failing to conduct an energy audit are assessed fines of USD 1,600-USD 7,800. Incumbents of those entities are also penalized with smaller fines for the failure to have the audit done. All energy auditors need to be incorporated in a self-regulated body to which they should regularly make insurance and membership payments. The self-regulated bodies are mandated to maintain a compensation fund in order to cover potential liability cases of their members.

The energy audit requirement spurred an almost exponential growth of the number of energy auditing companies. Almost every engineering company in Russia is now a member of one of these numerous self-regulated bodies. A recent roster of the Ministry of Energy lists 111 self-regulated entities[1] of auditors, which collectively account for about 5,000 energy auditors (both single-purpose legal entities, natural persons as well as subsidiaries of diversified holding companies).

The Ministry of Energy is in charge of the collection and analysis of energy certificates produced by the auditors.

Together with the passage of the EE Law, a number of seminal amendments were made to the Budget Code and the Public Procurement Law (also known in Russia as Law 94) that made EPC possible in the public sector. The Budget Code now directly specifies that public customers may enter into energy service contracts (which will be paid from energy savings only) for a term exceeding the current budgeting period[2]. There are no limitations on the maximum term of these contracts. No special budget appropriations are required for treasurers to pay for ESCO services as ESCO costs are to be covered from savings.

Law 94, which regulates procurement of all works, goods and services by public customers in Russia[3], was also amended as a result of the passage of the EE Law. Law 94 was expanded by Section 7.1 as well as by Government Decree No. 636 of August 18, 2010 (adopted in pursuance of the EE Law) which in great detail regulates procurement of EPC services.

Under Law 94, municipalities are free to place orders for energy services without any special authorization. Procurement itself can be done not only via an auction-type of tender (whereby bids are evaluated by their price only) but also through a 'competition' type of tender that allows selecting the contractor based on several parameters (with the price being just one of them):

- bidder's reference list and/or qualifications;
- project implementation timelines; and
- term or amount of the quality guarantee.

Under the procurement law, an ESCO is subject to a defined minimum penalty for failure to reach promised savings. This penalty is three times higher than that levied against the customer for a failure to make a timely payment for ESCO services.

The law also provides for a mechanism to adjust annual consumption figures in a post-installation period for changes in weather conditions, staff count, hours of operation or equipment loads. A model energy services contract designed by the Ministry of Economic

Development (MED) and freely available on its website contains formulas designed to perform the required calculations[4].

As regards the public sector, besides mandatory energy audits that the public agencies need to undertake before the end of 2012, the EE Law also contains a requirement that public agencies save 15 percent of energy by 2014 relative to their metered or (where metering was absent) contractual energy consumption in 2009. In order to make this requirement palpable, the EE Law mandates that treasurers cut funding of energy-related expenditure items by a certain percentage annually to a total 15 percent by 2014 in comparable conditions. The MED has yet to define the list of factors that public agencies may account for to offset increased energy use, as the case might be. This 15 percent energy savings requirement is a significant driver for EE and EPC for public sector clients.

Existing Legal Barriers

Although it contains many essential provisions, the legislation is not free from the obvious drawbacks, some of which deserve to be mentioned.

- The 15 percent cut of energy-related expenditure items is applied to all public agencies indiscriminately regardless of whether they entered into an energy performance contract. This means that public customers will have over the next three years 10-15 percent less money compared to what they paid for energy in 2009. This implies that there is going to be less money available to ESCOs. Therefore, in the current climate, only those projects are feasible where savings significantly exceed the planned 15 percent funding cuts.

- As are all other contractors under Law 94, ESCOs are subject to a generic requirement to present collateral for 10-30 percent value of the project in case the cost of the project exceeds EUR 1.25 million (USD 1.64 million). This collateral requirement makes sense in virtually all regular procurement contracts where contractors receive compensation soon after supply/installation of the goods in question to ensure replacement of faulty equipment. However, applying this requirement to ESCOs is redundant, as in the Russian model ESCOs will not receive the money they invest in the project back from the client after the installation is complete. ES-

COs will only receive compensation gradually; i.e., as the savings are being realized. Public treasurers simply have no other account than that for the utility bills to generate the needed cash flow to ESCOs. Therefore, ESCOs are materially interested in proper performance of equipment and the aforementioned collateral requirement only makes their life harder and payback periods for the client longer. Moreover, ESCOs are subject to fines for savings shortages.

- The issue of collateral is further complicated by the fact that the maximum price of the contract is set in a distorted way. Indeed, it is not set as minimum savings requested by the client and bid upwards as one would expect but as the client's energy bill costs in the baseline year times the tariff and times the term of the contract. Although bids are eventually evaluated based on ESCOs' energy savings proposals, the maximum price—rather than being purely based on savings requested by the client—is artificially increased, which demands a higher collateral needed from ESCOs.

- Payment to ESCOs under energy performance contracts can only be made from savings on a reduced amount of energy (and water) supplied to the client. This is despite the fact that other economies might result from the project such as savings achieved through reduced maintenance or staff costs, avoided one-time costs or optimized energy rates. Under current regulations, ESCOs will have to recover all their investment costs from energy and water savings only.

Current Legal Trends

A number of legislative initiatives to improve EPC regulations were initiated in 2011 by a range of actors, including the Russian Parliament (State Duma) and the Russian Energy Agency. The MED was planning to publish a revised model contract in early 2012. All this work was done against a backdrop of the Russian government deliberating a major revamping of the procurement law with both the MED and the Federal Anti-Monopoly Service presenting their draft laws. Each of the two drafts were to have different impacts on the EPC landscape if adopted in the proposed form. There is no doubt

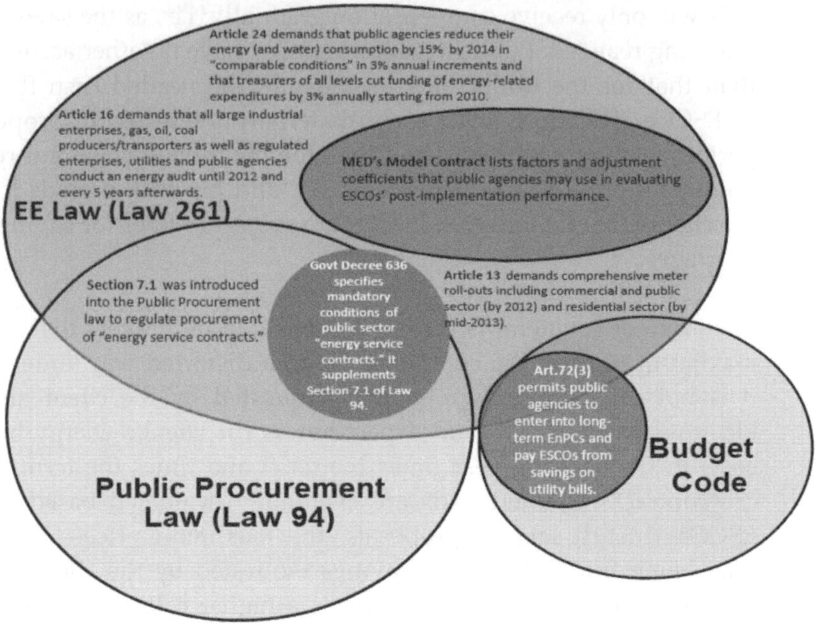

Figure 9-33: Government Actions Impacting Energy Efficiency

that the legal framework of the EPC approach in Russia will develop at a fast pace in the coming years.

MARKETS

The players of the Russian ESCO market are quite different in size and capabilities. There are no companies which dominate the Russian market as a whole.

Most ESCOs (companies for which EPC in a variety of forms is the core business) are small-size companies with a head-count of 10-60 employees. The origins of these ESCOs can be traced back to energy auditing, patented energy savings technology or turnkey engineering services. Due to lack of long-term funding and their small size, the projects in which these ESCOs are involved tend to be small (in the range of USD 10,000-USD 200,000) and focus on quick payback measures. These include lighting retrofits, heat metering (to reap financial savings for the client when making a switch from overestimated contractual volume to metered consumption), heat regulation, variable-frequency drives and cogeneration (including fuel switch projects) assisted with leasing schemes.

The mid-level slice of the market is represented by mid-size private sector companies of a variety of backgrounds (producers of metering equipment, IT companies, real estate developers) that make inroads to the EPC sector due to perceived low risk on projects with familiar clients and attractive internal rates of return (IRRs). Here, the investment volume (project size) tends to oscillate around USD 300,000 and USD 2,000,000.

On the other side of the spectrum are ESCO subsidiaries of a handful of major utilities or other state-owned enterprises being spun off to do EE projects both for parent companies and their existing clients—municipalities, cities and regions. Project volumes beyond the USD 3,000,000 mark are not uncommon in this segment.

There are several examples of real energy performance contracts in the industrial sector. Typically, Russian industrial companies are interested in improving their own energy efficiency and reducing their energy bills with their own means, but there are obvious limitations in using their own working capital for financing these projects. Under these conditions, ESCOs usually offer to work via shared savings schemes, leasing schemes and deferred-payment supply and install contracts and draw necessary financing by themselves. These schemes are favorable for industrial clients but have very clear limitations, mainly due to relatively high costs of financing. This narrows the market to only very attractive investment projects with high IRRs and short payback periods. Usually, these projects are quite short (up to three years) and rather small in volume relative to client revenues.

FACILITATORS

The Ministry of Economic Development through its Department of Tariffs, Infrastructure Reforms and EE, the Russian Energy Agency under the Ministry of Energy, as well as the Presidential Commission on Modernization and Technological Development of the Russian economy are actively involved in monitoring the market and (re)shaping the legal framework.

As recently as 2011, the Global Environment Facility provided a grant of USD 10 million to the EBRD to facilitate provision of technical assistance to both the regions tendering public sector EPC projects and the MED developing the enabling legal framework. The EBRD has committed to lend to active players in the sector, and it is expected that this will have a catalytic effect on the market.

EXPECTED TRENDS OF INDUSTRY DEVELOPMENTS

Having in mind the current state of the Russian ESCO market at the end of 2011 and major existing trends, one may anticipate the following opportunities for Russian ESCO market development:

- Russian ESCOs will widen the product range and start to provide full services to their clients. Energy performance contracts will become more popular.
- Consolidation of the market will decrease the number of existing associations and the number of market players. Consolidation between SROs will lead to the establishment of one to three federal-level associations. The most viable regional ESCOs will get a chance to become federal-level players. Small companies will quit the market or find a specific niche.
- Independent ESCOs will likely be acquired by energy equipment manufacturers.
- Special financing system(s) to fund ESCO activities will be established with the active participation of international and Russian development banks.
- Financial institutions will initiate the creation of ESCO industry funding standards (creditworthiness, criteria for project selection, requirements for collateral, rules of refinancing the loans, etc.).
- Consolidated SROs and the largest ESCOs will take active part in the development of industry standards like M&V standards, investment grade audit standards, etc. Along with the standardization of the market, the programs for the certification of energy efficiency professionals will be developed.

The following options to create a sustainable system for EPC financing can be envisioned in Russia:

- direct financing to municipalities, regional administrations, or ESCOs via development banks or commercial banks;

- creation of a dedicated financing facility which would purchase the receivables under energy performance contracts from ESCOs (these receivables could be securitized); and
- creation of a dedicated credit line for Russian commercial banks to refinance the loans to ESCOs (shared savings scheme) or ESCO clients (guaranteed savings scheme).

The selection of option(s) will define the future ability to raise capital for EPC projects in Russia.

Slovakia

Marcel Lauko

ACTIVITIES

The start of energy service contracting in Slovakia can be attributed to conditions in the early 1990s. With the rising cost of energy and the availability of efficiency technologies, ESCO projects have become much more commonplace. The term ESCO is increasingly known among potential clients looking to upgrade building systems that are either outdated or need to be replaced as well as among potential clients. Who wish to improve the efficiency of district energy plants. The real market for ESCOs started to develop in the mid-1990s.

There are several types of ESPs in the market (energy companies or utilities, ESCOs, energy agencies, consultancy and engineering companies, equipment providers). However, in the narrow sense, we define an ESCO as a company offering a comprehensive energy service for the implementation of energy projects at client sites. Thus, the ESCO must provide for project identification, implementation and financing as well as ongoing energy management and plant operation services. Pure ESCOs are able to provide EPC, together with or without ESC and other energy services. Definitions for ESCOs vary but the common understanding is that they are usually differentiated from other energy companies in mainly three ways: ESCOs guarantee energy savings, their payment is related to the level of energy savings achieved and they usually provide financing for the project.

Independent, specialist ESCOs in Slovakia vary in size from small privately owned companies operating at local or regional level to large companies operating nationally or internationally. They account for an increasing share of the total ESCO market and have the provision of energy service contracting as their sole (or main) business. They may meet the requirement for project finance from their own resources or (more usually) through an agreement with a bank or finance house.

Although EPC in Slovakia is quite well known, it is not a widely used form of ensuring energy efficiency in most sectors of the economy. As the relative energy intensity of the Slovak economy is significantly higher than the European average, there is a substantial market for EPC applications. However, there are many barriers to EPC market development, especially in the public sector. Although the EPC concept is not widely applied in Slovakia, the basic legislation exists and creates, in general, conditions for the development of ESCO activities based on this type of energy services.

The state of the EPC/energy services market on the demand side provides valuable information with respect to the situation on the supply side. There are not many ESCOs active in Slovakia. In fact, in Slovakia, there are currently no ESCOs for which EPC-based services would be the sole or main business.

According to publicly available information, approximately 30 companies consider themselves to be ESCOs. As most of them provide services in the private sector, only limited information on their activities is available. Based on available information relative to several projects, the majority of ESCOs are specialized in projects involving energy delivery contracting, focused on heat production/distribution system reconstructions and fuel switching. From information on several projects implemented in Slovakia, it appears that only some of these companies have completed genuine EPC projects.

At the present time, concerning EDC and EPC, the following companies have achieved a certain level of experience:

- **Siemens, s.r.o.**

  Siemens Building Technologies provides a complete offering of technical infrastructures for security, comfort and efficiency in buildings. It also provides innovative products, systems as well as total solutions and services in Slovakia.

- **COFELY, a.s.**
  Cofely is the energy and environmental efficiency services company of GDF SUEZ for businesses and local authorities. Cofely is the leader in energy and environmental efficiency services, designing and implementing solutions to help businesses and public authorities make better use of energy, while reducing environmental impacts.

- **Intech Slovakia s.r.o. company**
  Intech provides services in the utilization of energy efficiency and renewable energy sources. It also supplies Czech-made CHP units ("Tedom" type). Intech delivers assistance services for the preparation of energy audits and feasibility studies. Its service called "Energy Efficiency Management" can be considered as typical EPC. It allows consumers to invest in energy efficiency without engaging their own financial resources. This service was awarded the Slovak Gold quality certificate in April 2004.

- **Johnson Controls**
  Slovak Johnson Controls International spol. s.r.o. is the daughter company (100%) of Johnson Controls Holding Company, Inc. Several divisions of Johnson Controls have their headquarters in Žilina, Lučenec and Trenčín.

- **Dalkia, a.s. company**
  Its daughter companies are operating district heating systems in the regions (towns). Some of them are renting energy facilities from the municipality on the basis of long-term contracts, which makes these companies energy contractors obviously regulated by the Regulatory Office for Network Industries.

CONTRACTS

There are two main types of ESCO contracts used in Slovakia.

1. **Energy Delivery Contracting (EDC)**
   In delivery contracting, a so-called ESCO invests in facilities used for energy conversion at a client's premises. Usually, the ESCO installs a combined heating and power system in a public building

and provides the necessary fuels and services, including O&M of the installed systems. The settlement of accounts is based on the delivered heat and electricity volumes.

2. **Energy Performance Contracting (EPC)**
While in delivery contracting the focus is on energy supply, in the case of an EPC project, the ESCO aims at reducing energy requirements. Such a reduction can be realized through all kinds of methods to increase efficiency and reduce energy costs. In EPC, the contractor's remuneration is based on the cost savings achieved. An important aspect of EPC is the financing of the investments via the guaranteed cost savings achieved thanks to the ESCO through improved energy efficiency under the terms of the contract.

LEGAL FRAMEWORK

The only legislative document mentioning energy services (not specifically EPC) is **Act No. 476/2008 Coll. on rational use of energy** (transposition of Directive 2006/32/EC of the European Parliament and of the Council of April 5, 2006 on energy end-use efficiency and energy services and repealing Council Directive 93/76/EEC). Paragraph 10 of this act sets the legal frameworks on energy services. According to the law, energy services must help reach rational use of energy and profit. Energy services have to be based on contracts that may cover the following activities:

- energy analyses and audits;
- project implementation focused on rational use of energy;
- O&M of energy devices;
- monitoring and evaluation of energy devices;
- provision of fuels/energy;
- delivery of energy devices.

Paragraph 11, letters g) and n), assign to the Ministry of Economy the duty of preparing template contracts for providing energy services, unfortunately without a specified deadline.

Another legislation which affects energy services, especially EPC, is the philosophy of regulating prices of heat from district heating systems, which are reflected in edicts of the Regulatory Office for Network Industries. In the event that a district heating system operator manages to lower input costs, the price of heat must decrease as well. This philosophy of regulation prevents district heating systems and ESCOs from collaborating, thus making it unprofitable.

MARKETS

The main non-industrial target sectors for ESCOs are described below.

- **Public Buildings**
  Municipalities and regional authorities (Higher Territorial Units) are operating different types of buildings; e.g., office facilities, social buildings and sports buildings. Municipalities have considerable problems with operating costs. The current technical state in public buildings can be described as critical. The biggest potential for energy savings is in the public building sector (school facilities and health service sector). The most cost-intensive energy systems and installations are heating, lighting, hot water and HVAC systems.

- **District Heating Systems**
  Nearly every town or village, which has one or more apartment buildings, also has a district heating system. District heating systems provide hot water or steam for heating systems. The scale and structure of district heating systems depend mainly on locality and end-use demands. The majority of these systems were built more than 15 years ago and no radical reconstruction and rationalization have been performed until recently. The main barriers to the application of these methods include non-clear property rights as well as too many end-users and customers.

- **Public Lighting**
  Public lighting systems are installed in small villages or large cities. Problems with their operation occur almost everywhere. The prevailing part of public lighting systems is in poor technical condition, which in turn leads to very high operating costs. Electricity consumption is especially expensive in the municipal budget. Owners of these systems are municipalities, although they are

sometimes rented to private companies. Based on energy audits conducted for public lighting systems, the simple payback period is more or less ten years, which shows that the potential for EPC or EDC is very high.

FACILITATORS

In Slovakia, there is currently no association helping the ESCO sector with its activities (such an association has been set up recently in the Czech Republic). There is a state energy agency (Slovak Innovation and Energy Agency, SIEA) which is a contributory organization of the Ministry of Economy is responsible for tasks in the field of energy efficiency, including conditions for the development of energy services (required by EU or national legislation).

The SIEA. inter alia, is responsible for the certification of energy auditors on behalf of the government. Some of them provide broader services including energy consultancy, energy management or even the installation of energy-related equipment and systems.

There are no official statistics about the number and type of energy consultants and energy service project facilitators. Since most of them provide services in the private sector, only a limited amount of information on their activities is available.

GOVERNMENT ACTIONS

The same general legislation for all business companies including ESCOs applies in Slovakia. The legislative framework and legal conditions that affect ESCOs, or energy activities in particular, are not only unclear but also changing, thereby making local legal advice a necessity. While there has been considerable harmonization in recent years, the legal and regulatory climate will remain unfamiliar to most foreign companies. Furthermore, the drive towards accession and transposition of the EU legislation has led to a rapidly changing legal environment, which makes some legal requirements very unclear. Therefore, there is no substitute for obtaining local legal advice on these specific issues as well as on more general aspects such as employment, taxation, health and safety.

Status of ESCOs

The status of ESCOs is not detailed in any Slovak legislative document. An ESCO is a standard business company. Each business company can carry out ESCO-like activities. Energy services are not regulated activities. According to the local legislative framework, any company offering services in the energy sector (supplying heat, electricity) needs to have a license to undertake business activities in the energy sector in compliance with the Act on Energy No. 656/2004 Coll.

Tendering Procedures

The awarding of a contract to provide services or goods to a public sector client must be conducted through public tender, making proactive business development by the ESCO less attractive. Specific tendering requirements apply to the procurement of goods, services and public works by, or on behalf of, a public body, or using public funds. These requirements are laid out in the Act on Public Procurement (No. 25/2006 Coll.) and subsequent amendments.

Energy Services

Basic legislation exists and creates general conditions for the development of ESCO activities, including energy services. However, general legislation in the field of energy efficiency and energy services is not supportive. In particular, there are still missing template contracts for providing energy services, as required by EU and national legislation. In addition, there are no financial instruments or subsidy schemes, which could support energy service projects.

Slovenia

Alexandra Waldmann

ACTIVITIES

Slovenia is by nature a small market. With just over two million inhabitants and a little more than 20,000 km^2, it is comparable in size to the US state of Vermont. The country's GDP per capita in 2009 amounted to USD 24,417. Slovenia has been a member of the EU since 2004 and joined the EURO zone in 2007.

With regard to EPC, while the market has been seeing activity, it still has to be characterized as an immature market. Although one could say that there is no market altogether, there are still a few ongoing projects. Some of the reasons for this situation are related to the lack of adapted legislative environment, which is only recently changing, and to the confusion within the market about options and opportunities related to energy efficiency projects, which inhibit demand.

Activities on capacity building and project development started in the late 1990s with the TRANSFORM program of the German government geared towards the democratic and economic development of advanced transition countries, which were formerly part of the Soviet Union or Yugoslavia. TRANSFORM provided consultation on contracting in the public sector; i.e., the development of market-based instruments for energy efficiency and EPC. The first energy performance contract was the result of efforts under such program. It was signed in November 2001 in the municipality of Kranj[1].

Projects have been implemented as an amalgamation of supply- and demand-side refurbishments and retrofits. As is the case in other countries across Europe, street lighting retrofit projects have recently become attractive in the public sector, largely due to legislation demanding a reduction of light pollution as well as available financial support. Despite an energy savings target of 15 percent for Slovenia's public sector, more projects have been initiated in the industrial sector. In turn, it is to be noted that the housing sector has seen fewer projects brought to fruition. The public sector is lagging behind and in the health sector, no projects based on a full EPC model are available.

A frequent project core is boiler house refurbishment. Many boiler houses (industry, industrial zones and multi-story dwellings) are in dire need of refurbishment, being aged and inefficient. Cogeneration, based on biomass as well as highly efficient gas cogeneration units play, an increasingly greater role here. These projects have opened the door for new market players.

For a long time, there was only one proper ESCO along with a second (supply-side) company which had a focus on ESC with performance and savings guarantees. Companies present on the market can be grouped into: (a) one ESCO; (b) three to five companies active with ESCO-type services, and (3) another three to five companies which have the potential to enter the market relatively quickly. For the most part, companies have a background in energy supply and network O&M (district heating, other energy supply) while a smaller proportion boast expertise in equipment manufacturing, equipment O&M and lighting technologies. Another category consists of engineering companies that have experience in engineering design, project implementation and commissioning, and which are not involved on the financing side of projects.

Utilities are expanding their services sector, based on legal obligations to provide energy efficiency services to their customers, with some explicitly gearing up to provide ESCO projects. Strategies to achieve this are either know-how acquisition (companies with the needed know-how) or in-house know-how building. Both have been seen over the 2010-2011 period. The only real ESCO is now owned by Slovenia's largest supplier of oil and of other energy products.

Financing EPC projects in Slovenia faces similar issues as in other EU countries. Banks on the Slovenian market are not keen on lending to small ESCOs, especially since the "credit crunch." High-risk premiums by banks increase project costs for ESCOs and customers alike. Banks lack an understanding of the key principles of EPC, including that the guarantee on savings is actually the equivalent of future cash flow. Financial instruments that could alleviate difficulties, such as forfeiting or even mezzanine funding, are not available or have not been tested yet. In return, projects are looking for public funding. The Slovenian market generally seems to be overly dependent on national and EU subsidy programs as well as on financial incentives, such as structural funds or local grants for addressing energy efficiency needs. This tendency has also been shown in an analysis of energy efficiency policy and its effectiveness in Slovenia[4].

There is finally an urgent need to provide practical support (technical and legal) to procurement officers, building owners and project developers in order to help them use the EPC concept. Badly prepared tender procedures based on lack of know-how have increased project risks for bidders. Policy needs to provide accompanying guidelines as well as to support capacity building. Appropriate financial instruments for a market the size of Slovenia is critically needed to stimulate the use of EPC within the country.

CONTRACTS

While both the guaranteed savings and shared savings models can be found, a mix of models is also not uncommon. Costs are nevertheless based on a fixed energy price (by contract), whereby the risk of energy price fluctuations remains with the customer. Furthermore, the ESC model, in combination with performance and savings guarantees, is applied. In fact, because of the nature of companies offering ESCO services, supply has been part of contracts, probably in most cases.

While many, if not most, contracts contain some kind of a performance and/or savings guarantee, the majority of projects are technology-focused. They also lack the system view and cross-cutting measures that make EPC most attractive.

LEGAL FRAMEWORK

The Slovenian national activities are activities of an EU member state and, as such, are based on common EU goals and framework legislation. EU legislation today favors EPC as an important market-based instrument to increase building energy performance more explicitly than it did a few years ago.

For the private sector, the key framework is energy and environmental legislation (supply & demand, renewable energy & energy efficiency, climate change) as well as energy prices (currently still lower than EU average). All of which set the stage for (potential) demand and supply of ESCO services. Public sector activities are bound to follow procurement and budgetary regulations specific to the public sector.

Looking at the public sector, the existing energy act and related decrees are in principle setting a framework in favor of ESCO activities by not prohibiting them. However, a combination of more stringent energy savings targets for the public sector, mandatory energy strategies at municipal level, mandatory energy bookkeeping and a lack of resources, both capacity-wise and financially, clearly support the use of external know-how and financing. Since there is no EPC law as such, the legislation that rules EPC projects are the concession law and the law for PPPs. Additionally, the law on municipal financing, the energy act, and building codes form other parts of the framework for EPC.

Restrictions for the public sector to apply EPC stem from strictly enforced debt ceilings, a limitation of contracts to four years (with a possible extension, but not guaranteed) and a lack of procedural guidelines. Presently, municipalities cannot have a contracted debt in the current fiscal year, including existing debt, that exceeds 20 percent of realized revenues in the year prior to the year of contracting such debt. It also provides that repayment of the principal and interest in any year of repayment does not exceed 5 percent of realized revenues in the year prior to the year of contracting the debt. The Law on Execution of the State Budget defined EPC contracts as not counting towards this debt ceiling. However, the relevant part was scrapped only one budget year later by the Ministry of Finance. This means for EPC implementation that each project faces the risk of being considered towards debt by the Ministry of Finance on a case-by-case basis. As a result, readiness to enter into such projects at municipal level is limited.

The Public-Private Partnership Law, while being very specific on procedures for the establishment of PPPs, does not include service contracts and profit sharing contracts, which again leaves EPC open to interpretation on a case-by-case basis.

Green public procurement, generally a potential vehicle for more energy efficiency in the public sector, is not yet being applied. An action plan for the 2009-2012 period has been published, but lacks visible (accountable) implementation. An accompanying regulation has been drafted, but is not in force.

In 2011, a draft for a revision of the Energy Law was published. The lengthy draft law states that the municipal sector can implement its energy efficiency goals via service contracts with external partners and that contracts being repaid via savings—despite being considered a debt—will not be counted towards the debt ceiling of a municipality. The law also refers to (relies for its statement on) a Ministry of Finance recent review of such contracts. The fate of this law is still unknown.

MARKETS

The largest short-term potential for EPC to develop a market (demand) seems to be in the public sector. The Slovenian public sector currently has an estimated energy consumption of 1,850 GWh annually[2]. The country's 2008 NEEAP set a goal of 95 GWh savings to be achieved via investments in building energy performance improvements, 247 GWh via investments in efficient heating and cooling equipment, and 339 GWh via investments in efficient use of electricity.

MUSH sector-related buildings total about seven million square meters of useful floor area[3]. One-third of those buildings (erected before 1980) consume about two-thirds of the energy, with an estimated savings potential of 30 percent. Given the above, a rather rough estimate provides for a possible theoretical initial investment volume of EUR 250-300 million (USD 329-394 million) at current average energy prices, contract terms estimated at 15 years and approximately 50 percent of the buildings deemed suitable for an EPC project. Fully aware of the non-scientific approach to that number, one has to start somewhere since no market player or facilitator has provided any kind of estimates thus far; hence, a daring first step needs to be taken.

However, there is a clear challenge ahead in the short term. While the technical potential is available, capacity and awareness on options and opportunities are still lacking, as is the willingness of potential clients to do things in a new way. A request for expression of interest in EPC among Slovenia's 210 municipalities not long ago resulted in only eight responses.

The provision of hands-on guidelines on energy-efficient products and energy efficiency services targeted at procurement officers as well as work on innovative financing solutions for smaller ESCOs would work towards a positive development.

FACILITATORS

Institute Jozef Štefan has been involved from the beginning in projects and capacity development activities revolving around EPC in Slovenia via its Center for Energy Efficiency. In recent years, a growing number of energy agencies established by, or with the support of, municipalities as well as under the EU IEE program and its *ManagEnergy* Initiative have provided for another basis for facilitation of customer-side capacity building. Support for both project development and other services, such as measurement and verification of projects, is provided through these agencies. For the public sector, energy bookkeeping is mandatory, as are energy concepts (both areas of energy agencies' activities). While two agencies are currently most active in the field and have acquired substantial knowledge, two others are close to following suit.

GOVERNMENT ACTIONS

The framework conditions for EPC are slowly improving, and the legal and financial supporting framework is visible.

Financial Support

- Feed-in tariff which supports, and is attractive for, the renovation and upgrade of energy supply systems towards highly-efficient cogeneration units or renewable-based generation ;
- Since 2005, the local "Eco Fund" (Eko Sklad) is operative (grants, preferential loans); e.g., co-financing of lighting upgrades (focusing on business sectors, SMEs);
- Availability of structural funds (national co-funding secured) ; and
- Lack of technical assistance support for project preparation.

Legal Support

- Mandatory energy efficiency action for energy supply companies;
- Green public procurement, the revised law (2011) will gear (once in force) public procurement towards more energy efficiency ; and
- Regulation on street lighting, building codes, energy law.

South Africa

AZ Dalgleish, LJ Grobler

ACTIVITIES

Eskom, South Africa's national electric utility, formally recognized DSM in 1992 when integrated electricity planning (IEP) was first introduced. Recognizing that South Africa might run out of capacity by 2006, the first DSM plan was produced in 1994. In this plan, the role of DSM was established and a wide range of DSM opportunities and alternatives available to Eskom were identified.

Early on, Eskom viewed the use of the ESCO model as an effective DSM tool. In the mid-1990s, two weeks of intensive ESCO training for the utility was delivered by Dr. Shirley Hansen. In September 2002, the DSM Fund was approved. The year 2003 was spent mainly preparing the DSM business model and operations, developing customer awareness and education campaigns as well as setting up the ESCO industry.

A National Energy Efficiency Agency (NEEA) was established

in 2006 with the broad mandate to promote energy efficiency across the South African economy. This Agency is responsible for supporting energy efficiency projects for the public sector and targeted industrial end-uses (in the residential sector as well as in public and commercial buildings). The NEEA is also responsible for the accreditation and development of an ESCO industry in the country. The NEEA faces a daunting challenge in scaling up its organizational capacity to undertake these responsibilities.

Since 2008, evolutionary trends in Eskom Integrated Demand Management (IDM) have involved expanding funding models offered for DSM projects through the ESCO model to also include a standard offer option as well as a standard product option.

Evolvement outside the utility mainly includes legislation drafted by the South African government to promote energy efficiency in South Africa by publishing regulations 12I in 2009 and 12L in 2011. Regulation 12I concerns the qualification of green-fields and brown-fields as industrial policy projects through the promotion of sustainable energy efficiency, local production and job creation. Regulation 12L allows tax incentives for businesses that can demonstrate sustainable energy performance improvements through energy efficiency.

Barriers

In the past, cumbersome procedures for evaluating and approving project funding requests have delayed projects by months or even years. Eskom IDM has, therefore, designed the Standard Offer and Standard Product financing models to remove such barriers.

The approval timeframe for the Standard Offer and Standard Product models has been drastically reduced. Standard Offer projects can be approved in less than eight weeks and Standard Product initiatives in less than one week. Before the introduction of these two additional financing models, the EEDSM Fund primarily relied on ESCOs to conduct energy efficiency activities. However, there are now relatively few strong ESCOs remaining.

There is no mechanism for assisting the energy auditing and prefeasibility study process, which means that ESCOs must assume considerable risks and time delays before projects can access financial support. As there is little or no commercial lending for energy efficiency, ESCOs have nowhere else to go. Some energy efficiency markets, such as SMEs and municipalities, are virtually untapped.

Enabling Factors

Probably the most important positive enabling factor for the South African ESCO industry is that it has a strong backing from the national utility, in addition to the targets for demand reduction set by the DME for 2012 and 2025. Both these organizations realize that for these targets to be met implementation arrangements need to be improved.

Although, initially, project approval timeframes were a barrier, the new funding models for ESCOs (Standard Product and Standard Offer) that make project approval much quicker have become an enabling factor. Standard Product projects could be approved in a week while, in the case of initiatives led under the Standard Offer model, this could be in less than two months.

Another major enabling factor is the high tariff increases experienced in South Africa. In 2010, Eskom was approved for a three-year price increase (25 percent per year). In effect, the price of electricity nearly doubled over a three-year period, which makes the business case for investing in energy efficiency much more attractive.

Other enabling factors for the ESCO industry are standards such as SATS 50010, the recently published international energy management standard ISO 50001 and the legislation for tax incentives with respect to measureable and sustainable energy efficiency improvements in industry. It is interesting to note that a requirement of EEDSM programs is that they be subject to a full M&V campaign based on the IPMVP.

The expectation is that, over the next three to five years, the current ESCO program will not be as intensely utility-driven as it is at the moment. The expectation is that the role of ESCOs will move towards delivering energy services with focus on value added to industry.

Three factors stand out as having the potential to drastically affect the development of ESCOs in South Africa:

1. The approved tariff increases of 25 percent for three years resulting in electricity prices almost doubling over the same period, which could lead to bigger industries becoming their own ESCOs. The ESCO concept, in its traditional role, is expected to be mostly active in the public sector.
2. Legislation drafted by the South African government in Regulation 12L allows tax incentives for businesses that can demonstrate sustainable energy performance improvements through energy efficiency.

3. Carbon taxes, which could be required of businesses that cannot demonstrate improvements in terms of energy performance.

CONTRACTS

The three funding models described below are currently offered by Eskom IDM for EEDSM projects.

Eskom Improved Standard Offer

The Standard Offer is a mechanism used by Eskom for acquiring demand-side savings under which Eskom is to pay for verified energy savings using a pre-determined and pre-published rate in c/kWh for the implementation of an approved technology. Any energy user (customer), project developer or ESCO that can deliver verifiable energy savings, from 50 kW to 5 MW, can propose projects and, if successful, is paid the fixed amount per kWh over a period of three years. Achieved savings is verified by an authorized, independent M&V organization.

Standard Product

The Standard Product (SP) program is a mechanism designed to provide specific rebates for efficiency improvements derived from the implementation of approved technologies. Standard products are solutions designed to act as replacements for less energy-efficient technologies. Participation in the SP program requires no formal contract—only a formal commitment by the customer. The project approval turnaround timeframe has been streamlined in order to create the capacity to implement small and medium-sized projects. In order to accelerate energy efficiency projects, the SP project approval timeframe tpicall6y takes no more than four weeks.

ESCO model

ESCOs, which are accredited by Eskom, operate by establishing a three-way partnership between themselves, Eskom and the customer. Their combined knowledge of DSM technologies and programs is used to determine the best way to obtain results at customer premises.

There are three contractual arrangements in all EEDSM projects. These are summarized below.

The Eskom IDM contract with the client

This contract governs the relationship between the customer and Eskom IDM, and is the first completed contract in any DSM project.

The Eskom NEC contract with the ESCO

Whereas the DSM contract is negotiated first, the NEC contract defines the relationship between the ESCO and Eskom IDM. It sets out performance, payments and penalties.

The ESCO-client contract (maintenance)

An important changeover point is the end of the performance assessment period. This is usually three months after completion and it is during this time that the ESCO must prove that the project has met its contractual objectives. During that period, the ESCO is liable for penalties due to underperformance. However, as soon as the project is performing, or over-performing, the liability shifts to the client to maintain the project at design performance.

MARKETS

The South African Association of ESCOs (SAAEs) was created in 2003 to assist in the creation of a sound ESCO industry. The association holds regular forum meetings at which matters of mutual interest are discussed. These meetings also promote a code of conduct to which members must adhere. The SAAEs' latest list includes 85 members.

All members of the SAAEs must be registered with Eskom IDM. There are currently around 400 ESCOs registered with Eskom IDM, of which only 20 percent have submitted a project proposal since 2010. A full list of registered ESCOs is available from the Eskom IDM website. This includes ESCOs and other entities such as redistributors.

Growth in the number of ESCOs has not been important. The number of vendors selling energy-saving products, devices and equipment has shown substantial growth.

Eskom IDM requires that EEDSM projects be subject to a full M&V campaign based on the IPMVP. Energy Audit, which is independently under the Performance Assurance section in the Eskom Assurance and Forensic department, is managing the M&V program. Cumulative verified DSM savings from the inception in 2005 to March 31, 2010 under the EEDSM program amounted to 2,118 MW. Over that same period, a total of 11,755 GWh was saved. This resulted in CO_2 emission savings of 11,755,000 tons.

The ESCO business in South Africa focuses on all market segments, although it is rather fragmented between the residential, commercial and industrial sectors. There is currently a big drive towards energy efficiency in commercial buildings. The largest impact in the industrial segment has been made in the gold mining sector, with special attention given to underground lighting, water pumping, hot water systems, fridge plants and rock winders.

The measure most frequently pursued in the commercial segment is efficient lighting. ESCOs most frequently focus on hot water load control and efficient lighting in the residential segment. An extensive solar water heating (SWH) program was also launched in 2009. The program is currently slowing down while its counterpart heat pump program is gaining momentum.

Program types that will be carried out over the next three years and that have specific associated targets include lighting, HVAC systems, SWH, heat pumps, demand reduction, compressed air, industrial process optimization and low-flow showerheads.

South Korea

Christopher Seeley

ACTIVITIES

South Korea has had one of the most long-standing ESCO industries in Asia with ESCO-specific activities traced back to the very early 1990s. This industry began growing as a result of various government programs intended to support the development of the ESCO industry. This began with specific low interest loans that ESCOs could access in order to develop their business and offer ESCO services to their customers. This resulted in a steady growth in the number of ESCOs in South Korea since the mid-1990s.

There has been a balance in presence in the market between the largest multinational ESCOs and their locally established counterparts which have also seen a share in the market over the past five years. The market has benefited from the continued demand for ESCO services from international companies. In addition, US public facilities have formally requested energy efficiency projects to be undertaken by means of energy performance contracts.

There has been increased interest overall as well as a push for improved energy efficiency, predominantly led by the government but also well adopted by the private sector (especially large conglomerates). This continues to generate an attraction on the part of ESCOs, which have experienced a steady growth in South Korea, currently estimated at around 150.

With the adoption of green growth as a main focus of the South Korean government (both at the national and Seoul City levels), energy efficiency activities, including ESCO services, have become part of a nationwide aspiration. This has resulted in an increasing business opportunity for ESCOs. However, with the market being so well established and one of the most mature and active in Asia, the competition is very tough. This means that every year a number of ESCOs fail as well.

CONTRACTS

Although both traditional types of ESCO contracts are available and used in the South Korean market, it is dominated by shared savings contracts. The general market assumption is that close to 98 percent of the contracts signed with ESCOs are based upon a shared savings approach. This is not surprising considering that ESCOs have traditionally had access to preferential funds from the government with the specific intention that they would be used to develop and implement these types of contracts.

This percentage has been declining in recent years, with increased willingness, and even preference, on the part of clients to enter into performance guarantees as opposed to shared savings contracts. This is the result of several factors. Firstly, for those clients who have had good experiences with energy efficiency projects, the investment is seen as far less risky than is the case of the initial project. Hence, they are more willing to self-finance these projects while taking advantage of a performance guarantee provided by the ESCO. On the other hand, there will be other clients for whom the shared savings model has resulted in bad experiences, which may have brought about a conflict in post-implementation measurement and verification activities. Consequently, as these clients still have an energy efficiency mandate, they will elect to embark on performance guarantees in place of shared savings contracts. The largest multinational ESCOs seem to be leading the way in recommending this type of contract, with many of the international clients implementing EPC projects. Although still few in number, there have recently been power purchase agreement (PPA) contracts provided by ESCOs in South Korea. This may also see some growth as a model for achieving both improved energy efficiency as well as energy cost reductions for clients.

LEGAL FRAMEWORK

The current legal framework reflects a mature ESCO market and continues to present an environment that can support the respective industry. There is a strong legal infrastructure set up to facilitate the business practices of the various types of organizations that make up the industry whether they are ESCOs needing to formally register (ESCO business status), international providers of financing, specifically for ESCO projects, or building owners who need to understand the legal context of a specific performance contract. There is an abundance of experts, who can provide advice relative to the ESCO industry coming from educational institutions and/or private firms.

MARKETS

The industrial sector remains the most valuable for ESCOs in terms of business opportunities. The majority of ESCO projects have traditionally been within this sector with the commercial building sector still a rather distant secondary market. ESCO projects appear to be well spread out across different sub-sectors within industry.

Most projects address the main areas for energy efficiency; i.e., process improvement, cogeneration and tri-generation, waste heat recovery and generation of power on site. Improved lighting efficiency has been a very common measure in almost all energy efficiency projects across both the industrial and commercial sectors. Furthermore, this is likely to continue to be a viable component in these projects with light-emitting diodes (LEDs) being manufactured locally.

The average range with respect to the size of investment in ESCO projects appears to be rather wide. There is a very stable market for small ESCO projects (USD 250,000 to USD 350,000) in both the industrial and commercial sectors. It is still uncommon to come across a project that exceeds USD 1.5 million. Large projects are usually undertaken in phases where the investment is divided into a number of stages and over a period of up to three to four years. This is especially common in shared savings contracts where the ESCO is unable to take on a large investment due to limitations in both availability of, and access to, liquidity. However, US government facilities in South Korea are the exception. Indeed, large energy efficiency retrofits have been undertaken through performance contracts and have well exceeded the USD 2 million amount.

Although there are ESCO projects carried out over eight- to ten-year terms, from both the ESCO and client perspective, it is evident that the preference in the market is between five and seven years. This is considered to be above average contract length compared with other countries in the region that prefer terms of between three and five years. Perhaps this is a reflection of the maturity of the market with a lower perceived risk and government loan programs that allow for long-term contracts to be executed (with long-term repayments).

FACILITATORS

There has been a good presence of facilitators that continue to promote the use of ESCOs as a means of achieving improved energy efficiency in South Korea. The Korean Association of ESCOs (KAESCO) was established in 1999 and has remained active both domestically as well as internationally. It continues to work on developing the recognition of the ESCO industry among building owners and operators across the country. The association is also organizing conferences, training programs and workshops to that end. It also actively participates in ESCO activities all over the world, with partnership activities and collaboration with the National Association of Energy Service Companies (NAESCO) and its neighbors from Japan, the Japan Association of Energy Service Companies (JAESCO), and China, the Energy Management Company Association (EMCA). The Korean Energy Management Corporation continues to actively engage in initiatives locally and internationally to promote and facilitate energy efficiency projects, although not solely through the use of ESCOs and performance contracts.

GOVERNMENT ACTIONS

For the past couple of decades, the national government has played a significant role in the development of the ESCO industry. It not only continues to launch programs as well as policies to encourage the increased implementation of energy efficiency projects in both large facilities (industrial sector) and buildings (commercial sector) but it has continued to provide support to the ESCO industry. This has ranged from direct financial incentive programs (such as the Energy Conservation Fund which has provided tax refunds, low interest loans and other financing vehicles for ESCO projects) to indirect initiatives that support the development and training of ESCOs across South Korea.

At local government level, the Seoul Metropolitan Government (SMG) has followed in the national government's footsteps, setting up various direct and indirect initiatives to support improved energy efficiency within the capital city, once again through mechanisms such as low interest finance. Perhaps one of the best methods to do this was to set an example. In 2008, the SMG embarked on a retrofit program targeting all of its public buildings using EPC as delivery model. This was undertaken in phases over a couple of years and demonstrated to the private sector the use of ESCOs as a viable option for achieving improved levels of efficiency in a number of buildings. The private sector then followed this lead with several commercial buildings in Seoul being retrofitted using ESCOs.

Spain

Jose Guerra

ACTIVITIES

The emergence and development of energy services in Spain cannot be explained without analyzing the evolution of the Institute for Energy Diversification and Saving (IDAE), which is a public entity created in 1974 as a center for energy studies. In 1984, the IDAE was renamed, and started to drive audits, energy diagnoses, advice on energy savings and diversification to administer funds supporting national energy policy actions. In 1986, it started functioning as a public company, which developed projects in the market to promote energy saving and diversification, acting as the first ESCO reference in the Spanish market.

Throughout the 1990s and early 21st century, the IDAE signed a large number of energy services contracts with several companies in order to execute projects in the fields of energy saving, energy efficiency and cogeneration[1].

Table 9-18: Energy Services Contracts Signed in the 1990s and Early 2000s

	Energy Saving	CHP
Years	*IDAE* investment M€*	*IDAE* investment M€*
1998	9.3	3.4
1999	9.5	4.6
2000	6.2	6.2
2001	8.2	4.1
2002	1.8	5.1
2003	3.1	4.5
2004	3.2	1.0

Source: IDAE Annual Reports. Own analysis

In 2003, the IDAE established the Spanish Strategy on Energy Efficiency and Savings and launched the first Energy Efficiency Plan for the 2005-2007 period. Through such an initiative, the IDAE assumed a clear leadership in the coordination and promotion of plans through the Spanish Regions (Comunidades Autónomas) and energy service associations. In collaboration with one of the most important ESCO associations in Spain[2], the Association of Maintenance Companies (AMI), a member of the European Federation of Intelligent Energy Efficiency Services (EFFIES), a model of an energy services contract was published in 2005. It focused on the public administration (contract involving supply and services) as a boost measure to promote the ESCO industry in Spain. This contract stated the basis for all relevant changes in the Spanish law regarding energy service contracting by public entities. This ultimately led to the law on public-private partnerships, published in 2007, and other subsequent decrees.

Table 9-19: Planned Investment

	Planned Private Investment M€	Planned Public Investment M€	% Public Help
Plan 2005-2007	7.926	7.29	9%
Plan 2008-2012	22.184	2.844	13%
Plan 2011-2020	81.980	9.742	12%

Source: IDAE

As the IDAE became more active as a national energy agency and reduced its participation in the ESCO market, some private ESCOs started to attack that market. However, there was a critical lack of knowledge on the part of those ESCOs about the definition of the service, risk management, methodologies, savings contracts and M&V protocols.

At the present time in Spain, there is a wide range of ESCOs in operation. More than 451 companies are registered as ESCOs on the IDAE website (http://www.idae.es). However, just a few met the necessary requirements to be recognized as an ESCO[3] in accordance with European Directive 32/2006.

Among the traditionally large European ESCOs, such as Dalkia and Elyo-Cofely, we can find ESCOs that belong to traditional companies in the industrial and tertiary sectors for the O&M of their installations. Among others:

- those mainly associated with the Association of Maintenance and Energy Service Companies (AMI)[4], like Acciona, Applus, Clece, Eulen, EMTE, Ferroser, etc.;
- recently established ESCOs owned by Spanish utilities: Endesa, Iberdrola, Gas Natural Fenosa or HC-EDP;
- ESCOs belonging to technology vendors of efficient products or control systems mainly associated with Association of Energy Service companies (ANESE)[5], such as Schneider, Siemens, Honeywell, Johnson Controls, or engineering or consulting companies associated with Association of Energy Efficiency Companies (A3E)[6].

The Joint Research Center of the European Commission estimates the current market size in Spain to be around EUR 100 million (USD

131 million), but it is to be noted that this data underestimates the actual size of the market. Private companies estimate that the potential business for ESCOs to be in the range of EUR 1.4-4 billion (USD 1.8-5.2 billion).

CONTRACTS

Historically, the most extended energy efficiency contract in Spain has been the energy supply contract (often identified as a "chauffage" model), mainly in buildings with central heating systems, for the replacement of inefficient coal-fired boilers or fuel boilers. This kind of solution continues to be the dominant energy efficiency contract in Spain, to the detriment of the shared savings or guaranteed savings models. Only after 2008, with the translation[7] of the IPMVP and the certification of professionals in M&V did the other models of EPC begin to be implemented in Spain.

In the industrial sector, for the cogeneration projects implemented during that period, the most common contractual approach used has been the BOOT.

As of today, energy supply contracts are the predominant model used by Spanish ESCOs in the tertiary sector (mainly for residential buildings with central heating) as well as in tertiary private sectors for hotels, offices, sports facilities, elderly homes and hospitals. However, solid information is poorly collected or not available.

LEGAL FRAMEWORK

The three energy objectives for 2020 are: (i) 20 percent of energy efficiency over the business-as-usual (BAU) scenario; (ii) a 20 percent contribution of renewable energy; and (iii) a 20 percent reduction in GHG emissions. In Spain, the EU climate change strategy goals have been included in the Spanish Law for Sustainable Economy in 2010.

The following are interesting initiatives that further focus on the promotion of renewable technologies under energy services contracts:

- GEOTCASA: financing geothermal installations in buildings;
- SOLCASA: financing solar thermal installations in buildings;

- BIOMCASA: financing biomass-fired facilities in buildings;
- GIT: financing large thermal installations using renewable sources in buildings.

MARKETS

An interesting aspect of Plan 2000 ESE is that it includes public lighting in several municipalities, which represents a large potential market to develop energy services in Spain. A few tender processes that ran through 2011 seem to clarify different options for the procurement process in this market, where around 9 million old and poorly efficient luminaries were installed in 8,300 municipalities. So far, three different approaches for procurement have been inventoried with respect to three municipalities:

- Soto del Real: "Dalkia Model, 4P" contract in a competitive dialogue tender process where three competitors were selected from a list of pre-candidates, which initially competed for the best technical solution and the best economic offer.
- Alcorcon: "Dalkia Model, 4P" contract where the technical solution is defined by the municipality in a non-binding energy audit. The best economic offer wins.
- Salobre: Energy performance contract with an initial predefined technical solution.

An initiative set up by the Diputació de Barcelona in the context of the Covenant of Mayors. It negotiated a loan of EUR 500 million (USD 657 million) from the European Investment Bank and two other commercial banks for implementing energy efficiency and renewable energy projects in Barcelona's municipalities. During the 2010-2011 period, the Diputació conducted audits in a large number of buildings, installations and public lighting systems and it was expected that municipalities would start the tender process for public procurement in 2011 and 2012.

In the tertiary sector, the Spanish tourist industry is one of the most dynamic and experienced sectors with the fastest growth in recent decades and representing more than 10 percent of the gross domestic product. Many hotels in Spain, mainly those that belong to large corporations, have implemented energy efficiency measures and some of them have implemented the first energy performance contracts in the country[8].

BARRIERS TO OVERCOME; RISK MANAGEMENT FOR THE DEPLOYMENT OF ESCOs

The development of energy services in Spain requires overcoming a set of barriers that are not very different from those that exist in other markets, which are already developed. Thus, in financial, legal and institutional terms, the most significant barriers can be found below.

Financial

- Lack of credit guarantee mechanisms
- Low activity of existing lending institutions in energy efficiency
- Weak lending institutions

Legal

- Lack of clear and transparent ownership rights within the existing legal framework
- Weak legislative structures that prevent existing laws from being enforced
- Absence of mandatory medium-term budget forecasting, which prevents long-term service agreements and deters involvement of ESCOs

Institutional

- Underdeveloped ESCO market
- Inadequate managerial and technical expertise at municipal level in designing and implementing bankable EE projects
- Inadequacy of information about the financial markets and services available to regional and city administrations
- Lack of best practices to show potential benefits to customers

CASE STUDIES

Supply and Goods Mixed Contract: City of Alcorcón[9]

Description

The Alcorcon City Council convened in September 2010 and set up a mixed supply and goods procurement procedure for the management, comprehensive maintenance and repair of exterior lighting facilities, allocating an amount of EUR 1.58 million (USD 2 million) per year (including VAT) for a ten-year fixed period targeting ESCOs.

The specifications of the procurement procedure were:

- The City Council transfers the use and operation of outdoor lighting facilities to an ESCO for the time needed by the ESCO to undertake its improvement plan (adjustment of the lighting installations REEIAE) and recover the investment by means of obtained savings during operation.
- This implies ensuring a comprehensive conservation of the abovementioned lighting facilities by the ESCO.
- In contrast, for that period, the City Council will pay the ESCO the costs it incurred to date concerning its current lighting service: the cost of electricity consumption, estimated at EUR 1,000,000/year (USD 1,314,940), and maintenance operations management related to the facilities, estimated at a rate of EUR 42 per light/year (USD 55); i.e., EUR 580,000/year (USD 762,665). In the particular case of the Alcorcón City Council, the maximum budget reached EUR 1.58 million/year (USD 2.08 million).

The contract, published in January 2011, was awarded to ETRALUX, whose financial offer was:

- For management, energy supply, maintenance and conservation: EUR 1,500,993.49/year (USD 1,973,720) (down 5%) during the ten years of the contract period. With this amount, ETRALUX took over the costs of the street lighting electricity facilities in the municipality of Alcorcón along with their management, maintenance and conservation protocols.
- According to the procurement process, and at its own risk, ETRALUX offered an improvement plan for the renovation of the outdoor lighting facilities with an amount of EUR 3,039,397.18 (USD 3,996,620) (VAT not included).

Sweden

Therése Utsi

ACTIVITIES

The 1980s saw the development of performance contracting in Sweden. Unfortunately, when energy prices dropped drastically, many ESCOs went bankrupt since their contracts did not protect them from such fluctuations. This early background tarnished the reputation of the industry. Despite this setback, the EPC concept came back and even expanded during the 1990s. In the early 2000s, there were approximately three or four energy producers and six equipment manufacturers or consulting/installation firms providing different types of offers that could be linked to EPC, even though no company really specialized in that field. The level of activity varied from a few to hundreds of projects which targeted markets, such as the building sector.

In the last ten years, the ESCO market in Sweden has been dominated by the public sector and has gone from dormant to being more mature. Approximately 60 EPC projects have been carried out on the Swedish market since the beginning of 2000. Nevertheless, during that decade, there have been short pauses in the market when very few RFPs were launched.

CONTRACTS

While the format of the legal framework has not changed since the beginning of the year 2000, the content and evaluation criteria of RFPs have been modified a great deal during the last decade, most likely as a result of a maturing market. A negotiated procedure based on a guaranteed savings approach was the model preferred by the public sector.

Figure 9-34 shows the current accepted structure for procuring an ESCO in Sweden. Phase 1 ensures that the facility owner and the ESCO plan the upcoming project together based on specific needs, goals and prerequisites. The facility owner is free to negotiate the details and terms during Phase 1 since it has not yet committed to proceed with Phase 2.

The evaluation criteria have also evolved based on market changes.

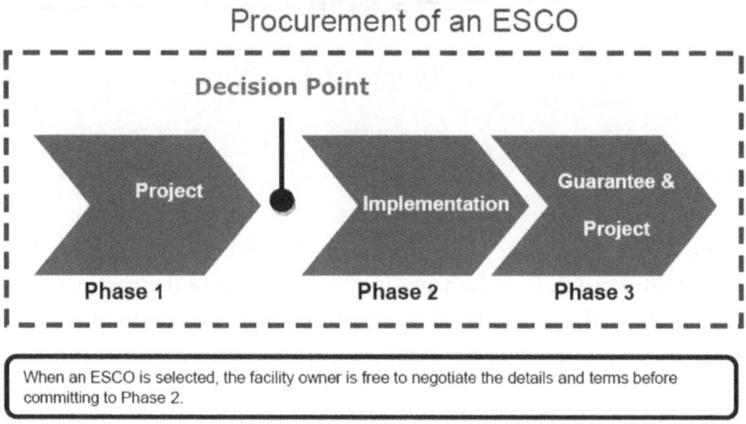

Figure 9-34: Procurement of an ESCO

The trend has been that the price of Phase 2 is becoming more relevant and EPC projects are increasingly commoditized.

It is also of great importance to remember that the result of project development (Phase 1) is based on the energy audit that the energy controllers convey and what they can deliver in terms of time and knowledge. In the most recent procurements in the Swedish market, this evaluation point has been downsized in favor of price and other quantitative parameters. Hence, another risk with a commoditized service is that these types of projects can easily turn into a traditional investigation and energy report instead of a tailored energy-saving project that will maximize results.

MARKETS

The dominant ESCO market is in the public sector: municipalities, public housing and county councils. In the Swedish market, there are less than ten active ESCOs, with three dominating the market.

FACILITATORS

The Swedish energy agency has supported a national project that resulted in a number of templates and checklists on how facility owners can draft their RFPs for EPC projects. This material dates back to 2006 and is still communicated to the public.

Over the last couple of years, various seminars and road shows have been arranged across the country by different stakeholders in the market to educate public facility owners on how to procure an ESCO. This most likely has benefited the market in terms of getting the EPC concept accepted. It helped develop basic documents for the launch of RFPs and allowed building relationships between ESCOs and their clients. On the other hand, the risk with the use of pre-set forms and checklists is that it is taken for granted that every single facility owner has the same specific needs and prerequisites when implementing a project. The outcome of the projects tends to be standardized and the format of the RFP might jeopardize the main purpose with this type of collaboration, which is to meet the specific needs of each single facility owner.

According to reference stories in Sweden, one crucial factor of successful EPC projects is the level of commitment and the right amount of allocated resources from the client. This applies to all levels within the customer organization and throughout the entire project—from the decision to issue the RFP to the validation of the project's initial purpose.

GOVERNMENT ACTIONS

When the EPC market was fairly new in Sweden, the Swedish government launched a substantial grant for public facility owners that encouraged investments in energy-saving projects. All investments that lowered the MWh consumption received a 30 percent grant. This economic parameter made it easier for decision makers to motivate a large investment and benefited the emergence of an EPC market, which was quite new at the time.

In 2002, the European Commission legislated (EU Directive 2002/91/EC) that all buildings in Europe had to be energy certified. This directive means that audits had to be performed to measure the energy performance of buildings whenever they were built, sold or rented. Certificates had to be renewed every five years and buildings had to meet minimum performance standards. This helped the topic of energy savings rise on the agenda of many facility managers, thereby triggering the development of a stronger EPC market in Sweden.

Switzerland

Jean-Marc Zgraggen

ACTIVITIES

ESCO activities are still underdeveloped in Switzerland. The ESCO concept is actually either unknown or partially known by energy players. According to energy experts, calls for tenders for EPC are almost nonexistent in Switzerland. The only one reported to date was made by the International Airport of Geneva in 2010. Also noteworthy, the national energy policy, although containing numerous tools to improve energy efficiency does not actually mention ESCOs as potential actors to activate energy savings. However, signs of ESCO-like activities are beginning to emerge in the country.

On the other hand, ESC is quite well developed in Switzerland, mainly in the form of district heating. The national ESC association called Swiss contracting[1] reports to have about 30 members which perform contracting and 30 qualified partners for conception, building, implementation and execution of energy supply contracts. The model promoted by this association is based on outsourcing the planning, financing, installation and operation of power generation facilities. This kind of ESC is an interesting tool to improve production-side efficiency, although it does not give incentives to the contractor to reduce the load on the demand side since its remuneration is tied to the quantity of final energy sold. What is more, a customer switching from an old fuel boiler to a brand new district heating system will not always see a reduction on the energy bill. The added value of doing it may be found, for instance, in no longer having to manage O&M tasks. In some cases, other factors can strongly attenuate future fluctuations of primary energy prices, such as the use of the open space left by removing the fuel tank (if any) or the price structure of the kWh bought. Based on these elements, it is not clear whether this kind of activity should be integrated within the range of services offered by an ESCO. The risk is to get a confused picture of the role of an ESCO, which is seen today by the few Swiss actors as mainly aimed at working on the demand side by providing energy savings. For these reasons, we think that such ESC activity should not be included under the ESCO label in order to keep and strengthen a clear distinction between ESCO-like companies selling energy savings through EPC and companies selling primary or final energy through ESC.

A few actors in Switzerland are beginning to offer EPC to their customers. The great majority of them are multinational companies, usually specializing in HVAC equipment and associated services. These companies possess the specific know-how required to build such contracts and are recognized as main actors on the international ESCO market. Three companies promoting EPC in Switzerland today are: Siemens, Schneider Electric and Johnson Controls. For some of them though, EPC is still seen as a prospective offer and the companies have not yet integrated in their Swiss offices the specific resources able to carry out such contracts. Those resources are available from foreign offices if necessary. Other such companies may also promote EPC, but they are not as visible as the three above.

Other quite active in this market are:

- Services Industriels de Genève (SIG), the Geneva energy utility. Currently, SIG seems to be the only utility offering EPC to its customers. SIG also took a leading position in the measurement and verification process by organizing the first IPMVP training in the country with the help of EVO[2]. The IPMVP is a worldwide reference tool for evaluating energy savings. This training is now given twice a year, contributing this way to the diffusion of the IPMVP through energy specialists. Finally, SIG helped introduce the IPMVP within the local legal framework as the reference method to be used for energy savings evaluation in order to comply with the energy law requirements.

- Alpiq, one of the largest energy utilities in Switzerland, seems to offer some kind of energy efficiency services as well, but it is not clear whether a commitment to energy savings has been made.

- Energho[3] is a non-profit association supported by the Swiss Federal Office of Energy. This association was set up in 2002 to provide energy savings to public buildings through the optimization of their HVAC systems. The energy consumption of public buildings is about 12.5 TWh per year. By the end of 2009, around 15 percent (1.9 TWh) of this target market was under the management of Energho. The cumulated measured savings achieved by the association since 2002 are 125 GWh of heat, 25 GWh of electricity and more than 800,000 m3 of water. This represents total gross savings of about CHF 20 million (USD 22 million) for consumers.

The association works in partnership with about 70 local engineering offices spread around the country. Energho does not invest in projects but only sells services aimed at improving energy efficiency. However, one of their products called energho©PLUS contains all the ingredients of a CPE with guaranteed savings. This specific contract represents about 33 percent of all 561 signed contracts listed by the association in 2010. Within the five-year term of the energho©PLUS contract, Energho has committed to reduce the global consumption of energy and water of any building by at least 10 percent at the end of the contract. A local and certified engineering office is subcontracted by Energho to carry out the contract. The mandate includes as well a transfer of know-how from the engineer to the O&M team through personal coaching along the five-year term. The customer is charged a fixed yearly subscription fee plus 20 percent of the energy savings achieved, which are distributed as follows: 12 percent for Energho and 8 percent for the engineer. In the case where the global reduction of 10 percent is not reached within the fifth year, Energho will repay the customer up to 50 percent of the subscription depending on the results achieved. The engineer will be charged 50 percent of the repaid amount.

The energy and water savings are evaluated using a methodology that is close to what is requested by the IPMVP. Firstly, consumptions of a reference year are identified. Then, baselines are set for electricity, heat and water based on a fit of measured values compared to external temperature. The resulting equations are reported within the contract and used along the years to estimate the savings compared to the reference year consumptions. The impact of changes in the static factors must also be evaluated and integrated in the estimation of savings. While this method is not as strongly defined as in the IPMVP, the spirit in which it has been developed is fundamentally the same.

In 2010, the global consumers' energy bill for combustibles and electricity (mobility not included) was CHF 17 billion[4] (USD 19 billion). A rough and conservative estimate of the ESCO market potential can be made by considering that, on average, ESCO projects can achieve 15 percent of savings on the global consumption. That amount saved represents an annual cash flow of CHF 2.5 billion (USD 2.7 billion) available to repay the projects and then available to the customers at the end of the contract. Assuming that, on average, the duration of an EPC agreement is between five and ten years, the present value of future cash flow represents an amount that is between CHF 9.5 and 15.4 billion (USD 10.4 and 16.8 billion) considering a 10 percent discount rate and a fixed price of energies. A more serious assessment of the ESCO market potential is of course needed in order to have a sharper picture of this potential, which obviously has not been fully activated as yet. However, this first estimate shows that there is certainly a real potential for the ESCO business in terms of cash flow available to finance energy efficiency projects.

LEGAL FRAMEWORK

Currently, there is no legal framework in Switzerland similar to European Directive 2006/32/EC that lays the foundation for energy services[5]. The legal framework is made of national and regional laws (cantons) that provide incentives to foster both energy savings and GHG emissions reductions. At national level, the following relevant points should be highlighted:

- An action plan of 15 energy efficiency measures was published in 2008[6]. A real emphasis was put on buildings, mobility and low-efficiency technologies. One specific measure concerns the sector of industry and services. It states that energy utilities should ensure to create incentives so as to encourage end-users to reduce their energy consumption or increase their energy efficiency. Within the framework of the energy policy for 2050[7] currently under consultation, a new action plan has been proposed. A great number of the 50 proposed actions are related to energy efficiency. The final version of this new energy policy should be set by mid-2012.
- Since 2009, the CO_2 Act levies a tax on combustibles to encourage large consumers to achieve energy savings[8]. In 2011, this tax was up to CHF 36 (USD 39) per ton of CO_2 leading to a CHF 600 million (USD 655 million) budget. Almost one-third of this budget was used to fund a program for building retrofits, the balance was redistributed to the population and companies. This law is currently under revision with some debates on whether carburants (mobility) should be included.

- Since early 2009, the production of electricity from renewable energy sources has been encouraged by a compensatory feed-in remuneration funded by a tax levied on consumed electricity[9]. In 2011, this tax was up to 0.45 Swiss cents per kWh leading to a CHF 256 million (USD 279 million) budget.

At regional level, each canton has its own legal framework. However, municipalities and cantons tend to align their energy policy strategies on the objectives set by the 2,000-watt concept[10]. Almost all building constructions and retrofits follow the recommendations made by the Society of Engineers and Architects (SIA)[11] on minimal requirements for energy consumption.

The Fukushima tragedy clearly triggered the end of nuclear activity in Switzerland. Following this event, the government made the decision to turn off all of its five nuclear plants by 2034. These power plants are currently providing up to 40 percent of the production capacity. This decision has naturally placed the energy efficiency potential under the spotlight as one of the pillars of the new energy policy that must be set up.

FACILITATORS

At national level, there are two main actors driving the surge in energy efficiency: the Swiss Federal Office of Energy[12] and the Energy Agency for Economy[13]. At the present time, there is no ESCO association in Switzerland.

The Swiss Federal Office of Energy is a government office whose main target is to create the necessary conditions for a country level energy supply that is sufficient, stable, diversified, economically viable and sustainable. Furthermore, the Swiss Federal Office of Energy promotes rational use of energy, enhanced renewable energy production capacity and reduced GHG emissions. Finally, it also encourages and manages research activities related to energy at national level.

The AEnEc is an agency founded in 1999 by the main economic associations in reaction to the introduction of the CO_2 Act. This agency, working in close collaboration with federal agencies, provides services to help industries economically reduce their GHG emissions through energy efficiency. If eligible, a company can be exempt from the CO_2 tax if a commitment is made with the AEnEc in terms of CO_2 reductions and the goal is reached at the end of the given period (a couple of years).

At regional level, each canton has its own energy office in charge of applying national and regional energy laws. They meet regularly in order to coordinate and harmonize their actions, and are good local drivers for energy efficiency because of their proximity to consumers. There is currently no public fund dedicated to investing in ESCO projects. However, a private equity fund called SUSI[14] partners has recently announced that it would invest in EPC projects where energy savings are guaranteed by a technological partner.

GOVERNMENT ACTIONS

At national level, there are three main programs driving the improvement of energy efficiency: the National Buildings program[15], ProKilowatt[16] and the Swiss Climate Foundation[17]. At some point, they can have a positive influence on the development of ESCOs, as they make energy efficiency projects more attractive.

The National Buildings program is a ten-year program launched in January 2010 to reduce building energy consumption and associated CO_2 emissions. The government expects from this program a reduction of CO2 emissions of 35 to 52 million tons. The operating mode is to give financial incentives in order to trigger retrofit actions. This program is funded by the CO_2 Act, which contributes up to CHF 200 million (USD 218 million) per year (one-third of the CO_2 tax budget), and by the cantons (regions), which will in turn add between CHF 80 and 100 million (USD 87 and 109 million) per year. The program's main target is the energy retrofit of building envelopes, but it also gives incentives to boost renewable energies, heat recovery systems and HVAC retrofits. By the end of its first year of existence, about 26,000 funding requests for envelope retrofits had been accepted, leading to CHF 204 million (USD 223 million) of engaged incentives for about 5 million square meters to be retrofitted. About three-quarters of these requests concern fossil fuel heated buildings.

ProKilowatt is an incentive tool launched in 2010 by the Swiss Federal Office of Energy to improve electric efficiency specifically. This tool gives financial incentives to electric efficient projects or efficiency programs through a public call for tenders. The selection of the projects is made upon a cost/efficiency ratio—a maximum of electricity saved for a minimum of invested money. A contribution of 5 percent (maximum) of the compensatory feed-in remuneration tax is used to fund this tool. A first procurement call of CHF 9 million (USD 10 million) was made in 2010 and the budget for 2011 has been raised to CHF 15 million (USD 16 million). ProKilowatt finances up to 20 percent of the investment if the return on investment is below five years and up to 40 percent for a return on investment below nine years.

The Swiss Climate Foundation is a positive collateral effort resulting from the passage of the CO_2 Act. This incentive tool has been set up by the private sector with the objective of helping SMEs that are proactive in their approach to reduce CO_2 emissions. The basic mechanism is that all companies are taxed through their consumption of fossil fuels, save for the mobility sector as carburants are not taxed. As mentioned before, almost two-thirds of taxes collected is redistributed to the population and companies. The amount given to a company is proportional to its wage bill. As a result, certain companies, such as services for instance that do not have a big energy bill but which employ many people, happen to recover more money than what they spend on CO_2 taxes. The Swiss Climate Foundation then collects this "extra earning" among partner companies in order to fund energy efficiency projects or innovative solutions. If an SME makes a commitment in terms of CO_2 reductions with the Energy Agency for Economy (AEnEc, see below), the foundation will also take in half of the annual subscription fee. In 2010, the foundation received CHF 5.3 million (USD 5.8 million) (based on the 2008-2010 period) and paid CHF 1.1 million (USD 1.2 million) in incentives. A total of 13 projects were funded and 90 SMEs were helped to commit with the AEnEc.

Thailand

Arthit Vechakij, Ruamlarp Anantasanta,
Morakot Gerddang, Nawarat Tritipros

ACTIVITIES

The ESCO business in Thailand originated in cooperation with the Department of Alternative Energy Development and Efficiency (DEDE) under the Ministry of Energy and the Electricity Generating Authority of Thailand (EGAT) in 1999. This effort was the result of growing awareness of the effect of the energy crisis on the country's long-term and sustainable energy utilization. It intentionally promoted and supported energy savings and energy efficiency through ESCO services. An ESCO pilot project was implemented as a case study to analyze the strengths, weaknesses and barriers in developing a suitable ESCO business for Thailand.

Resulting from the success of the ESCO pilot project, some companies were created to qualify as ESCOs. These ESCOs drew the attention of industrial and commercial entrepreneurs through the performance of energy-saving projects as ESCO service solutions. The trend of ESCO growth since 2002 has shown that the number of ESCO projects has slightly yet consistently increased measured by the amount of EPC agreements and investments.

At present, there are 39 companies registered as ESCOs under the Institute of Industrial Energy of the FTI. Each ESCO has various energy-saving solutions/technologies under ESCO service-type standards. The classifications are guaranteed savings and shared savings.

The ESCO business in Thailand has grown because of such factors as continuously rising energy costs, successfully implemented pilot energy-saving projects, financial institutions' required performance guarantees before granting loans as well as government agencies' support and subsidy programs for energy-saving projects. Many clients have high confidence in investing in energy-saving projects through ESCO services.

The current ESCO market in Thailand indicates that the ESCO business is gradually picking up momentum. The relentless promotional activities carried out by the FTI over the past five years have allowed countless entrepreneurs to better understand the ESCO concept and energy-saving projects. Although Thailand continues to learn from the ESCO pilot project, some problems and obstacles in the development of the ESCO business remain, including a misunderstanding of the energy consumption profiles in their factories, and the selection of the wrong size of equipment, or worse, wrong technologies. These make energy-saving projects fail.

If clients could have enough information and acceptance of the ESCO concept, they would be in a better position to maximize the benefits of energy-saving projects. Moreover, the energy conservation and ESCO business in Thailand will be more successful. This is why government agencies try to promote and support the ESCO business by arranging activities and launching special campaigns in order to help entrepreneurs understand and access ESCO information easily.

CONTRACTS

There are two types of EPC agreements currently in use in Thailand, guaranteed savings and shared savings contracts. The guaranteed savings approach is more popular than is the shared savings concept because there are only a few qualified clients willing to engage in shared savings.

LEGAL FRAMEWORK

Due to the current legal framework in Thailand, clients and ESCOs have to conform to all terms and obligations set out in signed EPC agreements. Proven examples indicate that ESCOs have paid a number of customers for deficit savings in amounts ranging from THB 100,000 to THB 800,000 (USD 3,220 to USD 25,750). In some cases, the arbitration rules of the Thai Arbitration Institute, the Office of the Judiciary and the Ministry of Justice were executed where customers and ESCOs could not amicably negotiate. ESCOs were able to recover and get compensation as stipulated in the signed EPC agreement. These cases have led many entrepreneurs to trust and believe that risks can be mitigated, thereby increasing the benefits of the ESCO business.

However, the approval of M&V as proof of savings remains difficult to get from customers. In some cases, it takes a very long time to come into resolving M&V issues despite ongoing operation phases.

MARKETS

ESCO services can be categorized according to type or size. Based on market demand for ESCO service solutions in Thailand, the industrial sector is the prominent group. In 2002, the investment cost of energy projects under ESCO service solutions for this sector was THB 45 million (USD 1.4 million) and it jumped rapidly to THB 825 million (USD 26.5 million) in 2009.

If EPC agreements are considered (small and medium EE projects versus large EE projects), it will indicate that the number of small and medium EE projects are outstanding. This means that entrepreneurs are interested in ESCO projects even if they are relatively small.

For small and medium projects, the focus is on low-pressure boiler replacements, chiller replacements, ozone for laundry, heat pumps, automatic demand management systems and energy management systems. The investment cost is about THB 5-40 million (USD 161,000-1.3 million) per project. On the other hand, large projects include cogeneration power plants, waste heat recovery, waste-to-energy projects and renewable energy power plants. These projects require very high investments averaging between THB 100-1,000 million (USD 3.2-32.2 million) per project. The gap in investment cost between small and medium projects and large projects is huge. Therefore, the number of small and medium EPC projects is higher than the number of large EPC projects.

FACILITATORS

The Ministry of Energy, through the DEDE and the Institute of Industrial Energy under the FTI, the Board of Investment (BOI), financial institutions like the Export and Import Bank (EXIM), Kasikorn Bank PLC, CIMB PLC and Krung Thai Bank PLC as well as other government agencies such as EGAT, the Provincial Electricity Authority (PEA) and PTT PLC all act as key facilitators for the ESCO business in Thailand.

Although these organizations do not gain direct benefits from the ESCO service solution, they still promote the ESCO business. The latter grows gradually in Thailand. Even though there are some barriers to ESCO business development and implementation, entrepreneurs are confident that they will get total benefits from projects implemented under the ESCO service solution.

GOVERNMENT ACTIONS

The following are government actions supporting the growth of the ESCO business:

The Institute of Industrial Energy under the FTI (DEDE sponsorship)
Registered ESCOs

The objective is to stimulate the registration of real ESCOs as there are many energy consulting firms, equipment manufacturers, suppliers or contractors that claim to be ESCOs while they cannot provide real ESCO services even though they do not meet obligations to entrepreneurs or do not execute actual EPC agreements. Registration can confirm to entrepreneurs and financial institutions that their selected ESCO provides actual energy efficiency services.

After encouraging many ESCOs to register, the FTI started to change the conditions and process of registration in order to screen and pick up qualified ESCOs only.

ESCO Information Center

The objectives of the ESCO Information Center are to:

- become the center of information linking ESCOs, entrepreneurs and financial institutions;
- ensure that entrepreneurs can access all information related to the EPC concept and the benefits they can obtain in using the approach;
- build a stronger relationship with financial institutions in terms of loan approval for energy-saving projects under the ESCO service solution;
- review ESCO business operation performance and collect data in order to analyze problems which obstruct the process of the ESCO business and try to solve such problems;
- collect information about past experiences with ESCO business development for planning structure, policy and implementation with a view to establishing an ESCO association in Thailand;
- produce a guideline for expanding ESCO projects by defining the long-term promotional plan, which will result in a sustainable ESCO service solution.

Annual Awards

ESCO Excellence Award (annual)

The objective is to award the ESCO which has best followed the concept of an actual ESCO. This award will let entrepreneurs know that if they want to perform an energy efficiency project under the ESCO service solution, they should consider these ESCOs first, or at least select an ESCO which possesses these qualifications.

ESCO Project Award (annual)

The objective is to award the entrepreneurs, who have fully and completely operated energy-saving projects through the ESCO service solution. This award will let other entrepreneurs know that if they want to carry out an energy-saving project, they should proceed with an ESCO as the latter will really guarantee the success of project performance. The measure of reward looks at the savings, which result from the M&V plan, to which both parties will have agreed.

Thailand ESCO Fair & Regional ESCO Fairs (annual)

The objective is to stimulate investments in energy-saving and energy efficiency improvement projects in Thailand.

ESCO Matching Seminar

The objective is to arrange the business market for entrepreneurs and ESCOs to meet and enter into business deals.

ESCO-Bank Networking Seminar (bi-monthly)

The objective is to establish a good relationship and expand the network between ESCOs and Thailand's leading financial institutions.

Department of Alternative Energy Development and Efficiency under the DEDE

Energy Fund for Investing in Energy-Saving and Alternative Energy Projects (ESCO fund)

The ESCO fund will provide investment capital to entrepreneurs who have a potential project but still need investment budget. The ESCO fund will join as a project shareholder with an investment portion not exceeding 50 percent and a co-investment budget not over THB 50 million (USD 1.6 million) per project. The required return rate is very low compared with private funds (10 percent on average). There is also an equipment leasing scheme available (less than THB 10 million or USD 0.32 million). In addition to a credit guarantee facility and technical support assistance which were announced but never actually implemented.

Low Interest Rate Loans for Energy-Saving and Alternative Energy Projects (revolving fund with low interest rate)

The objective is to stimulate investments in energy-saving and alternative energy development projects and increase confidence in, and familiarity with, financial institutions that join the loan approval process under such projects. The DEDE will support loans of THB 50 million (USD 1.6 million) per project. This charging loan interest rate is lower than normal interest rates which financial institutions charge to entrepreneurs. The maximum interest rate is expected to be only 4 percent per year. Since 2005-2009, two rounds of 2,000 MB (USD 64 million) each were lent into energy efficiency, alternative energy and renewable energy projects.

Investment Support and Promotion to the ESCO Business by the BOI

The objective is to engage real investments in the ESCO business. The ESCO business is a special business to which the BOI pays attention because it supports energy-saving and energy efficiency improvement projects in Thailand. Furthermore, the ESCO business contributes to the success of the BOI because ESCOs manage and administrate energy-saving projects committed to the EPC concept.

The BOI leads many investment campaigns targeting entrepreneurs who wish to invest in energy-saving, energy efficiency improvement and alternative energy development projects.

- **Tax Incentive on Proven Energy-Saving Projects**

  Government fund will subsidize tax savings based on performance and impose a limit of 2 MB (USD 64,000) per year per company. Entrepreneurs can claim energy savings as expense (tax deduction).

- **Promote Investments in Energy Efficiency Management and the Renewable Energy Sector**

  The BOI privileges activities relevant to energy conservation, such as solar cell manufacturing or manufacturing of equipment and tools related to energy savings or renewable energy. Moreover, the renewable energy generation business is included in such activities; e.g., the production of alcohol or fuel from agricultural by-product/waste, renewable energy generation (electrical/steam).

  Investment Promotion under Section 7.1 (in the case of setting up a new company): entrepreneurs can claim 100 percent of corporate income tax exemption for eight operating years and 100 percent of import duty exemption.

  Investment Promotion under Section 2/2010 (in the case of

an existing company): entrepreneurs can claim corporate income tax exemption up to 70 percent of total investment costs for three operating years and 100 percent of import duty exemption.

Other Government Agencies, such as EGAT, the PEA, the Energy Policy and Planning Office (EPPO) under the Ministry of Energy, the National Innovation Association (NIA) and Leading Financial Institutions

These government agencies support and promote the ESCO business in many ways. For example, they buy electricity from renewable power generation projects at high rates and sell natural gas to natural cogeneration projects at low prices. Moreover, they offer a subsidy to energy-saving and alternative energy development projects to which new innovative technology is applied. The Demand-Side Management by Bidding Mechanism program was created in order to convince entrepreneurs to manage and reduce the demand of energy consumption in their factories. The winning project will receive money as a reward (USD 0.05 per kilowatt-hour for the reduction of electricity consumption or USD 2.34 per 1 MMBTU for the reduction of thermal consumption).

All in all, these agencies try to encourage entrepreneurs to invest in the ESCO service solution as they are a significant part of ESCO success. Furthermore, financial institutions launched the special loan for energy-saving and alternative energy development projects through the ESCO service solution. This campaign supports ESCO business growth as financial institutions will provide special loan interest rates and conditions for projects implemented using the services of an ESCO.

Tunisia

Jalel Chabchoub

ACTIVITIES

In Tunisia, ESCOs are considered as one of the most important components to the development of a sustainable market for energy efficiency. The Energy Efficiency Project in the Industrial Sector (PEEI), which is financed by the GEF and the World Bank, has brought major support to ESCO development. The five-year pilot phase initiated in 2005 focused on the implementation of 125 projects targeting large and medium-sized enterprises and emphasizing the development of the ESCO approach.

The first ESCO in Tunisia was established in 1999, through an initiative led by Econoler and three different banks, with the support of the national utility (STEG). That ESCO, called Société Tunisienne de Gérance Énergétique (STGE), operated as the only ESCO in Tunisia for a period of five years. Since then many actions have been accomplished both on the part of the private sector and the government for the development of the concept. The first legal recognition of ESCO activities appeared in a related energy efficiency law enacted in 2004.

Currently, around ten ESCOs are accredited by the National Agency for Energy Conservation (ANME) and entitled to offer EPC. Most of the accredited ESCOs are engineering firms, one is an energy provider and three are equipment providers.

MARKETS

The 29 EPC agreements signed by ESCOs between January 2006 and December 2010 totaled an investment of USD 16 million into which ESCOs participated in the amount of USD 1.17 million. In addition to the existing incentives, they have benefited from an additional support of about USD 1.14 million as the projects were carried out by ESCOs.

It should be noted that for all EPC agreements signed, the contribution of ESCOs to the project financing scheme was limited to honorary (fees) contributions since most ESCOs did not have enough financial resources to invest. Most projects were implemented in the industrial sector through the financial support offered by the government and various stakeholders.

FACILITATORS

The above-mentioned achievements for support to ESCOs were delivered mainly through the ANME. This agency was established in 1985 as a non-administrative public entity under the authority of the Ministry of Industry and Technology. The mission of ANME consists in implementing state policy in the field of energy conservation, renewable

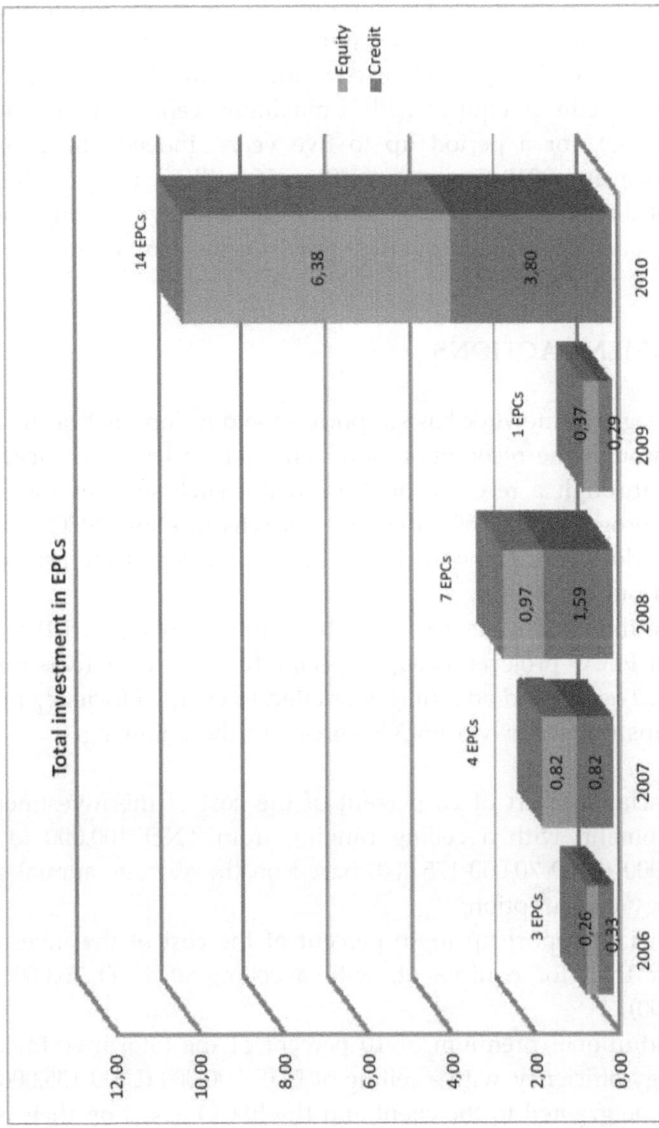

Figure 9-35: Total Investment in Energy Performance Contracts

energies and energy substitution. It has acted as the government vehicle to support the development of EPC in Tunisia since the mid-2000s.

In 2010, strong progress in terms of project implementation was due essentially to the guarantee fund and to the efforts put in place by the PEEI to promote EPC and encourage ESCOs. This guarantee fund helped ESCOs with a guarantee covering up to 75 percent of ESCO investment (credit or equity) with a maximum ceiling of EUR 400,000 (USD 527,000), for a period up to five years. Indeed, the guarantee disbursed during 2010 was about USD 2.53 million, bringing the total amount of disbursement to USD 3.66 million and representing approximately 98 percent of the budget assigned to this component.

GOVERNMENT ACTIONS

The legal framework has supported the development of the ESCO market through the recognition and definition of ESCO services since 2004 and through a recommendation to use such services for energy efficiency programs. In 2005, the terms of reference for ESCOs were issued, including conditions for ESCO accreditation and the activities to be carried out by ESCOs.

Since then, many decrees have been issued to support the development of ESCO projects through specific financial incentives for EPC initiatives. The aid and advantages granted to energy efficiency projects in the industrial sector via ESCOs consist of the following:

- financial support of 20 percent of the cost of the investment for equipment, with a ceiling ranging from TND 100,000 to TND 250,000 (USD 70,000-175,000) based on the average annual global energy consumption;
- financial support up to 70 percent of the cost of the investment, other than for equipment, with a ceiling of TND 70,000 (USD 50,000);
- an additional premium of 10 percent of the total investment for energy efficiency, with a ceiling of EUR 100,000 (USD 135,000) that may be granted to the client and the ESCO based on their contribution to energy efficiency project investment;
- additional support of 10 percent is attributed to the ESCO, under the same above ceiling, based on funding;

- bank guarantee of 75 percent of the credits granted by financial institutions to the clients and the ESCO, with a ceiling of USD 400,000—this guarantee is assured by the guarantee fund SOTUGAR for energy efficiency projects in the industrial sector;
- guarantee for ESCO credits: this is a guarantee for ESCO investments and unresolved payments covering 75 percent of the outstanding debt with a ceiling of USD 540,000;
- funding ESCO investments through an equity fund up to 15 percent of the total investment;
- VAT exemption;
- payment of the minimum customs duties for imported energy-efficient equipment with no equivalent equipment manufactured locally.

Furthermore, the PEEI provided support for ESCO development by reducing barriers to EPC implementation and by providing additional assistance (financial and awareness) to promote energy savings in the industrial sector and help the ESCO market. By the end of 2010, 487 energy efficiency and natural gas substitution projects in the industrial sector had received approval for existing financial incentives.

To achieve the PEEI objectives, a total fund of USD 8.5 million was allocated to finance the three main components:

1. investment support to energy efficiency projects through an additional grant of 10 percent of the project investment;
2. establishment of a guarantee fund to facilitate ESCO development;
3. technical assistance for capacity building of market stakeholders (industrial sector, technical centers, ESCOs, consultants, financial institutions, government institutions, utilities, etc.) and enhanced awareness for the potential of energy efficiency.

Within the PEEI, specific working groups have been established to promote ESCOs and the EPC concept as well as to increase awareness in industry, financial institutions and engineering firms through targeted workshops and seminars. In addition to ESCO training, technical assistance for project development and the preparation of energy performance contracts, additional support included helping ESCOs in their financial packaging and submission of project applications to financial institutions.

In addition to the above, credit lines have been developed with the agreement of the government in order to drive the development of EPC projects and foster ESCO activities, including:

- World Bank Credit Line for Energy Efficiency and Cogeneration in the Industrial Sector: USD 55 million managed by three local banks dedicated to the industrial sector for energy efficiency and cogeneration projects, up to 15 years with a five-year grace period.

- French Development Agency (AFD) Credit Line: EUR 40 million (USD 52 million) for two components which include both funding and technical assistance for the implementation of energy efficiency projects. Approved projects can benefit from a maximum credit of EUR 5 million (USD 6.6 million) up to 12 years, with a two-year grace period.

Turkey

Prof. Dr. Esin Okay
Prof. Dr. Nesrin Okay
Prof. Dr. Ugur Akman

ACTIVITIES

Officially, Turkish ESCOs emerged together with the country's Energy Efficiency Law (EEL) in 2007. However, Turkey was not foreign to ESCO-like activities. In the 1980s, under the guidance of the Turkish Electricity Authority, an assessment of the country's renewable energy potential for power generation and energy audits for several important energy-intensive industrial establishments were carried out with a US-based company. Consequently, an Energy Conservation Measures division was set up under the General Directorate of Electrical Power Resources Survey and Development Administration (EIE, www.eie.gov.tr, established in 1935) and the National Energy Conservation Center (NECC), established in 1993 is attached to the Ministry of Energy and Natural Resources).

In the 1990s, the EIE cooperated with the Japan International Cooperation Agency (JICA), through a Japanese grant, to develop a training course on energy efficiency management of industrial facilities and buildings. After 1995, although there were no laws and regulations referring to ESCOs, the EIE started training engineers as energy managers. Various international organizations such as the United Nations Industrial Development Organization (UNIDO), the World Bank, the EC and JICA have supported EIE/NECC projects aiming to gain energy efficiency in the industry, transportation and residential sectors.

In 1997, the International Energy Agency (IEA) recommended that the Turkish government implement some policy measures and, consequently, the government issued a decree requiring energy-intensive industries to establish an "energy management unit" or to employ an "energy manager." During that period, USAID supported Turkey in developing the appropriate laws and regulations, and provided training support to launch industrial energy audits and ESCOs. In 1999, National Britannia Energy Management Services (a UK-based firm) performed energy audits for several factories, hospitals, hotels and malls. Those activities were the first ESCO-like activities in Turkey, and were done without any support in the absence of laws and regulations.

The Turkish economy experienced crises in 1991, 1994 and 1999, all leading to the severe crisis of 2001, which was felt predominantly in the banking and finance sectors. These economic instabilities along with the high inflation rate as well as the absence of relevant laws and regulations prohibited the healthy birth of the Turkish ESCO industry.

In a 2005 report, the IEA mentioned that ESCOs were not active in Turkey. However, although there were no certified ESCOs in Turkey, energy efficiency activities, particularly for the industrial sector, had already been carried out since 1995. There were several large-scale (but mostly small-to-medium-scale) private Turkish energy companies that conducted energy efficiency-related business in the pharmaceuticals, chemicals, automotive, agriculture, paint, food and beverages, airport, hospital and building sectors. The financing mechanisms of these projects, which are mostly small-scale, have predominantly involved end-user financing.

One of the most critical difficulties that local companies face in Turkey is capital inadequacy and, therefore, they are unable to act as market makers. Since 2005, international partners have seen the potential of a stable economy and good indicators to invest, free from country risk. Consequently, they are more willing to invest in this promising sector. Although Turkey's current account deficit is the fourth in the world, mainly caused by energy dependence, foreign direct investment continues to be interested in Turkey's strongly growing economic capacity.

The setback composition of the Turkish ESCO market is a bit complicated, made up of macro-/micro-economic and financial risks coming through both domestic and global means. Unfortunately, market performance is bound to market-size problems. The country's current low interest rate and low inflation potential due to a small debt-to-GNP ratio unlocks the potential for growth in this sector. However, large debt problems of developed countries and a worldwide fear of associated risks may block the foreign direct investment flow. A flow that is necessary for market progress as well as for the reduction of the country's current account deficit. In order to maintain a lower credit risk of financing the ESCO market, it is a must to align with international ESCOs with a view to seeking possible business partnerships and securing TPF from international finance organizations (beyond the World Bank) such as IFC, the Environmental Enterprises Assistance Fund (EEAF) and the Renewable Energy Equity Fund (REEF).

Despite the banking sector's high liquidity and profit, lending volumes remain depressed. This is primarily because of the risk-averse approach of the new banking system as well as the private sector and its capital inadequacy. The Turkish private sector (consisting of 99 percent SMEs) has maintained the effects of weaknesses and inadequacies for years, providing insight on the limits to firm growth and sustainable investment. SMEs, being small in size, cannot bear the entire financial risk of projects. Therefore, only large-sized firms are on the platform to establish an ESCO market. If they come together to form joint ventures with domestic firms or established foreign ESCOs the result might change. Additionally, for secured loans, it is important to assess the end-user payment default risk, the ESCO non-performance risk and the bank loan-repayment default risk. These risks must be appropriately addressed in order to ensure successful projects.

Beyond such financial and risk-related issues, there are no technical-/engineering-related barriers since many Turkish private companies have already proved their qualifications, especially with large-scale housing and industrial construction projects successfully completed abroad.

CONTRACTS

In Turkey, financing mechanisms used by ESCOs are not yet publicly available. In our view, currently only four to eight out of the 38 EIE-listed ESCOs may be considered as chief players, which may possibly employ shared savings as a financing mechanism. For the remaining ESCOs, the guaranteed savings mechanism looks more promising at first. On the client side, about 99 percent of the Turkish private sector consists of SMEs. However, SMEs as clients are not financially very stable. Indeed, most of them have foreign exchange liabilities under the floating currency regime of the Turkish economy, and, consequently, they are not accustomed to risk management methods.

A few bank advertisements for energy project financing campaigns give a clue that the ESCO financing mechanism in Turkey may develop into a scheme as depicted in the figure below. If the EPC firm (which may be conscripted by the bank according to some bank flyers) is also a licensed ESCO, then the project illustrated in the figure collapses into the guaranteed savings mechanism.

In this context, some private Turkish banks have been able to obtain financing conveniences from the EBRD and the Clean Technology Fund to grant lending for energy efficiency improvements in

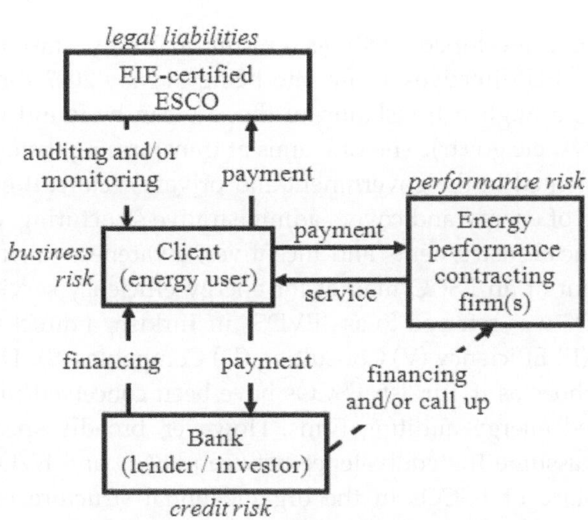

Figure 9-36: Forecast of Financing Mechanism Developing in the Turkish ESCO Market

residences, SMEs, manufacturing sectors and commercial buildings as well as for renewable energy projects. The take-up of this program has been encouraging. About EUR 150 million (USD 197 million) of the proceeds provided by the European Investment Bank and the EBRD to private Turkish banks will be used for financing the mid-side renewable energy and energy efficiency projects under the Mid-size Sustainable Energy Financing Facility (MidSEFF) framework. This funding is provided through an innovative financial scheme based on diversified payment rights (DPR) securitization. Some private Turkish banks have had a long-standing relationship with the European Investment Bank in terms of SME lending and new facility MidSEFF expands upon the area of cooperation between the European Investment Bank and banks. The credit package comprises two 5- and 15-year fixed-term payments, and was given as part of the EBRD's USD 200 million Turkey Sustainable Energy Financial Facility (TurSEFF) fund, which has been allocated to Turkish banks. The credit will be used for small household owners, energy efficiency projects for SMEs, energy efficiency supplier investments and renewable energy projects.

LEGAL FRAMEWORK

The EEL, developed partly as a result of Turkey's task of complying with the EU directives, came into being in May 2007 through Law No. 5627. The English translation of the law can be found on the EIE website (www.eie.gov.tr). The EEL aims at transforming the energy policies implemented in the government and private sectors. It exploits the efficient use of energy and covers administrative structuring, energy auditing, financial instruments and incentives, awareness raising and the establishment of an ESCO market for energy efficiency services. In the EEL, "ESCO" was referred to as "EVDS" in Turkish, a direct translation for Energy (E) Efficiency (V) Consulting (D) Companies (S). This Turkish acronym echoes as if Turkish ESCOs have been conceived primarily as state-licensed energy auditing firms. However, broadly speaking, one may safely assume the equivalence between ESCO and EVDS.

The place of ESCOs in the organizational structure of Turkey's EEL is depicted in Figure 9-37. The main organizations responsible for energy efficiency policies and activities are the Ministry of Energy and Natural Resources (MENR), the EIE and the NECC. The MENR is responsible for formulation of policies and supervision of their implementation within the context of national energy policies while the EIE/NECC is responsible for implementation and coordination of energy efficiency programs. The EIE/NECC carries out training, energy auditing, drafting of legislation and public awareness promotion activities for enhancing energy efficiency in all end-use sectors. The Energy Conservation Coordination Board (ECCB), the composition of which is given in Figure 9-37, under the auspices of the EIE/NECC, is responsible for motivating public awareness. The EIE shares its authorization in energy manager training and certification only with universities and the chambers of electrical and mechanical engineers The EIE and these institutions are responsible for ESCO training and provide laboratory support for programs to be offered to ESCOs for the training of energy managers in the industrial and building sectors. With the adoption of the EEL, only certified companies will be considered as ESCOs authorized to conduct energy efficiency auditing, training and consulting activities in the industrial and residential sectors.

As shown in Figure 9-37, in the EEL of Turkey, "universities" and "chambers of profession" are expected to provide laboratory support and training to ESCOs. The Chamber of Electrical Engineers and the Chamber of Mechanical Engineers received their authorizations from the EIE only recently. Among the 105 state universities of Turkey, only Gazi University applied and was approved by the EIE in February 2011. We could not find information on whether any other universities had applied to become ESCO-related centers. In our opinion, the number of such universities will not exceed three to five in the near future.

The companies certified as ESCOs are regularly listed on the EIE website (www.eie.gov.tr/duyurular/EV/yetki_belgesi_EVD/yetkili_sirketler.html). The EIE issues ESCO certificates in two categories: industrial and residential. Considering the sectors targeted by ESCOs around the world, in Turkey, the commercial and industrial sectors together can be lumped as "industrial," and the municipal, building and residential sectors together can be lumped as "residential" in practice.

MARKETS

Figure 9-38 depicts the six-monthly development of the Turkish ESCO market in terms of the number of companies that have been ap-

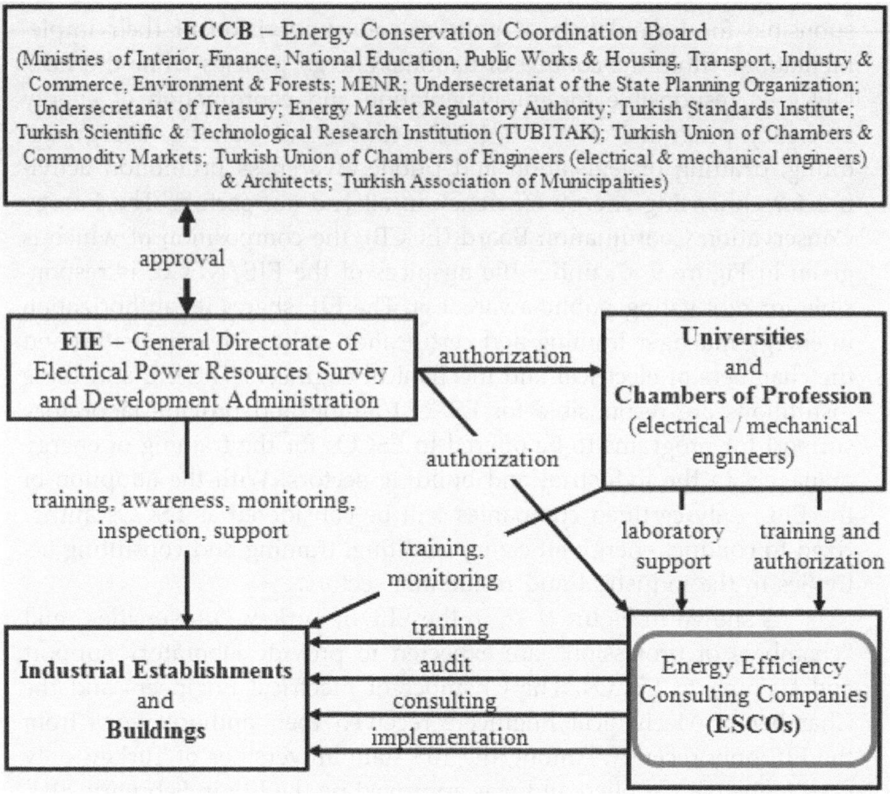

Figure 9-37: The Place of ESCOs in the Organizational Structure of Turkey's EEL

proved as ESCOs by the EIE. As of October 2011, there were 38 ESCOs on the EIE's list. Eight of them had received ESCO certificates that were valid for industrial activities only and 15 of them had received ESCO certificates that were valid for residential activities only. The number of ESCOs certified for both industrial and residential activities amounted to 15. From Figure 9-38, it can be concluded that, as of October 2011, the percentages of the sectors targeted by ESCOs were 43 percent (industrial) and 57 percent (residential). The growth in the number of ESCOs licensed for residential activities only was higher than the growth in the number of ESCOs licensed for industrial activities only. In EIE has recently ceased issuing ESCO licenses to candidate companies due to ongoing changes in the "Regulation on Increasing Efficiency in the Use of Energy Resources and Energy" issued by the MENR. The EIE states

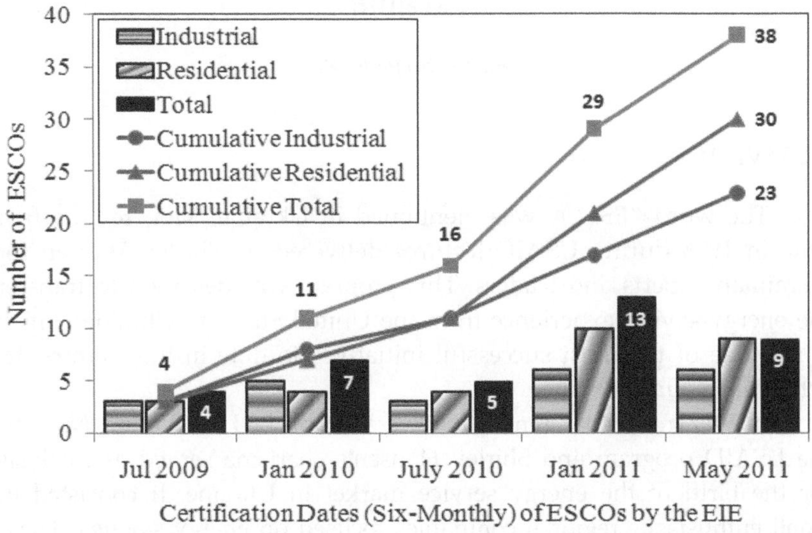

Figure 9-38: Evolution of the Turkish ESCO Market

that ESCO licensing will resume right after the approval of regulatory modifications.

The current value of energy projects carried out by EIE-listed ESCOs is not publicly available. As stated previously, ESCO-like business, particularly for the industrial sector, has been conducted since 1995 by several private energy companies. We anticipate that energy projects in Turkey will continue to be undertaken by private energy companies, whether certified as ESCOs or not. However, formal auditing and monitoring tasks will be conducted primarily by EIE-licensed official ESCOs. Although there are 38 EIE-listed ESCOs today, only between four and eight of them may currently be considered as chief players in the EPC market.

FACILITATORS

Currently, there is no ESCO association in Turkey. On the state side, the EIE/NECC presently acts as the only facilitator to help develop EPC in Turkey.

Ukraine

Vasily Stepanenko

ACTIVITIES

The word "ESCO" was mentioned in Ukraine, Kiev, for the first time in 1996 during USAID lectures delivered by Shirley Hansen for Ukrainian experts and leaders. This program was designed to transfer the energy saving experience from the United States to Ukraine, which is still one of the most successful initiatives to date in the country. Its echo stills sounds today.

In 1997, more than ten ESCOs were established in Ukraine. Namely, the USAID program and Shirley Hansen's charisma served as catalysts for the birth of the energy service market in Ukraine. It consisted of small enthusiastic regional companies focused on energy savings. These companies were typically small engineering firms. However, the lack of working capital, a sustained demand for energy efficiency and potential customers' confidence did not even provide for medium-sized businesses. All of these companies had an annual turnover lower than USD 200,000. Their main market was the Ukrainian industry, which needed professional advice to reduce energy costs during this transition period.

In 1998, UkrESCO, the state ESCO, was founded. It was established as part of the implementation of the Credit Agreement between Ukraine and the European Bank for Reconstruction and Development (EBRD). Special Ukrainian law was adopted to create UkrESCO. According to the Credit Agreement, UkrESCO received a EUR 20 million (USD 26 million) loan under the government guarantee, which was successfully implemented over six years.

In 1999, the first five ESCOs established the Association of Energy Service Companies in Ukraine. It was the first attempt to establish cooperation between private ESCOs, the state, the banking community and businesses. Despite a good start and a few successful joint projects, the rapid expansion of AESKO members to 37 companies caused a loss of control. In five years, the Association of Ukrainian ESCOs ceased to exist and, by common consent, was eliminated.

From 2000 to 2005, the industry and energy sectors were privatized, which renewed the interest in energy efficiency projects on the part of owners and investors, and led to the new ESCO business development cycle. Although the actual business volume increased only slightly, the quality of the projects changed. More capital-intensive medium-term projects replaced low-cost energy efficiency initiatives. ESCO contracts grew in value and their number increased. This period can be called the birth of the real energy-saving market.

Customers' pragmatism and exactingness to operation results were increased over that period. During that period, the natural selection process for the ESCO market began. A number of companies went bankrupt and small ESCOs merged into larger companies. Competition among ESCOs increased and small/weak companies lost their independence or left the market. The total number of such ESCOs in the country had risen to about 30-40 by 2005.

From 2005 to 2008, the Ukrainian ESCO business initiated the road to recovery, which resulted from increased natural gas prices and a rapid growth of overall rates for all energy types. The low-cost classic approach to energy saving lost its attractiveness with average tariffs increasing by 25 to 45 percent per year. In this period, there was a clear need for significant energy-saving projects geared toward growing gas prices. New methods, financing schemes and technologies contributed to increased requirements for ESCOs, with their number rising to 70-80 companies. Energy efficiency projects worth millions of dollars in investments were implemented during this period in Ukraine.

Furthermore, the energy efficiency scale-up was the most rapid in energy-intensive sectors (metallurgy and mining industry, food and chemical industries, building and public sectors).

The atmosphere of the industrial growth and the investment inflow fixed into the modernization of main industrial funds on the background of increasing energy prices brought about new quality standards in the management of capital-intensive energy efficiency projects. The Institute for Energy Saving and Environment, led by the former Ukrainian Energy Minister, established a new type of ESCO, financial-industrial group, Donbas's Industrial Union, which developed energy efficiency projects for its plants. The institute was functionally similar to an ESCO and implemented large projects for more energy-efficient metallurgical plants.

The global financial crisis of 2008-2010 destroyed the growing energy service market in Ukraine. Almost all large and medium projects were interrupted, investors' money went out of the country, the construction of thousands of buildings stopped and energy efficiency projects were frozen or eliminated. As a result, many companies including

ESCOs went bankrupt. In 2010, the number of ESCOs, which prevailed after the crisis, did not exceed 20 companies.

The ESCO market underwent a transformation and the focus shifted from the industrial sector to the municipal energy sector and public construction. Due to the six-fold increase in natural gas prices and the rise in heating tariffs over the previous five years, the demand for ESCO services grew significantly and continues to escalate.

LEGAL FRAMEWORK

The post-crisis period in 2011 saw economic growth as well as a stronger demand for energy-efficient upgrades in the industry and energy sectors of Ukraine. However, nowadays, the highest expectations in energy efficiency are related to heating, water supply and public construction. The situation is becoming more and more critical in this sphere and low energy efficiency levels in these sectors has transformed it into a political issue for the government.

In 2009, another attempt to unite ESCOs under the Association of Ukrainian energy auditing companies proved unsuccessful and, today, only two companies and a handful of individual members remain.

However, in 2011, the Ukrainian energy service market started to blossom along with the number of energy-saving companies. The industrial sector has continued to recover, and the introduction of a "green tariff" for electricity is contributing to the rapid growth of the renewable energy and biofuel markets.

Today, the greatest expectations are in the termomodernization market of residential and public buildings. A directive from the EPBD was observed in Ukraine and the Parliament is preparing the new Ukrainian Law on "energy efficiency of buildings." The new USAID project entitled "Reform of Municipal Heat Supply" could be a strong catalyst for change in the way power is supplied to Ukrainian cities. This could mark the beginning of the era of municipal energy planning. In essence, this project provides the basis for new ESCO markets (heating and water supply for residential and public sector buildings for hundreds of Ukrainian cities). Ukrainian ESCOs are involved in this project and are looking into the international energy-efficient modernization experience of urban households. They have already trained several dozen members of Ukrainian city authorities in energy planning.

The coming into force of the international standard for energy management (ISO 50001) should contribute to the development of the ESCO market in Ukraine. The standard will significantly change the quality of control systems for thousands of Ukrainian enterprises and hundreds of cities over the coming five years. The ESCO market could drive these changes.

Significant positive changes have taken place in the education sector. Several Ukrainian institutions of higher education have been producing energy managers and energy auditors since 2006. This profession has become popular in the labor market and the demand for skilled professionals continues to go up.

MARKETS

Ukraine is still among the most energy-inefficient countries in the world. Millions of old engines as well as hundreds of thousands of energy-inefficient pumps, boilers and compressors are still in operation in industry today. As almost all residential and public buildings are cold, termomodernization could be a great way to address this issue. Only seven percent of industrial assets were impacted by energy-efficient modernization over the last 15 years. Only three percent of utility assets had been upgraded by 2011. The Ukrainian energy service market is much less developed in contrast with the EU and US markets, especially in qualitative terms.

Energy audits constitute the basic service provided by Ukrainian ESCOs. There is a great demand for express and demonstration audits in Ukraine. Their cost is low but they describe energy losses as a system and objectively assess the whole in money terms. Over 120 companies, which offer these types of energy audits, have already been registered. These companies are small. In fact, often only one employee works in these companies.

Today, comprehensive or full energy audits are also developing, which is more useful for customers. Typically, this type of audit is implemented when ownership changes. There are about 30 companies able to carry out full energy audits professionally.

A more complex investment grade energy audit is performed before the modernization starts. This type of audit requires a team of highly qualified specialists and sometimes special, expensive equipment. Investment projects and programs, business plans and technical specifications for detailed design are the results of investment audits. Energy modernization strategies for industry and municipalities, municipal energy plans, and heat and water supply schemes for cities and regions have also appeared in recent years. There are less than ten ESCOs in Ukraine which can carry out all these new projects. The industrial and municipal heating system sectors are the dominant markets for ESCOs. ESCO municipalities have started to emerge in Dnipropetrovsk, Rivne and Kherson.

Third-party financing still represents a new type of service, which is yet to be mastered by Ukrainian ESCOs. Furthermore, to date, the EPC approach still has not settled in the Ukrainian reality. Today, existing Ukrainian economic and contractual rights reject energy performance contracts. Bank rates are escalating and risks are high—it is currently impossible to get a return on investment costs in energy efficiency projects.

Direct service contracts, more rarely trade credit or leasing, are the most common type of ESCO contracts today. Risk insurance is never used and insurance companies know very little about ESCOs. Guaranteed savings contracts are extremely unpopular.

The legal basis for promoting energy efficiency in the country is taking shape very slowly. Frequent government changes in Ukraine have significantly reduced municipal and public management qualifications. The fierce political environment in force is not conducive to a successful power relay. The strategic planning has not developed completely. The annual budgeting and planning system does not contribute to the development of the ESCO market. The most critical sectors are municipal services and public buildings.

The legal and regulatory framework, which has been inherited from the centrally planned economy period, has not changed over the last 15 years. These old economic relations represent the major barrier to attracting investments in these sectors, despite the ever-increasing efficiency of termomodernization projects.

FACILITATORS

The development of the Ukrainian energy service market lags behind that of markets in Eastern Europe as a consequence of the long-term public policy stagnation in energy efficiency. In the Ukrainian energy service market, there are no intermediaries such as organizations, associations or regional and sectorial energy agencies, which are all characteristic of the markets in developed countries. Furthermore, there is neither long-term partnership nor support, which reduces the efficiency and effectiveness of the ESCO market in Ukraine.

The change is coming slowly but inexorably. Today, the Ukrainian ESCOs form the face of the business of energy efficiency in Ukraine as well as determine the professionalism and competence level that defines the industry. The future belongs to them.

United Kingdom

Anees Iqbal

ACTIVITIES

The growth of EPC, which in the UK is better known by the term contract energy management (CEM), originated just prior to the first oil price shock of the early 1970s. It focused initially not so much on efficient use of energy rather than on providing cost savings in the production of energy (usually steam or hot water). The term ESCO was not widely known in the UK at the time and the concept in the present form evolved much later.

The first companies offering ESCO-type services in the UK were essentially boiler house operating and management companies initiated by the National Coal Board, under the chairmanship of the Liberal peer Lord Derek Ezra. One of the first such companies owned by the National Coal Board was Associated Heat Service (AHS). Many other similar companies came on the scene and offered very similar services of providing operation and manning (using the "Milk Round" concept) of central coal-fired boiler houses. In other words, the services provided were entirely supply-side energy management, and savings to the clients were largely generated by a reduction in operation and manning costs. Associated Heat Service was later acquired by French company "Compagnie Générale de Chauffe" which itself was a subsidiary of "Compagnie Générale des Eaux."

It was not until 1984 when Shell UK launched its own contract energy management subsidiary EMSTAR (Energy Management Services Technology and Resources) that true demand-side energy management services came on the horizon. Up until 1984, Shell UK had been trying out this concept from a small unit within the organization called Heating Management Service. Under this, Shell invested its own capital and technical expertise to modernize the heating plant in multi-dwelling buildings, installing new modular boilers and controls, clearing up the backlog of maintenance and changing the fuel (which very often involved converting oil-fired boilers to gas). This essentially proved to the market that the concept was not another vehicle of the oil company to sell more oil. In addition, the clients were offered guaranteed savings in their fuel bill. Although this business was new in the UK, Shell as a company had similar businesses in other parts of the world, notably in Holland and the US. The business model for EMSTAR was largely based on Shell's activities. Shell was already operating in the US under the name Scallop Thermal Management Inc. (STM). STM concluded a major hospital demonstration contract in Philadelphia, which at the time was its flagship project.

Very soon after the launch of EMSTAR, British Petroleum also launched its own almost identical activity under the name BP Energy. Thus, until recently, the demand-side energy management business was dominated by the two oil giants in the UK—Shell and British Petroleum.

As of today, the 11 members of the ESTA CEM subgroup are now the major ESCOs in the UK. However, a number of smaller companies that might call themselves ESCOs also exist serving mainly the smaller end of the market; e.g., single commercial premises, private dwellings, etc. Most of these smaller end clients will have an annual fuel bill well below GBP 50,000 (USD 78,725) per annum. These new smaller companies are usually formed by consultants who may have managed to acquire some financing to help their clients implement their recommendations. We estimate that around 15 percent of all smaller buildings are actually owned and operated by larger companies (bank branches, pubs, etc.). The combined energy cost in these organizations is substantial. However, they may not provide the full breadth of service, such as guarantees and/or long-term O&M. On the other hand, some O&M companies can also extend their services to include aspects of a traditional ESCO (such as access to finance and performance guarantees) and may call themselves ESCOs. Some of these smaller ESCOs are now being taken over by larger companies. This trend of mergers and acquisitions is likely to continue and further changes may take place in the coming months and years.

CONTRACTS

In the UK, there is no single contract model in exclusive use. Shared savings, guaranteed savings and the heat service models are all used depending upon the type of client and project concerned.

UK ESCOs use a wide variety of methods to finance their investments. Most ESCOs have access to pools of capital from private sector lenders or from their own organizations. ESCOs are able to structure the loan to be on or off balance sheet, depending on the tax situation and other considerations of the client.

Companies such as Dalkia frequently use the technique of "undisclosed third-party financing." Under this arrangement, the ESCO has an agreement with a preferred bank or a finance house whereby there is no separate contract for finance that is signed between the customer and the bank, it is rather implicit in the ESCO agreement. In addition, the financier remains in the background. It is only in the event of the client defaulting that the financier steps in.

LEGAL FRAMEWORK

The concept of contract energy management flourished in the UK but initially only in the private sector. There were a considerable number of barriers in its path for it to be accepted in the public sector. The HM Treasury initially labeled CEM as an unconventional financing method and as such saw it as a backdoor means of circumventing the government's capital restrictions in the public sector. It was even labeled illegal for use in the National Health Service as it seemed to violate a major clause in the National Health Service Act of 1977.

The CEM industry fought a long and hard battle with the HM Government Treasury, which ultimately saw the merit of CEM and permitted its use subject to detailed option appraisals demonstrating the "Value for Money" criteria. Today of course, the Treasury not only approves the use of CEM but encourages all public sector bodies to consider private finance under its Private Financing Initiative (PFI) before approaching the state to fund major capital-led programs.

However, ESCO/CEM still has a 'rip off' image which stems from the apparent similarity with energy tariff consultants' style of contracts (negotiate lower utility rates and make a killing). In addition, the general attitude to the PFI is very negative—in the press and on TV. Observers generally decry the PFI as very expensive for the public sector. Whether this is right or wrong, the negative publicity degrades what could be achieved through private finance under the ESCO/CEM heading.

Other barriers in the public sector that remained were competitive quantitative tendering. While supply-side energy services were relatively easy to evaluate on a competitive basis by comparing the cost of steam, the wider demand-side services were impossible to offer and evaluate on a competitive basis. Novel ways of tendering such as "qualitative tendering" are being promoted by the industry to overcome this difficulty. Another way being approached in the UK and elsewhere is for clients to appoint consultants to develop a detailed specification for demand-side energy management and on the basis of which competitive bids are sought. The winning bidder is then asked to include the fees of the consultants as part of the investment. While this procedure works to a certain degree, the novel approaches of various CEM companies cannot be used by clients.

The legal and contractual documents, negotiations and personnel transfers that are often required means that the transaction cost for a project is high. ESCOs are therefore careful about when and where they tender and the contract process is protracted—often up to two years. The lack of common contract terms is also a barrier. Overall, this makes tendering for smaller and medium-sized projects very costly and restricts the market for users spending less than GBP 50,000 (USD 78,700) per annum. Although larger ESCOs would say GBP 500,000 (USD 787,000).

Even though the ESCO model is now a mature concept and has been used in the UK for more than 25 years, there is still an important lack of awareness and knowledge in the market about the concept and about how to use it in order to extract maximum benefits.

MARKETS

UK ESCO projects may be classified under three main categories:

1. Demand-side refurbishments/retrofits. This service normally of-

fers the provision of finance and performance guarantees. These include projects related to (i) building envelope and hot water distribution improvements; (ii) controls; (iii) efficient lighting; (iv) boiler decentralization; (v) energy recovery; (vi) routine & breakdown maintenance; and (vii) fuel purchase and management.

2. Supply-side retrofits/refurbishments includes: (i) boiler house retrofits; (ii) fuel switch; (iii) improved hot water and steam distribution systems; (iv) controls and insulation; (v) medium-scale CHP; (vi) fuel purchase and management; (vii) routine and breakdown maintenance; and (viii) finance and performance guarantees. There is also a wide scope for better skilled operation and management (a contract to run a facility without necessarily investing in new hardware). The potential for savings was clearly demonstrated in the public sector as display energy certificates (DECs) got underway on October 1, 2008. There is likely to be considerably more scope for achieving savings from better management, operation and maintenance of sites than we currently perceive.

3. New buildings: This is a new area for larger ESCOs and for those in consortiums with monitoring and evaluation (M&E)/facilities management (FM) contractors. This new business for ESCOs came about with the launch of the Government of UK's Private Finance Initiative for Public Estate. This category includes provision of construction finance, turnkey contracting, operation and maintenance as well as total facilities management where required (catering, gardening, decoration, etc.).

More recently, we have identified ESCOs in three further categories different from the above.

1. Community ESCO: This is where a local authority or housing association creates a dedicated ESCO. The entity is called an ESCO but it is not offering services to the wider market.

2. Domestic ESCO: This is generally the utility which offers household services. Domestic ESCOs are in a good position to progress as they have contacts with millions of households and are used for mass marketing—most Energy Services Trade Association (ESTA) members are not experienced in mass marketing.

3. FM-style ESCO: These are ESTA members offering a commercial service across all non-domestic sectors. FM-based ESCOs have not been totally successful and a new government initiative called the "Green Deal" (see below) is an attempt to make the desired effect in the marketplace.

We also have to be careful about the definition for "energy services." Many companies offer energy services such as a bureau service to collect consumption data in order to analyze and present it. However, we would not recognize these as ESCOs.

Unlike in Canada, the UK CEM industry has evolved in a very different format. The public sector was not the first market sector to embrace the CEM concept. As mentioned earlier, the Treasury placed many restrictions and the industry has had to fight a long and hard battle to get CEM accepted in the public sector.

It is only a recent development that the public sector has become properly involved with the industry through the Government of UK's PFI and this has continued in many ways so that a private-public "partnership" has developed. Under this PFI partnership, a significant number of new build projects are likely to be executed in the coming months and years.

The health and hotel sectors have traditionally been a lucrative market for the CEM industry. Both feed and sleep people in large numbers and the 24/7 demands for reliable energy services together with shortage of investment capital lend itself ideally to the CEM business model. This is a sector where the CEM industry has found a natural fit.

The same can be said for the multi-dwelling residential sector—early CEM contracts in the UK were largely implemented in this sector. Educational and commercial establishments are also sectors where the CEM industry has been highly successful.

The industrial sector is still a large and lucrative market for the CEM industry, although it has remained flat in terms of market growth. This sector can be broken down into three main components:

1. demand for reliable and cost-effective heat, power and compressed air;
2. space heating and hot water demands for industrial building envelopes;
3. industrial process side.

The CEM industry has been admirably successful in the first two but has by and large shied away from getting involved in the processes of their industrial clients. Moreover, CEM companies in the UK are wary of direct involvement in social housing (where they have to get tenants to pay). Generally, they require guarantees from local authorities or housing associations.

FACILITATORS

As such, there is no UK Association of ESCOs. However, the industry in the UK is represented as a CEM subgroup within the Energy Services and Technology Association (formerly Energy Systems Trade Association). The CEM subgroup was formed in 1987. The Association encourages orderly growth of the industry through accreditation, support and advice to both CEM companies and customers. Current ESTA memberships, which include equipment suppliers, utilities and consultants in addition to CEM companies, stand at 117 companies, of which CEM subgroup members total 11.

GOVERNMENT ACTIONS

The UK ESCO industry has, unfortunately, not enjoyed the same strong support from the government as it has in many other countries. The UK government is a strong believer in letting the "market forces do the talking." However, as stated earlier, the government now does encourage the public sector clientele to consider the use of private finance as a priority over state finance, provided that it offers a better value for money. UK ESCOs have had to fight hard to win large public sector contracts.

On the other hand, the Green Deal program will be a key government action of the present Coalition Government for improving the energy efficiency of buildings in the UK. It is a new market framework. It is based on a key principle that some energy efficiency-related changes to properties pay for themselves, in effect, through resulting savings on fuel bills. The Green Deal will create a new financing mechanism to allow a range of energy efficiency measures (loft insulation or heating controls) to be implemented in people's homes and businesses at no upfront cost. Although the Green Deal will not start until late 2012, there is much anticipation around the types of installations it may fund.

Government intends that the Green Deal UK will revolutionize the energy efficiency of British properties. It is establishing a framework to enable private firms to offer consumers energy efficiency improvements to their homes, community spaces and businesses at no upfront cost. It also intends to recoup payments through a charge in installments on the energy bill, thus favoring the development of the energy efficiency market.

Uruguay

Marcelo Gonzalez

ACTIVITIES

At the beginning of 2000, engineering firms were offering energy consultancy services, and charging for their services from a portion of the savings generated after the proposed measures had been implemented. Still no one spoke of ESCOs at the time as there was no financing available for project implementation—neither from ESCOs themselves nor from the financial market. There was therefore no talk of performance contracts and little understanding of what the concept was.

In 2004, the Uruguay Energy Efficiency Project (UEEP) was developed by the World Bank through support of a GEF grant. As investments in energy efficiency projects were very limited in the country, it was proposed within this project to address the market barriers through the development of a utility-based ESCO (USCO) and private ESCOs.

The national Uruguay Electricity Company (UTE) developed an ESCO operation within its ranks, and the EPC concept was launched. Under this initiative, UTE-USCO developed, implemented and financed energy savings investments using the EPC shared savings concept. The program was designed for USCO to act as a Super ESCO, meaning that it would support the development of local ESCOs and provide them with financing for their own project implementation.

The components of the UEEP program related to the development of the EPC concept included the following:

- support for the creation and development of USCO, as mentioned above;
- support for the development of private sector ESCOs through training, operation, support and financing mechanisms; and
- creation of the Uruguay Energy Efficiency Fund (FUEE) to support private sector ESCOs in the implementation of energy efficiency projects.

From its inception, UEEP management has established a register of existing ESCOs and other companies interested in becoming ESCOs. The first registration was voluntary with no related conditions to that effect.

The FUEE fund was created as a guarantee fund that worked through financial intermediation institutions. FUEE is managed by the National Development Corporation (NDC), an entity dependent upon the central government. So far, two commercial banks (BROU and BANDES) have signed agreements to participate in this financing system for energy efficiency projects.

This funding scheme was launched in December 2008 and has so far had little movement, as the whole process is quite cumbersome.

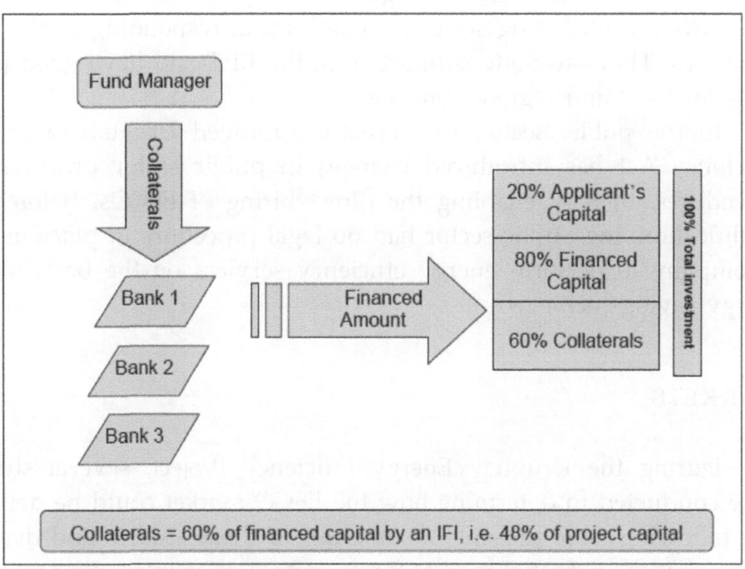

Figure 9-39: Basic Operating Scheme

Improvements are underway to make it a flexible and a more attractive tool for both applicants and financial institutions.

CONTRACTS

At this time, the shared savings model is the one that is best known. It is based on its use through the UEEP World Bank program.

In the private sector, the developing EPC market has yet to adapt a performance contract model that adequately satisfies the Uruguayan legal and commercial framework.

LEGAL FRAMEWORK

Uruguay has defined its energy policy through the use of energy efficiency as one of its key strategies. Actions have been implemented in this regard in order to stimulate the development of a sustainable ESCO market. To support this trend, the Energy Efficiency Act addressed procurement and contracting mechanisms for public institutions, introduced tax exemptions and supported the implementation of adapted financing mechanisms for energy efficiency projects. Despite the small size of the potential Uruguayan energy efficiency market, there is already an interesting set of companies that responding to the new initiatives. They are quite supportive of the EPC and have good prospects for local and regional business.

In the public sector, the recently approved Uruguayan Energy Efficiency Act has introduced changes in public sector procurement procedures, thereby enabling the direct hiring of ESCOs. Before this modification, the public sector had no legal procedure in place to hire a company to perform energy efficiency services on the basis of the energy savings achieved.

MARKETS

During the Uruguay Energy Efficiency Project, several studies were conducted to determine how the ESCO market could be defined. The following table presents a summary of those studies and demonstrates that the potential market for projects is quite interesting as it has not been tapped up to now.

Table 9-20: Distribution of the ESCO Market by Sector

Sectors	Number of Potential ESCO Clients	Total Energy Costs (USDx10^3)
Industrial, Agriculture	383	350
Commercial	226	35
Public Sector	201	50
Public Lighting	44	1,3
Water Works	33	13,3
Residential	16	3,4
Total	**903**	**453**

FACILITATORS

The Energy Efficiency Act empowered the Ministry of Industry, Energy and Mining (MIEM) for the creation of the Energy Efficiency Unit (EEU). This unit under the National Department of Energy and Nuclear Technology (DNETN) is responsible for supporting the development of EPC in the country.

GOVERNMENT ACTIONS

The government has undertaken the following actions in order to promote the ESCO market.

The Energy Efficiency Act was approved in September 2009. The many aspects of this law have given focus to the development of an EPC market in the country. As the following articles demonstrate:

Article 4: Creation of the Energy Efficiency Unit under the Ministry of Energy and Mine (MIEM).

Article 16: Development of an Energy Efficiency Certificate Market to help finance energy efficiency projects.

Article 17: Creation of the Uruguayan Saving and Energy Efficiency Trust (FUAEE) in order to provide financing for technical assistance in energy efficiency, promote energy efficiency at the national level, finance investment projects in energy efficiency, foster energy efficiency research and development and act as a contingency fund in the context of the energy sector crisis.

Article 21: Inclusion in the Financial Administration of Accounting text, which regulates how public sector contracts can develop direct contracting in the case of EPC with potential contractors (ESCOs).

Through these initiatives, it was possible for ESCOs registered through the MIEM to be paid out of the savings generated by a project, thus enabling the use of EPC in Uruguay. This act is still in the process of being fully implemented.

Through Decree 354/009 the government has decided to implement an incentive for the development of ESCOs through the application of tax exemptions. In August 2009, a new decree of the Law on the Promotion and Protection of Investment (which encourages investments in Uruguay through tax exemptions) extended the benefits under the law to services offered by ESCOs registered with the MIEM and designated as Category A.

The **Institutional Energy Plan** has addressed the need to generate specific energy efficiency plans in the public sector. There is also a proviso that these agencies may make direct contract with companies by linking payments to performance. The Plan is expected to generate new opportunities for ESCO business.

United States of America

Donald Gilligan
Terry E. Singer

ACTIVITIES

The history of EPC in the US can be usefully divided into stages.

- **Initial stage (pre-1986)**—ESCOs grew out of the efforts made by Scallop Thermal, a Division of Royal Dutch Shell, as discussed in Chapter 1.

- **Oil price drop (1986)**—The embryonic ESCO industry almost died when the price of oil dropped in the late 1980s. At the time, all projects were using the shared savings model (a pre-determined split of the financial savings), which worked well when prices were stable or escalating. The 1986 price drop protracted the paybacks, often to periods longer than the contract. As a result, the US ESCOs shifted to guaranteeing the amount of energy saved and the guaranteed savings model emerged.

- **Public sector focus**—The initial focus of guaranteed savings was on the public sector market where the US tax-exempt financing made the model more attractive.

- **Expanding EPC (1987-1993)**—A boost to the industry came from utility programs, especially in response to DSM and integrated resource plans (IRPs). ESCOs bid to provide the kW or kWh savings, delivered turnkey projects to large industrial and institutional customers and financed the projects themselves.

- **Success and consolidation (1994-2002)**—Successful experience with EPC, documented in studies by the Lawrence Berkeley National Laboratory (LBNL) and the National Association of Energy Service Companies (NAESCO), encouraged the federal and state governments to promote the use of EPC in their own facilities as well as those of private sector building owners and operators. The implementation of the IPMVP, which provided standard methods for documenting project savings, gave commercial lenders the confidence to begin financing EPC projects on a large scale.

- **Pause and then fast growth (2003-present)**—The collapse of Enron, the suspension of the federal energy savings performance contracting (ESPC) program and uncertainty about the deregulation of the electric utility industry caused a US slowdown in the growth of EPC from 2002 to 2004. EPC is now growing at more than 20 percent per year, driven by increasing and volatile energy prices, federal and state energy savings mandates, the continuing lack of capital and maintenance budgets for institutional and federal facilities, and the growing awareness of the need for large-scale action to limit GHG emissions of which energy efficiency is the cornerstone.

The US ESCO market continues to grow at a much faster rate than the overall US economy. The National Association of Energy Service Companies (NAESCO) estimated in its 2009 industry report that the total revenues for the US ESCO industry in 2011 would be between USD 6 and USD 7 billion.

Several factors are holding back the growth of the EPC market in the US, including:

- **M&V Limitations**—New systems are required to make the calculation of project energy savings more understandable to non-technical policy-makers depending on energy efficiency to meet public policy goals, such as energy savings and GHG reduction mandates. While M&V is viewed as a necessity, it is also seen as a cost burden to a project. Without major expenditures, the accuracy is only +/- 10-20 percent.

- **Shortage of Skilled Personnel**—ESCOs, utilities, state regulatory agencies and **customers** are struggling to find the skilled engineering and technical personnel required to implement large-scale energy efficiency and renewable energy programs, and to operate and maintain energy efficiency and renewable energy technologies.

- **Specific Market Barriers**—Each of the major EPC market segments suffers from its own constraints.
    - **The federal and MUSH markets** are hindered by landlord agency and financial control bureaucracies that too often resist large-scale program implementation in the face of executive and legislative mandates.
    - **The commercial real estate market** is hindered by the refusal of building owners to encumber their buildings with the debt required to finance comprehensive EPC projects, or by tenant problems.
    - **The industrial market** is hindered by the policy of most American manufacturing companies, resulting in project payback requirements of typically less than two years, which in turn preclude comprehensive EPC projects employing multiple technologies. Top management frequently fails to understand that financing energy efficiency out of avoided utility costs is not the same as requiring new budget allocations. Fears

related to the loss of intellectual property creates management reluctance as well.

MARKETS

The heavy concentration of the industry on government and institutional projects has not changed in the last few years, with almost 90 percent of projects implemented for federal, state and local government agencies and institutions. US ESCOs anticipate continued strong growth during the next few years, driven by increasing requirements among government and institutional customers for energy savings and facility modernization and a modest, but growing, interest among private sector commercial building owners.

The MUSH market and the federal market account for about 80 percent of the total EPC projects in the US. Commercial building projects represent about 9 percent of the projects, industrial projects represent about 6 percent, and the balance consists of residential and public housing projects.

The discussion below presents the growth factors that are driving the expansion of the ESCO industry in key market segments and the issues that the ESCO industry faces in those segments.

Federal Government

US federal government facilities—civilian and military—are subject to very aggressive energy use reduction mandates. By 2015, the entire portfolio of federal facilities is expected to reduce its energy use by 30 percent from a 2003 baseline. This goal is being achieved for new construction and major renovation projects, but existing buildings are not yet on the improvement glide path required to meet the goal. By 2015, new construction and major renovation projects must reduce their fossil fuel usage by 65 percent. By 2030, fossil fuel use must be eliminated for new and substantially renovated buildings.

These energy savings mandates, however, are not accompanied by increased capital construction budgets. For civilian agencies, capital budgets for the next five years will be minimized. Military capital budgets are also being cut, but not as severely as civilian agency budgets. The combination of the aggressive mandates and the diminishing capital budgets means that the only way that federal agencies will meet the mandates is through the increased use of EPC (referred to as ESPC by the US federal government).

The scope of ESCO projects will also have to change to meet the mandates. The type of projects that ESCOs have delivered for the past decade, which upgraded equipment to reduce facility energy use by 20-35 percent, will not be sufficient. If we assume that half of federal facilities will undertake retrofits, these retrofits must average 60 percent energy use reductions to meet the mandate of 30 percent reduction across the whole portfolio of facilities.

As a result, federal civilian agencies and the military services are now developing, on an accelerated basis, programs that will demonstrate the technical, economic and business feasibility of "Deep Retrofit" projects. These projects will combine unprecedented thermal upgrades of building envelopes, equipment replacements on single buildings, cogeneration or CHP as well as renewable energy generation (solar, wind and biomass) on campuses or groups of buildings. The ESCO industry agrees with government scientists that the technologies required to achieve the mandated savings are currently available. However, the economics and the business structure (contract terms and tenure, risk sharing, savings verification) are still predicated on the enabling legislation and the agency guidance created in response to current practices.

State and Local Government

In the US, the development and implementation of EPC projects in state and local government facilities is governed by state laws, which allow EPC projects exemptions from the traditional state public construction laws and subject them to alternative procurement criteria. Most states require that any public facility construction project above a minimal level (USD 25,000) be procured using a process in which contractors submit fixed-price bids on detailed scopes of work. This traditional procurement method has proven to be ineffective for energy efficiency projects because most state and local government agencies lack the expertise to design a project for bidding and lack the capital to pay the design and bid administration costs. As a result, state laws in nearly every state permit government agencies to procure energy efficiency projects using competitive selection processes that emphasize an ESCO's qualifications and track record. The ESCO is also required to provide some indicative pricing information demonstrating that the project produces enough savings, guaranteed by the ESCO, in order for the facility to meet its debt service and be able to repay the full capital cost.

In contrast to the federal government market, many state and local government projects are driven by the need for capital improvements rather than aggressive energy savings mandates. State and local government agencies have typically been starved for capital construction funds for the past decade, and so they use EPC agreements to replace critical building components like lighting and heating systems. The appeal of EPC is that it does not raise the costs for the government agency, but simply re-directs operating budget funds that are currently spent on wasted energy into a stream of payments that amortize the capital cost of the improvements. Funding for projects is provided not by the ESCOs but by a robust competitive private marketplace of banks and specialty project finance companies at very attractive rates, typically 2-4 percent in the current market.

Because the legislation, which enables state and local government EPC projects is state legislation, EPC programs are usually organized and promoted by state energy offices, which offer government agencies assistance that ranges from standardized project processes and documents to low-cost loans from state revolving loan funds. The state energy offices are in turn supported by the US Department of Energy (DOE), which provides a full range of technical assistance services through a national network of consultants, including the National Association of State Energy Officials (NASEO), the Energy Services Coalition (ESC) and NAESCO.

The DOE State Energy Program (SEP) got a huge boost in funding from the American Reinvestment and Recovery Act (ARRA), which was the federal government's stimulus program to counteract the 2008 recession, with funding levels rising from USD 50 million per year to about USD 3 billion per year from 2009 to 2012. The ARRA funding will not be replaced and the SEP program is dropping back to its pre-ARRA funding. To sustain the ability of the energy office to continue to operate and promote the state EPC program with reduced budgets, many states are now levying small fees on EPC projects.

During 2011, the steady growth of state and local EPC programs has been interrupted in several states. Twenty-nine of the 50 US states elected new governors in 2010, and many of these governors were elected with mandates to cut the size of government and eliminate public programs. The new governors were also faced with serious budget deficits that ranged from 10 to 40 percent. In response, cuts were made in many state energy office staffs and state EPC program staffs.

LEGAL FRAMEWORK AND GOVERNMENT ACTIONS

Two developments outside the ESCO industry and outside federal and state energy agencies as well may have significant impact on the ESCO industry in the near future.

Dodd-Frank Act

In 2010, the US Congress enacted legislation that was designed to prevent a repeat of the 2008 financial crisis. Part of the legislation mandates that the Securities and Exchange Commission (SEC) promulgate regulations that would require that any person or company that is providing financial advice to a municipality register with the SEC as a "Municipal Financial Adviser." Under the proposed rule, ESCOs and many of their employees would be required to register and report to the SEC, comply with unknown and potentially numerous and costly new rules as well as assume the potential risk of SEC enforcement liability and undefined fiduciary liability. The proposed rule could render some public building retrofits uneconomic and lead some ESCOs to leave the municipal building business altogether, costing significant jobs and resulting in lost energy savings.

FASB 13 Modifications

At present, third-party financing agreements, which fund many energy performance contracts is accounted for as an operating lease and thus presented separately from the balance sheets of building owners. The Financial Accounting Standards Board (FASB) has proposed a new interpretation of "lease" that would essentially eliminate the category of operating leases and cause many energy performance contracts to be inappropriately capitalized as liabilities for building owners, potentially causing many ESPC customers to exceed allowable debt levels under existing debt agreements and otherwise adversely impacting their balance sheets.

The proposed modifications to FASB 13 would discourage the use of energy performance contracts. The modifications would have the perverse outcome of preventing building owners from improving their long-term financial positions by reducing energy-related operating expenses and enhancing the value of their building stock.

To avoid this result and provide appropriate treatment of energy performance contracts, the FASB was requested to consider clarifying the final language of the modification to FASB 13 in order to ensure that energy performance contracts are wholly or (depending on the terms of those contracts) partially excluded from the definition of a "lease." This clarification is appropriate from an accounting perspective because an energy performance contract is fundamentally distinct from ordinary leases of real property or equipment in the following ways:

- An energy performance contract serves as a vehicle for carrying out a suite of permanent improvements and for providing long-term energy-related services at a customer's facility.
- Although an EPC customer commits to make a series of payments over time under a TPF agreement, payments are tied to the *performance* of the systems installed by the ESCO. Under most energy performance contracts, when contractually guaranteed energy savings are not sufficient to repay the financing, the amount of the payment shortfall is paid by the ESCO. This type of repayment mechanism demonstrates that the financing agreement is linked to the service rather than the equipment.
- Energy performance contracts may also include separate payments from the ESPC customer for operating, maintenance and energy monitoring services. These, too, are not payments for the use of property, but are instead payments for energy services.

An energy performance contract provides a no-risk method for the customer to essentially dedicate a portion of its operating expenses—a reduction in its energy bills—to cover the cost of energy services and equipment provided by an ESCO, in exchange for *guaranteed* reductions in overall energy bills during and after the energy performance contract. As such, it is logical and consistent with good accounting practice to treat the energy performance contract as a service agreement or operating expense, rather than a capital lease.

About 70% of ESCO projects are performance-based, and another 25 percent of projects are implemented using design/build or engineering, procurement and construction (energy performance contracts).

By dollar volume, ESCO projects are largely comprised of a mix of energy efficiency technologies (73 percent), renewable technologies (10 percent) and distributed generation or CHP (6 percent). The balance of ESCO revenues is derived from consulting and planning services.

Vietnam

Jalel Chabchoub
Phuong Hoang Kim

ACTIVITIES

Vietnam has given high priority to the development of the ESCO industry to ensure successful achievement of its targets in energy efficiency. The first effort of the country was made under the World Bank-financed demand-side management/energy efficiency project, which was implemented during the 2003-2010 period. The latter also encompassed the Commercial Energy Efficiency Pilot (CEEP) program. The CEEP program was designed to develop and test appropriate business models and mechanisms as well as to catalyze a sustainable energy service market to support energy efficiency investments in Vietnam. Through the technical and financial support provided under the Promoting Energy Conservation in Small and Medium Enterprises (PECSME) project, and during the five years that the program was in force (2006-2010), approximately 500 energy efficiency projects were implemented with the participation of 25 energy service provider (ESPs). These ESPs expected to converge toward an ESCO business and were willing to offer EPC to customers.

The Ministry of Science and Technology (MOST) and the Ministry of Industry and Trade (MOIT) have been involved in many initiatives to support the initial development of the ESCO market. Assistance was provided through training sessions and energy efficiency projects mainly funded by ADB, the World Bank, UNDP/GEF and bilateral donors.

The preliminary list of ESPs comprises about 50 companies including the 25 ESPs that were involved in supporting the PECSME project. However, only one of these ESPs made the decision to face the challenge and embark on EPC for implementing energy conservation projects in the industrial sector. The lack of financial capacity and often low willingness to shift from the ESP scheme to the ESCO business model explain, for the most part, the current situation in this regard.

In early 2011, MOIT, through World Bank support, prepared a strategic development program for the ESCO industry in Vietnam. In addition, the Vietnam National Energy Efficiency Program (VNEEP Phase II—2011-2015) includes a specific component focusing on expanding capacity building of ESPs, key market players and stakeholders. The target is to create an appropriate mechanism to deliver increased energy savings in the energy efficiency market with the main objective to convert ESPs to ESCOs.

CONTRACTS

The shared savings first-out EPC model was used for these pilot demonstration projects. The PECSME project subsidized service delivery (design, implementation, measurement and verification). The choice of contract type was mainly related to the lack of an adapted financing mechanism that would enable ESCOs to finance the full implementation of the project. The three EPC projects were carried out in food processing industrial facilities in the Ho Chi Minh and Can Tho areas with an average payback period of 12 months.

LEGAL FRAMEWORK

In the middle of 2010, the Vietnam National Assembly approved the Law on Energy Efficiency and Conservation under its energy efficiency program and targets. A number of decrees and regulations to implement the law were prepared. On March 29, 2011, Decree 21/2011/ND-CP was issued providing details and measures to implement the Law on Energy Efficiency and Conservation. The implementation of the Law falls under a number of ministries, provinces and organizations. MOIT acts as coordinator of the whole program and is also implementing several components under the latter. In particular, MOIT has set up the Energy Efficiency Office for program implementation.

The Law on Energy Efficiency and Conservation, constitutes a very important leverage to reinforce energy efficiency initiatives and create a dynamic market centered on energy efficiency services and EPC. This new law will reinforce actions related to energy efficiency, thereby creating more favorable conditions and tools to promote energy efficiency investments. The decrees and regulations for the implementation of the law will bring a better understanding of application modalities.

It is expected that the new law will provide an excellent basis for continued strengthening of SME and ESCO capacity for profitable energy efficiency activities.

It is anticipated that the government will reinforce its involvement and will thus likely support energy efficiency in less profitable sectors using the required tools and assistance mechanisms.

MARKETS

Despite all efforts, the ESCO business model has not yet taken off and only three EPC pilot projects have been implemented by an ESCO. This was done through the technical and financial support of the PECSME project between 2009 and 2010.

FACILITATORS

There are no well-established associations dealing with energy efficiency and promoting ESCO activities. At country level, the MOIT Energy Efficiency and Conservation Office plays the role of coordinator, promoter and implementing agency of government strategies and programs related to energy efficiency and conservation.

The Energy Conservation Centers support local people's committees in provinces/cities to implement national programs. They constitute the major partners for energy efficiency and conservation projects executed in Vietnam. They play an important part in performing and supporting energy efficiency activities. However, some ESPs consider these government centers as competitors and do not see them playing a catalytic role in ESCO and EPC development in Vietnam.

GOVERNMENT ACTIONS

In the documents drafted in July 2008 for the Climate Change Mitigation Support to the Vietnam energy efficiency program, it was indicated that a new energy efficiency fund would be established in Vietnam in accordance with the draft Law on Energy Efficiency and Conservation. The fund is expected to be available to both private and state-owned companies. While the size of the fund is not officially known, a proposal suggests a capacity of USD 20 million per year. The program would cover training sessions for audits, ESCO set-up, demonstration projects and investment assistance. However, the newly approved Law on Energy Efficiency and Conservation does not make any reference to a dedicated energy efficiency fund. Consequently, no dedicated energy efficiency financing mechanism or fund is currently available to promote energy efficiency or ESCO project implementation.

As experienced under the PECSME project, a number of incentive-based laws, policies and government programs could serve as a good starting point to accelerate SME adoption of energy efficiency measures and their implementation as part of related projects. Such measures include subsidies from the Vietnam Environment Protection Fund (VEPF) or the Local Science and Technology Fund as well as concessional borrowing rates and tax exemptions. In addition, the Loan Guarantee Fund established with the purpose of providing financial support (loan guarantee) to ESPs willing to implement energy conservation projects could be reinforced and used by ESCOs for energy efficiency project implementation.

Nevertheless, to stimulate the long-term commitment to energy efficiency, the government is expected to rely on voluntary agreements with industries. However, this does not seem to bring any noticeable results due to the lack of government commitment to put in place supporting mechanisms and accompanying incentives. The current conditions imply a low interest level in energy efficiency and, consequently, a limited environment for ESCO development.

Chapter 10

The Way Forward

As pioneers in the development of energy performance contracting (EPC) around the world, the authors find it most gratifying to see how much the industry has grown in recent years. Particularly heartwarming is the recognition EPC is receiving as a means to improve our energy efficiency use, spur economic growth, curb pollution and conserve our natural resources.

Documentation is growing that energy efficiency (EE) is the most cost-effective way to reduce environmental pollution. EPC is increasingly recognized as a valuable tool in our efforts to control climate change and achieve a healthier environment. In fact, it is becoming evident that EE is an extremely desirable way to finance our growing need for cleaner energy and achieve an ever more efficient economic environment. Further, energy service companies (ESCOs) make a natural marketing channel to combine both worlds.

Every country is unique. Reading the ESCO country reports offered in Chapter 9 underscores how very diverse EPC can be. A careful review of our contributing authors' thoughts on ESCO development in their respective countries, however, also points to certain aspects they have in common. This seems to be particularly true when we consider the prevailing barriers to EPC.

As we consider the way forward for ESCOs, we are repeatedly reminded that we can all learn from the experiences of others in the industry. As a result, the trends, which the industry shares across national borders, give us an important perspective and enable all of us to identify common barriers and share workable solutions.

The two dominant themes through the reports pertaining to the difficulties in the development of EPC are related to ESCO financing and the role of government. Before we discuss these two issues, there are several other matters that merit attention. Dominic Yin offers a good template for considering the issues before us when he lists those facing Hong Kong ESCOs, stating, "poor awareness, lack of experience in providers and prospects, a variety of complex legal and contractual issues, intransigence of the conventional procurement process and the host of challenges in any M&V attempts."

MEASUREMENT AND VERIFICATION

The need for, and the acceptance of, measurement and verification (M&V) is growing. In our research for *ESCOs Around the World*, declarations of the need for M&V appeared frequently. However, it was still a call for help.

Despite the "host of challenges" Yin mentions, M&V has now become an accepted and even vital part of ESCO work. There is an expressed desire to refine and extend its use, but whether or not M&V should be a component of EPC is no longer up for discussion. There is a call for more training and an obvious need for assistance across the industry in effectively gathering and interpreting the data.

There is clearly a growing consensus that there needs to be an internationally accepted protocol so that we are all using the same yardstick. It is very evident that the most universally accepted protocol is the International Performance Measurement and Verification Protocol (IPMVP).

Barriers

The most common M&V barriers mentioned are a) the cost of M&V; and b) the need for more training.

Lessons Learned

A uniform statement of support for M&V from various organizations within the international community, such as the EU, the UN and multilateral banks; e.g., the World Bank, would help cement this vital procedure as part of the ESCO industry. There is a concomitant need for demonstrable value and quality use of the process. Training is critical to realizing the goal. A procedure to certify M&V professionals that has some uniformity across national borders, such as the Efficiency Valuation Organization/Association of Energy Engineers Certified Measurement and Verification Professional program, would instill more confidence in the process.

CONTRACTS AND THE LEGAL FRAMEWORK

It is self-evident that a core component of EPC would include contracts. The nature of those contracts and the key components of the energy service agreement (ESA) are still evolving. The structure and usage of the contracts clearly varies with the type of financing mechanism; e.g., shared savings, guaranteed savings or chauffage. They also change considerably with the breadth and size of the project.

While there is clearly a need for contracts to comply with local laws and conditions, the basic elements and structure of the contracts seem to be reaching greater uniformity.

The use of a planning agreement (also referred to as a project development agreement) is increasingly relied upon. The industry is moving toward the common practice of using an agreement that assures that the ESCO gets some remuneration for its auditing services if the project does not go forward.

The major surprise in reviewing the reports was how few contributing authors mentioned contract enforcement. A small number of observations only were offered as to adjudication procedures. If the judicial system does not provide for contract enforcement, the "contract" is nothing more than a plan for implementation and does not offer conditions that bind either party.

James Wakaba, writing about ESCO conditions in Kenya, is one of the few who addressed enforcement issues. He summed it up observing, "The contractual context for the country is quite weak, with drawn contracts either unenforceable or, in the event of a breach, likely to result in protracted legal processes, which the fledgling ESCOs are ill-equipped to go through."

Dominic Yin, in his description of law enforcement in Hong Kong, reminds us that judicial procedures vary considerably, commenting, "Hong Kong follows common law, a heritage of its colonial past, and would typically look to Commonwealth countries such as the UK, Canada and Australia for legal precedents when so required. American, Japanese and European case law in the area of energy services contracts might be more difficult to call upon."

Barriers

National laws frequently impose a burden on contract language, making them far more complex than they need to be. In some parts of the world, legal guidance is required in structuring contracts. The extent to which contracts are enforced and adjudicated, and the degree to which this constitutes a barrier, is not clear.

Lessons Learned

It would greatly facilitate the growth of the ESCO industry if model contracts, or at least a framework, were made available, with the understanding that such documents would need to be modified to satisfy local laws and conditions. Investigation into the judicial system for contract enforcement would be beneficial. The results of this investigation could then be used to provide some model structure to countries seeking to support this aspect of EPC.

AUDITS

The sense of accomplishment conveyed by our contributors regarding the audits performed offers mixed signals. On one hand, the audits indicate that the knowledge about potential energy savings is being disseminated. It also is indicative of the revenue which potential ESCOs have before them as they prepare to become full-fledged ESCOs.

The ESCO industry has a long, painful history of doing audits that have not been paid for. Historically, they were regarded as a sales tool.

On the other hand, history repeatedly reminds us that audits do not save energy; people do. The information offered in an audit must be acted on if energy is to be saved. Unfortunately, such action typically costs money—money that audit recipients may not have. The obvious answer, germane to our discussion here, is interjecting ESCO services into the process, which can close the loop. In recognizing the value of ESCO services in this regard, however, it is critical to remember that EPC is a lot more than financing an audit.

Lessons Learned

In too many countries, the audit mill is churning out audits that go unattended. Some effort by those international organizations funding the audit should be made to be sure that there is follow-through on the audits as needed. It is also not clear if the audits being conducted are quality investment grade audits (IGAs). Follow-up efforts should ascertain if the energy audits meet IGA standards.

AWARENESS, ASSOCIATIONS AND COMMUNICATION

A review of the country reports shows that public awareness, associations and communications are frequently linked. Associations are often the means of establishing communications among ESCOs and promoting awareness in the customer base.

The vital role associations play in industry development is evident in the description of Austria's ESCO association, DECA, "Originally it was limited to EPC providers, but it has recently opened up to ESC providers and facilitators. Among other activities, DECA serves as an information hub, hosts a regular exchange of experiences for its members, organizes experts meetings, collects market data from its members and lobbies for the industry (www.deca.at)."

One of the strongest statements as to the need to educate customers came from Jan Bleyl and Monika Auer in their discussion of ESCO market development. They observed, "... efficiency markets need "**educated customers**" to demand (services) in the market. Furthermore, even the most "educated" customers will require independent facilitators/intermediaries to support them on their journey through this complex matter."

An awareness of the benefits of EE, and more particularly EPC, is not as prevalent as those in the industry would like. Several contributors lamented that customers still do not understand the need for EE, or the benefits of EPC.

While concerns about the environment have become a driver for EE and EPC, the very emphasis on the environment has typically shifted the focus away from EE to renewables and alternative energy. When the question is posed as to whether to invest in energy efficiency or renewables, the cost-effectiveness issue is seldom addressed.

Lessons Learned

The ESCO industry would benefit from the realization that EE can help pay for the more expensive renewables—and ESCOs offer an excellent model for delivering both in one package.

Concerted efforts to more effectively educate the public on the benefits and need for EE—and for EPC—can be made by trade associations and the government. The Lebanese Center for Energy Conservation, funded by UNDP, has launched a national campaign to promote audits and indirectly EE. The Lebanon report offers samples of the brochure used in this campaign.

FINANCING

Financing remains a pernicious and prevailing concern for ESCOs throughout most of the world. Pradeep Kumar, in describing the ESCO situation in India, reflects this concern observing, "Indian ESCOs face severe financial constraints and large international ESCOs are skeptical about investing in Indian public sector projects due to payment security risks and perceived/real risks about maintaining the sanctity of the legal contract."

Kumar also offers an insightful analysis of the banking sector, commenting, "Access to financing—most lenders and investors have a limited understanding of the dynamics of EPC business models and technical issues. Appraisal of ESCO projects requires different techniques and tools from those involved in the appraisal of projects designed to add to productive capacity. Financial institutions should recognize EE lending as a business strategy and adopt cash flow-based project evaluation for ESCO projects rather than asset/collateral-based lending."

Some countries have shown a preference to structure direct financing or co-financing for ESCOs or their clients in cases where financial institutions are not willing or able to meet the market demand. The Bulgarian Energy Efficiency fund (BgEEF) www.bgeef.com, set up with the support of the World Bank and the Austrian government, has been offering direct financing since 2005 as well as co-financing and partial guarantees, which address these financing barriers to energy efficiency projects. Other government financial support mechanisms may include partial risk guarantees, loan loss reserve funds, special purpose funds or interest credits. In these cases, initiatives focus on the elimination of the financing barrier to ESCO projects.

In addition to providing capacity building for financial institutions and ESCOs, governments may be able to support initial access to financing for ESCO projects through the development of guarantee funds. This mechanism can be designed to cover, in part, the needed guarantees to financial institutions. As examples:

- The Hungary Energy Efficiency Co-Financing Program (HEECP2) aimed at enhancing the financing capacity for EE projects of domestic Hungarian financial intermediaries; and

- In the mid-2000s, the Brazilian government created a partial guarantee facility for EE projects, called PROESCO, where the Brazilian National Development Bank (BNDES) shared up to 80 percent of the credit risk with the remaining 20 percent assumed by the intermediary bank.

As financing becomes available, there is evidence that the industry and its customers are becoming more knowledgeable regarding the benefits of the various EPC models. Sami Siltainsuu and Filip Medhammar observe, "As in the case with other Nordic countries, the guaranteed savings model is the most common in Norway. The shared savings model is almost non-existent. Not only is it due to the fact that it is riskier for vendors, but customers also understand that, due to the greater capital expense and risks involved, they get less with the same investment."

Barriers

Access to financing for EE is a frequent barrier in markets around the world. This becomes even more critical for EPC. Although there is usually sufficient funds available in the marketplace. The link between financial institutions and ESCOs and their project is often missing.

Reflective of the mounting frustration in some countries, Abdelmourhit Lahbabi describes the critical role of financing in Morocco by noting, "Despite the important number of good cost-effective projects and the promotion efforts deployed to attract financial institutions to join the ESCO tour de table, ADS Maroc could not secure the financing resources needed to carry out its ESCO business plan."

Factors which can limit the use of EPC include accounting practices and the way taxes are applied to the agreement by the different parties. Authorities often do not recognize EPC as a conditional sale where sales and income taxes are applied only when the transactions is actually made; i.e., when the payment is made by the end-user to the ESCO. Making sure that the tax authorities recognize EPC can frequently facilitate the initial phase of introducing the concept in a country.

Another important barrier to be addressed is the high transaction costs associated with developing EPC projects. Since the concept is new and quite different from what potential customers are familiar with, there can be a serious lack of customer confidence. In order to address such a barrier, some countries have developed national programs to promote the use of EPC, mainly in the public sector. These programs, like the Federal Energy Management Program (FEMP) in the US and the Federal Buildings Initiative (FBI) program in Canada, promote the concept and also support its use through training and subsidies to cover the initial transaction costs.

Lessons Learned

In addressing financial needs, governments can play a major role. Information and capacity building can offer important solutions to bankers to help them understand EPC and teach ESCOs how to deal with banks or financial balances. Such support can be offered by national energy agencies, energy ministries, dedicated non-governmental organizations (NGOs) or specialized consulting firms. Further, they can be structured to support government initiatives related to EE.

The transition between funding from public or international institutions in the form of loans, funds or subsidies to commercial banks is not easy. The former can usually give better conditions than the latter and might, therefore, compete with each other. It helps if a strategy is adopted to phase out publicly supported financing mechanisms as soon as the commercial banks are able, and willing, to engage in EPC by themselves. Good examples of such an approach is the International Finance Corporation (IFC) CEEF program (Commercializing Energy Efficiency Finance) and the Hungary Energy Efficiency Co-Financing Programme (HEECP). They not only provide credit guarantees to banks and lending institutions, but also offer technical support to the ESCO industry and the financial community. Under these programs, the financial institutions have, over time, demonstrated their ability to be able and willing to lend money to ESCOs on their own.

GOVERNMENT

Government actions revealed in some of the country reports illustrate the vital role which governments can play in fostering EE and the development of EPC. Other country reports indicate that governments, which fail to act, don't foresee the results of unintended consequences. Or, they don't appreciate the opportunity which incentives offer toward the growth of an ESCO industry.

A surprising number of countries seem to be waiting for the government to take some action. The lack of references to ESCOs in some legal systems is viewed with considerable apprehension among potential customers. In some countries, the potential industry, as well as its customers, is waiting for the government to establish guidelines and directives. In some instances, existing laws still make EPC exceedingly difficult and may even prohibit a private body from operating or managing a building's energy service infrastructure. Ironically, experience suggests that government action in some countries will be needed to build reliable and sustainable markets to save the ESCO industry.

Lessons Learned

Public procurement can be a very sensitive issue in EPC development. Since EPC is complex, it requires a deep knowledge of both the public procurement rules in a specific country as well as of the EPC concept itself. Unfortunately, such knowledge is rare in many institutions. To help overcome customer reluctance in such cases, governments can develop standardized procurement and other model documents. This approach also has the added benefit of reducing the cost of developing proposals by ESCOs.

Beyond the development of adapted procurement procedures, the country reports show that governments can also use their own buildings to stimulate the development of a strong EPC market. Such actions by themselves can create an initial market for start-up ESCOs. Finally, governments can use their own financing capacities to cover the needed project financing. In such cases, ESCOs are typically requested to compete based on their respective capabilities to develop the most interesting projects as long as they can offer the needed guarantees for the promised savings.

One of the first actions that a government can take is to review its own legal system and to identify statutes and procedures that could prevent the use of EPC between two consenting parties. Factors which can limit the use of EPC in such a sector include accounting practices and the way taxes are applied to the agreement by the different parties. Authorities do not recognize that EPC is a conditional sale where sales and income taxes are applied only when the transactions is actually made; i.e., when the payment is made by the end-user to the ESCO. Making sure that the tax authorities recognize the status of EPC can frequently facilitate the initial phase of introducing the concept in a country.

There is an evident need in many countries for governments to create a strategy for a sustainable solution that will enable ESCO access to financing for their projects. It would appear to be to the nation's benefit to have government address the development of a mature financial industry that understands and supports EPC. The major problems for banks with a nascent ESCO industry are mainly related to:

- the lack of knowledge about EPC and perceived risks;
- high initial costs and uncertainty about the creditworthiness of ESCOs and/or their clients; and
- limited understanding of EPC projects (not based on asset financing but more on future positive cash flows) generating a lack of internal capability to properly evaluate EPC offers.

CONCLUSIONS

Through our work, we have had the good fortune to exchange ESCO information with many of the prime actors of the EPC market around the world. We have learned much regarding the current status of EPC and on ESCO activities in the far corners of our globe. Evidence of ways to offer innovation in ways to use the EPC approach is certainly in place. It is our hope that *World ESCO Outlook* will help the readers understand how EPC has evolved over the last 30 years and gain ideas for future growth.

Probably the greatest finding in assembling and assessing these reports was the discovery of how rapidly the market is evolving and the many different ways EPC is used in all of these countries. In *ESCOs Around the World,* we stated, "We can say with confidence that EPC has become a clearly established industry with a far reach." That is even more true today.

While it will continue to grow and evolve, EPC will certainly change in character, but we expect that the basic elements of special services and guaranteed results will remain its hallmark. Concerns about climate change and our environment are apt to play a greater role in the use of EPC. We can imagine that the increase in energy costs over time will invite many end-users to inquire about the concept and find it attractive enough to request proposals from ESCOs.

We also believe that governments, while facing so many challenges related to energy security, the environment and budgets, will see the EPC advantage and make the necessary legal modifications to enable all public sector entities to use it more efficiently. And we fervently hope financial institutions will become more comfortable with , thus providing key support for local ESCO industry growth.

Finally, we anticipate that *World ESCO Outlook* will not only be a key industry benchmark, but will encourage the different stakeholders around the world to continue to promote and develop energy performance contracting. It is, after all, the best mechanisms to fund future energy and environmental challenges and to lead us into achieving a healthier sustainable planet.

Appendix A

Contributing Authors

Mr. Valentin ANDRIANOV (Russia)
JSC "ESCO Tyumenenergo"
andrianovvv@esco-te.ru

Mr. Valentin Andrianov holds an M.Sc. in nuclear engineering from the Moscow Engineering and Physics Institute, an M.Sc. in economics from the Higher School of Economics (Moscow) and an executive M.B.A. from the Stockholm School of Economics. He currently acts as the Director, Business Development, at JSC "ESCO Tyumenenergo." He has more than ten years of experience in the energy, municipal housing and utilities sectors. Mr. Andrianov has significant managerial experience in the Russian energy (electricity transmission and distribution) and energy equipment manufacturing sectors.

Currently, Mr. Andrianov is involved in the creation of an ESCO methodology in Russia, energy performance contracting (EPC) approaches and schemes to attract finance for performance contractors. Moreover, he is actively participating in the preparation of educational and training programs for Russian energy market professionals.

Mr. Ugur AKMAN (Turkey)
Dept. of Chemical Eng. Bogazici University
akman@boun.edu.tr

Prof. Ugur Akman obtained his B.Sc. and M.Sc. in chemical engineering from Bogazici University, Turkey, in 1985 and 1987, respectively, and his Ph.D. in chemical engineering from the University of South Florida, USA, in 1991. He is currently a Professor in the Department of Chemical Engineering at Bogazici University. His current research interests include computer-aided process and energy engineering, optimization and process design under uncertainty and risk, heat exchange networks, energy economics and portfolio optimization. He has been following the development of the EPC market in Turkey for many years.

Mr. ALKAM (Jordan)
Energy Management Services Int.
asalkam@ems-int.com

Mr. Alkam is currently CEO of Energy Management Services Int. (EMS), the first ESCO to have been created in Jordan, in addition to his capacity as General Manager of EMS in Jordan and Saudi Arabia. He has a B.Sc. in electrical engineering from the University of Jordan (1993).

Mr. Alkam has over 18 years of experience as an energy management and green building expert serving the government, industrial, commercial and residential sectors of the Middle East region. Mr. Alkam contributed to establishing and operating several ESCOs in Jordan, Lebanon, the United Arab Emirates, Egypt and Saudi Arabia from 1993 to 2011.

Mr. Carlos ALVAREZ (Colombia)
Colombian Energy Efficiency Council
info@cceecolombia.org

Mr. Carlos Antonio Alvarez Diaz holds a business administration degree from the Escuela de Administracion de Negocios in Bogota. He has also completed degrees in financial modeling from the Universidad de los Andes, in marketing and sales from the Autonomous University of Manizales (in association with the University of Cambridge) and in trusteeship from the Andes University.

He currently acts as an energy and finance consultant and advisor with over 25 years of experience. Since 2010, he has been the President of the Colombian Energy Efficiency Council (www.cceecolombia.org), which is following and supporting the development of the ESCO market in Columbia.

Mrs. Maria Cecilia AMARAL (Brazil)
Consultant
mcamaral@expertsmkt.com.br

Maria Cecilia Amaral is currently working as a Consultant Specialist in the energy efficiency sector. Previously, she worked for seven years as Executive Director of the National Association of Brazilian ESCOs (ABESCO) and was in charge of many projects with international organizations such as the Inter-American Development Bank (IDB), the World Bank, the United Nations Development Programme (UNDP) and the Canadian Government.

Mr. ANANTASANTA (Thailand)
Excellent Energy International Company Limited
ruamlarp@eei.co.th

Mr. Anantasanta has worked in the fields of project management and the consulting business since the beginning of his career. He joined Excellent Energy International Company Limited (EEI) in 2005 in marketing and business development BU. His role was to further extend the leading position of EEI in the Thailand ESCO industry by continuously acquiring new ESCO projects. He is also highly recognized by broad local and international agencies as a marketer and business developer in energy efficiency and the ESCO market. He is responsible for managing multiple projects with major sub-projects or interrelated projects as well as for ensuring integration, coordination, timing and consistency throughout the implementation process.

Mrs. Monika AUER (Austria)
Austrian Society for Environment & Technology
monika.auer@oegut.at

Mrs. Monika Auer studied history and political sciences with a focus on environmental politics. Since 1995, she has been an employee of the Austrian Society for Environment and Technology (ÖGUT) and has been sitting on the management board since 2006. For several years now, she has been leading an interdisciplinary working group consisting of major stakeholders in the energy performance contracting industry of Austria.

She has also issued publications with guidelines for municipalities as well as for the business, industry and housing sectors in addition to running the Austrian internet portal for energy performance contracting (www.contracting-portal.at). She carries out research projects for the development of contracting-based business models to be implemented by energy service providers. She has held numerous presentations on energy performance contracting. Since 2005, she has been managing the office of the umbrella association Energy Contracting Austria (DECA), a major player in energy performance contracting.

Dr. Paul BANNISTER (Australia)
Exergy Australia Pty Ltd.
paul@xgl.com.au

Dr. Paul Bannister is Managing Director of Exergy Australia Pty

Ltd. He has a Ph.D. from the Australian National University in the field of solar thermal power following from honors degrees in mathematics and physics from the University of Otago. He has spent his entire professional career in the field of energy management, with a specific focus on commercial sector buildings in Australia and New Zealand. He is considered one of the most knowledgeable experts on the energy performance contracting market in Australia.

Mr. Jan W. BLEYL (Austria,Germany)
Graz Energy agency
Bleyl@grazer-ea.at

Jan W. Bleyl has pursued studies in energy and process engineering as well as in energy economics at TU Berlin and Humboldt State University, California. Over the last 18 years, Jan has acquired knowledge and experience in a broad range of energy (efficiency) systems in national and international context. Since 1994, he has developed and managed the ESCO division of Berlin Energy Agency Ltd. and served as authorized manager (Prokurist).

In 2002, he founded "Energetic Solutions," an independent energy efficiency consultancy and joined Graz Energy Agency Ltd. as senior consultant. Since 2006, Mr. Bleyl leads Task XVI, the Energy Services task of the IEA's Demand-Side Management Implementing Agreement, which brings together ESCO experts from six countries around the world that join forces to advance ESCO models and markets.

Mr. Claudio CARPIO (Argentina)
Consultant
claucarpio@gmail.com

Mr. Claudio Carpio is a chemical and petroleum engineer as well as a specialist in the development of energy efficiency programs, especially in industry. He began his activities in energy efficiency in 1978 at a Repsol oil refinery in Argentina.

Between 1992 and 1999, he acted as Technical Coordinator of the Program of Rational Use of Energy between Argentina and the European Union (Energy Secretariat of Argentina). Between 2000 and 2002, he served in the Department of Sustainable Environment for the Inter-American Development Bank in Washington, DC (USA).

Finally, since 2002, he has been working as an independent consultant for various institutions and companies. He has been following the development of the ESCO market in Argentina.

Mr. Jalel CHABCHOUB (Tunisia)
Econoler
jchabchoub@econoler.com

Mr. Jalel Chabchoub is a senior energy efficiency expert who has been working at the international level for the last 12 years for Econoler. He has taken part in ESCO-related activities in Tunisia for several years as an employee of the first ESCO ever created in Tunisia and as a senior consultant working on the development of energy performance contracting for over eight years in several countries. His accomplishments include setting up ESCO market development strategies, producing ESCO business plans, delivering hands-on training and providing capacity building on the energy performance contracting approach and ESCO operations.

Dr. Braam DALGLEISH (South Africa)
Energy Cybernetics (Pty) Ltd.
mgiazd@puk.ac.za

Dr. Braam Dalgleish holds a Ph.D. in mechanical engineering from the North-West University of South Africa. He has authored and co-authored more than ten local and international articles on energy-related topics including on energy performance contracting. He is currently employed by Energy Cybernetics (Pty) Ltd. He has been involved in the energy field for the past 12 years. He has been granted both the Certified Energy Manager (CEM) and the Certified Measurement and Verification Professional (CMVP) accreditations by the Association of Energy Engineers (AEE). He has been actively involved in energy trainings, energy management systems, measurement and verification as well as energy cost-saving services to industry.

Dr. Alan DOUGLAS POOLE (Brazil)
Consultant
alan.douglas.poole@gmail.com

Dr. Alan Douglas Poole went to Cambridge University, in the UK, and took a bachelor's degree in agricultural sciences and biology in 1971. He joined the Energy Policy Project of the Ford Foundation in 1974, beginning a long career in addressing questions of energy supply and use. He has long studied the factors which influence energy consumption, including energy pricing, economic growth, technology change and policies. A primary focus over the past 15 years has been on how to establish and consolidate an energy efficiency services sector in the context of a developing country such as Brazil. He has worked with many stakeholders in this area and has prepared numerous reports on the energy performance contracting issue for Brazil. He has been part of several initiatives conducted by international organizations, such as the World Bank, to develop the ESCO market in the country.

Mr. Bruno DUPLESSIS (France)
Mines ParisTech
bruno.duplessis@mines-paristech.fr

Mr. Bruno Duplessis is an Assistant Professor at the Center for Energy and Processes, Mines ParisTech. Since 2004, his main research domain has dealt with demand-side management, more specifically the evaluation of energy efficiency services through the development of methodologies and tools for assessing energy savings. His work was rewarded in 2008 with a Ph.D. based upon a thesis titled "Implementation of Energy Performance Contracts for the Improvement of Chilled Water Plants." He is also the co-author of various papers and reports on the subject of energy performance contracting, namely for ADEME, the French Environment and Energy Management Agency, and for the Executive Agency for Competitiveness and Innovation (EACI) of the European Commission within the research program "Intelligent Energy—Europe."

Mr. Pierre EL KHOURY (Lebanon)
Lebanese Center for Energy Conservation
pierre.khoury@lcecp.org.lb

Mr. Pierre El Khoury holds a bachelor's degree in electrical engineering and a master's degree in engineering management, both from the American University of Beirut (AUB). He currently holds the position of Manager for the Lebanese Center for Energy Conservation (LCEC) hosted at the Ministry of Energy and Water and supported by the United Nations Development Programme (UNDP). His main duties include the management of the different aspects of the center including setting up strategies and programs related to the development of energy efficiency and renewable energy markets in Lebanon.

Mr. El Khoury also headed the LCEC team that developed the 2011-2015 National Energy Efficiency Action Plan (NEEAP) for Lebanon in December 2010, making it the first Arab country to develop its NEEAP. He also coordinated with the Central Bank of Lebanon towards the development of the National Energy Efficiency and Renewable Energy Action (NEEREA), a national financing mechanism to boost the work of ESCOs in the energy audit business.

Mrs. Jasmina FANJEK (Croatia)
HEP ESCO
jasmina.fanjek@hep.hr

Mrs. Jasmina Fanjek is one of the Croatian leading experts in energy efficiency with extensive experience in large and comprehensive investment projects related to the energy sector. In 1999, she took the challenge of joining the team that introduced and promoted energy efficiency in Croatia. As a member of the team, she started to work on the preparation of an innovative initiative of the World Bank to create the first ESCO in Croatia together with the World Bank. As a result of this project, a new utility-based ESCO was established in 2003, HEP ESCO Ltd. The long-term business experience that she gained as part of the start-up of the company and its management made her a recognized expert in the region and across Croatia. Since 2011, she has been working as an independent consultant for the region in order to contribute to the promotion of energy efficiency.

Mr. Donald GILLIGAN (USA)
National Association of Energy Service Companies
dgilligan@naesco.org

Donald Gilligan is the President of the National Association of Energy Service Companies (NAESCO), an organization whose ESCO members deliver about USD 5 billion of energy efficiency, demand response, renewable energy and distributed generation projects. He is responsible for coordinating NAESCO's federal and state advocacy activities as well as its relationships with other national and regional energy efficiency organizations. Mr. Gilligan has worked in the energy efficiency industry since 1975 as a consultant, entrepreneur and state government official. He is the author and co-author of a number of reports on energy efficiency and the growth of the ESCO industry that have been published by NAESCO and the Lawrence Berkeley National Laboratory. He is a graduate of Harvard College.

Mr. GO-ACO (Philippines)
Consultant
rgoaco@gmail.com

Mr. Go-Aco is a licensed mechanical engineer with a master's degree in business administration. He is also a Certified Energy Management trainer and a Certified Energy Manager (CEM) under the ASEAN Energy Management Scheme project. He has gained extensive expertise in energy efficiency, energy management and energy financing. Specifically, he has been involved in ESCO operations and in the implementation of energy efficiency financing through energy performance contracting in the Philippines for over ten years.

His proficiency in the ESCO business, energy efficiency and EPC applications has led him to work on key projects for different institutions such as the Asian Development Bank, the World Bank, the United Nations, the KfW Bank of Germany and the International Finance Corporation where he is a consultant for the Sustainable Energy Finance (SEF) program in the Philippines.

Mr. Andres GONZALEZ (Colombia)
Regeneracion Ltda
andres.gonzalez@regeneracion.com.co

Mr. Andres Gonzalez Hassig is General Manager and founding partner of Regeneracion Ltda (www.regeneracion.com.co). He graduated from Universidad de los Andes with a mechanical engineering degree in 2004 and specialized in industrial energy efficiency and power generation. He joined the Colombian Energy Efficiency Council (www.cceecolombia.org), an organization that has been following the development of the ESCO market for many years, where he holds a position on the board of directors as treasurer.

Mr. Marcelo GONZALEZ (Uruguay)
UTE—Administracion Nacional de Usinas y Trasmisiones Eléctricas
mugonzalez@ute.com.uy

Mr. Marcelo Gonzalez has been closely involved in the creation of the first ESCO in Uruguay called USCO, which was created by national Uruguayan electricity company UTE with the support of the World bank. He is involved in all ESCO projects implemented by USCO. Since its creation, USCO has implemented over 20 projects.

Dr. L.J. GROBLER (South Africa)
North-West University, Faculty of Engineering
LJ.Grobler@nwu.ac.za

Dr. L.J. Grobler is co-director of the company Energy Cybernetics CC. He holds CEM and CMVP credentials from the AEE. He is registered as a professional engineer (PrEng) with the Engineering Council of South Africa and received his Ph.D. at the University of Pretoria, South Africa.

He is president of the AEE and Chapter President of the Southern African Chapter of the AEE. He administers the CEM and CMVP training programs for the AEE in Southern Africa.

LJ is also a professor in mechanical engineering at the School for Mechanical Engineering of the North-West University. He has authored and/or co-authored more than 40 local and international articles on energy-related topics. He has also presented and co-presented more than 50 papers at local and international conferences and seminars on various energy topics. He has conducted more than 100 industrial and commercial energy audits. Through his extensive work for many years in the energy efficiency sector, he is one of the most knowledgeable experts in the market of energy performance contracting in South Africa.

Mr. Z'ev GROSS (Israel)
Consultant
gross.zev@gmail.com

Mr. Z'ev Gross has joined the Israel Ministry of National Infrastructures in 2002 and began his involvement in energy efficiency and the energy performance contracting methodology in 2003. He was named acting head of the Energy Efficiency Division of the Ministry in 2004 and officially took over the division in 2007.

Mr. Gross was responsible for managing the National Energy Efficiency program in Israel and was instrumental in promoting both regulatory and project-oriented innovations. He was instrumental in developing a national initiative to introduce energy performance contracting in Israel, which resulted in initial demonstration projects. He was also a key actor in launching many important projects involving different ministries in the country.

Mr. Gross left his post at the Ministry of National Infrastructures in April 2011 and is presently an independent consultant working in different sectors of the clean energy field including support for the launch of the most important EPC public project in the country.

Mr. Jose J. GUERRA ROMÁN (Spain)
Smarteec
jjguerra@unionfenosa.es

Mr. Jose J. Guerra Román holds an executive M.B.A. from the Universidad de Los Andes in Colombia. He is an industrial engineer from the Universidad de Oviedo in Spain and is in his 4th year of mechanical engineering at the University of Strathclyde in the UK.

As a key member of Natural Gas Fenosa, he was involved for six years in the development of the energy efficiency activities of this important energy utility in Spain. In this function, he started one of the first projects in Spain to operate exclusively under the energy performance contracting approach in the hotel sector using the International Performance Measurement and Verification Protocol (IPMVP) for measurement and verification of savings.

He also collaborates in several publications on energy efficiency and participates very actively in different forums to promote ESCOs, market barrier analyses and related solutions in Spain. Recently, he started a new professional challenge as a "smart energy efficiency adviser" while working for multilateral organizations and consulting firms.

Mr. Jesper Rohr HANSEN (Denmark)
Danish Building Research Institute
jer@sbi.dk

Mr. Jesper Rohr Hansen is a Ph.D. student at the Danish Building Research Institute (SBi), Department of Town, Housing and Property. He has an M.Sc. in sociology from University of Copenhagen. His research fields are organizational innovations in the public sector, focusing on leadership and collaborative dynamics as well as the energy efficiency sector in Denmark, including the use of energy performance contracting.

Mr. Eaton H. HAUGHTON (Jamaica)
Caribbean ESCO Ltd.
caribbean_escoltd@yahoo.com

Mr. Eaton H. Haughton is a professional engineer and the Managing Director of Econergy Engineering Services Ltd., Jamaica's first

ESCO. This company is a multifaceted regional leader in maintenance engineering, energy conservation and alternative energy, from which the ESCO Caribbean ESCO Ltd. has evolved.

Mr. Haughton has proven proficiency as a vocational training instructor and is an accepted member of engineering associations such as the American Society of Heating Refrigeration and Air Conditioning Engineers (ASHRAE), the Institute of Plant Engineers (Britain), the Association of Energy Engineers (AEE) and the Jamaica Solar Energy Association, as Charter President. He is one of the most active promoters of the use of energy performance contracting in Jamaica and in the Caribbean.

Mr. Albert HULSHOFF (Netherlands)
NL Agency
albert.hulshoff@agentschapnl.nl

Mr. Albert Hulshoff is a consultant with the NL Energy and Climate division of NL Agency. He is focused on expediting the development of the ESCO market in the Netherlands and is currently conducting a study together with market players on the feasibility of creating a Dutch ESCO platform. In addition to ESCOs, Albert's focus areas include the value of sustainable buildings, the Energy Performance of Buildings Directive (EPBD) and performance contracts.

Mr. Anees IQBAL (United Kingdom)
Maicon Associates Ltd.
AneesIq@aol.com

Mr. Anees Iqbal holds a master's degree in engineering (Southampton University, England). He is currently the head of Maicon Associates Ltd. and has over 40 years of experience in the energy industry, with over 20 in energy performance contracting. He was an Associate Director at Emstar, the Energy Management subsidiary of Shell UK and the first ESCO in the UK, which later became part of Dalkia, the utility arm of French multinational group Vivendi. Furthermore, he has been involved extensively in the promotion of ESCOs in the Czech Republic, Poland, Lithuania, Belarus and Romania.

Mr. Iqbal is currently active as a board member of the Efficiency Valuation Organization (EVO) and is also a member of EVO's Executive Council.

Dr. Jesper Ole JENSEN (Denmark)
Danish Building Research Institute
joj@sbi.dk

Dr. Jesper Ole Jensen has an M.Sc. in civil engineering from the Technical University of Denmark (DTU) and a Ph.D. from Aalborg University. He acts as a senior researcher at the Danish Building Research Institute (SBi), Department of Town, Housing and Property. He has been working as a researcher at SBi and DTU for several years in the field of sustainable housing and cities, from household consumption and sustainable buildings to sustainable urban development. His research themes embrace new types of collaborations shaped by new planning regimes, tools for sustainable buildings and cities, sustainable urban regeneration, facilities management and the use of energy performance contracting.

Dr. Phuong Hoang KIM (Vietnam)
Ministry of Industry and Trade—
Department of Science and Technology
kimph@moit.gov.vn

Dr. Phuong Hoang Kim (Ph.D. in engineering, MA) is the Deputy Director General of the Science and Technology Department of the Ministry of Industry and Trade in Vietnam. His main research interests have been in the field of energy and environment, in which he has over 15 years of experience. He is currently director of a wide range of national projects including the development of ESCOs in Vietnam.

Mr. Pradeep KUMAR (India)
Alliance to Save Energy
pradeep.ase@gmail.com

Mr. Pradeep Kumar has a bachelor's degree in mechanical engineering and is a post-graduate in energy engineering. He has more than 15 years of experience in the field of public and private sector energy efficiency, clean energy, sustainable development and climate change mitigation programs. He is one of the top energy efficiency professionals in projects that require facilitation of energy performance contracts, energy efficiency financing as well as measurement and verification of energy savings. He has been instrumental in providing strategic and technical advice to various governments and international development agencies in relation with the development of ESCOs in India. Finally, he has developed statewide municipal energy efficiency programs in India covering more than 200 municipalities and urban local bodies.

Dr. Abdelmourhit LAHBABI (Morocco)
ADS Maroc
adsmaroc@iam.net.ma

Dr. Abdelmourhit Lahbabi received his engineering degree in industrial processes from Institut National Polytechnique de Toulouse and his Ph.D. in chemical engineering from the University of California in Santa Barbara.

He is a certified engineer and has worked as an international energy and environment expert for numerous international organizations. He is the founder and President of ADS Maroc, created as the first ESCO in Morocco and the only one for a decade.

Dr. Lahbabi conducted more than 100 energy audits and gained extensive hands-on experience working on a wide range of energy efficiency projects. He was directly involved in the implementation of over 30 projects using energy performance contracting in Morocco. Dr. Lahbabi has more than 25 years of professional experience in energy efficiency and renewable energy projects.

Mr. Tamás LÁSZLÓ (Hungary)
Independent Entrepreneur
tamas_laszlo@chello.hu

Mr. Tamás László is one of the key Hungarian experts in the field of energy performance contracting. He has been involved in the early days of energy performance contracting in the country through his work at the Ministry of Industry and Trade where he acted as an officer for energy projects. Since then, he has issued several publications about the principles of third-party financing. He then joined EETEK—Energy Efficiency Technologies Ltd., located in Hungary, as the Head of the Cogeneration division. He was involved in the development and implementation of energy-saving measures based on long-term third-party financing contracts and energy performance contracting.

Mr. Marcel LAUKO (Slovakia)
Energy Centre Bratislava
lauko@ecb.sk

Mr. Marcel Lauko has 11 years of experience in the financial assessment of public sector policies. Over the last five years, he has been focused on policies related to energy efficiency and renewable energy sources. As a representative of Energy Centre Bratislava in several policy committees and working groups, he is responsible for drafting recommendations as well as for commenting all policy documents and legislation related to energy efficiency and renewable energy sources in Slovakia. His activities of interest are related to the promotion of energy performance contracting, especially in the public sector of Slovakia.

Mr. Peter LOVE (Canada)
Energy Services Association of Canada
peter@energyservicesassociation.ca

Mr. Peter Love has an M.B.A. and a B.A. from the University of Toronto and recently completed the Directors Education Program from the Institute of Corporate Directors. For many years, he was Ontario's first Chief Energy Conservation Officer with the Ontario Power Authority where his focus was on providing leadership in electricity conservation and advancing a conservation culture in Ontario. He is also a member of a number of corporate boards and teaches a 4th year course on energy efficiency as an Adjunct Professor at York University's Faculty of Environmental Studies. He currently acts as the President of the newly formed Energy Services Association of Canada, which is the Canadian ESCO association.

Mr. Janusz MAZUR, (Poland)
Przedsiębiorstwo Oszczędzania Energii ESCO sp. z o.o. (POE ESCO)
janusz.mazur@esco.krakow.pl

Mr. Janusz Mazur is a graduate of the Academy of Mining and Metallurgy in Cracow as well as of the School of Business of the University of Economics in Cracow. He is the President of Przedsiębiorstwo Oszczędzania Energii ESCO sp. z o.o. (POE ESCO), located in Cracow, Poland. The company was established in 2000 by the Municipal District Heating Company in Cracow (MPEC SA) at the initiative of the World Bank as one of the first active ESCOs in Poland. Before that, Janusz Mazur was Manager of the Bureau for Strategy and Promotion at MPEC SA—the sole owner of POE ESCO.

Mr. Filip MEDHAMMAR (Norway)
Schneider Electric Buildings Norway AS
filip.medhammar@schneider-electric.com

Mr. Filip Medhammar holds a master's degree in industrial engineering and management from the University of Linkoping, Sweden. He is a Business Developer for Schneider Electric and has been working in the ESCO business for just over two years. In 2009, he joined Schneider Electric and its International Trainee Program, under which he had the opportunity to study and experience the energy performance contracting market in Europe and North America. In that year, he was responsible for carrying through a comprehensive market analysis, revealing the potential of the ESCO business in Norway. Today, Mr. Medhammar is working as Business Developer in Oslo, spreading the message of the benefits of the ESCO model all over Norway.

Mrs. MELLADO (Chile)
Chilean Agency for Energy Efficiency
pmellado@acee.cl

Mrs. Mellado holds a master's degree in commercial engineering (economics) and a bachelor's degree in economics from the University of Chile. She has been Deputy Director of the Chilean Agency for Energy Efficiency (AChEE) since February 2011. She previously worked on the National Energy Efficiency program of the Ministry of Energy as Head of the Industry and Mining division and later for the Market Development division. In that capacity, she actively participated in the design and implementation of programs aimed at fostering the development of the energy efficiency market in various sectors of consumption, particularly related to energy services and energy performance contracting. Among her main achievements was her work with the Global Environment Facility (GEF) where she raised USD 5.5 million in financing for the implementation of projects in order to promote, strengthen and consolidate the energy efficiency market in industry, including through the use of ESCOs.

Dr. Remir MUKUMOV (Russia)
ESCO Tyumenenergo
mukumovre@esco-te.ru

Dr. Remir Mukumov holds a master's degree in economics from the Institute of Engineers of Civil Aviation (Kiev, Ukraine), a diploma with honors from the Russian National Academy of Foreign Trade, Moscow, and a Ph.D. from Tyumen State University. He currently acts as the CEO of JSC "ESCO Tyumenenergo." Dr. Mukumov has more than 15 years of professional experience in the Russian public sector at the municipal, regional and federal levels. Dr. Mukumov has experience in the implementation of energy audits and energy performance contracts in Russia. He actively participated in drafting municipal, regional and federal legislation on various aspects related to economic development, including energy efficiency issues. He supported public-private partnership initiatives in the areas of regional development and economic planning across Russia.

Dr. Chiharu MURAKOSHI (Japan)
Jyukankyo Research Institute Inc.
murakoshi@jyuri.co.jp

Dr. Chiharu Murakoshi, Ph.D., has been in charge of the secretariat of the Japan Association of Energy Service Companies (JAESCO) from 1999 to 2010. He has done research on energy demand, including analyses related to energy consumption, energy-saving and alternative energy technologies, energy efficiency policies, measures against global warming and the ESCO industry. In 1998, he served as committee member of the energy council of the Ministry of Economy, Trade and Industry. He successively held many seats as member of committees concerning energy efficiency standards and programs, energy efficiency regulations for the industrial and commercial sectors, building standards, labeling programs, feed-in tariffs and the ESCO industry.

Dr. Hidetoshi NAKAGAMI (Japan)
Jyukankyo Research Institute Inc.
hnakagami@jyuri.co.jp

Hidetoshi Nakagami, Ph.D., is the President of JAESCO. He founded the Jyukankyo Research Institute Inc. in 1973 and has served as its President from that time to the present. He is a certified and authorized first-class architect and builder and a registered consultant of the World Bank. He regularly lectures on energy situations and strategic options in Japan. He currently sits on the Advisory Committee for Energy Policy, Ministry of International Trade and Industry, the committee for Energy Consumption Labeling of Appliances for the Energy Conservation Center and the Study Committee of Technical Countermeasure for Global

Warming Prevention, Environment Agency of Japan.

Mrs. NIELSEN (Denmark)
Centre for Facilities Management—Realdania Research
sbni@man.dtu.dk

Mrs. Susanne Balslev Nielsen is Deputy of Centre for Facilities Management—Realdania Research and an associate professor at the Technical University of Denmark. She is a civil engineer, specialized in sustainable facilities management and has a Ph.D. in the transition of urban infrastructures. In 2010, she was awarded as "FM Researcher of the Year" by EuroFM. She has been following the development of energy performance contracting for many years in Denmark.

Dr. Esin OKAY (Turkey)
Dept. of Commercial Sciences Istanbul Commerce University
eokay@iticu.edu.tr

Professor Esin Okay received her B.A. in business administration from the Marmara University and completed her Ph.D. at the Banking and Finance Institution of the same university. She has been with the Department of Commercial Sciences at the Istanbul Commerce University as a faculty member since 2002. In recent years, she has focused on energy economics and finance, particularly on a systematic restructuring of the ESCO market in Turkey. Professor Okay is a member of the Professional Risk Managers' International Association (PRMIA).

Dr. Nesrin OKAY (Turkey)
Bogazici University
okay@boun.edu.tr

Dr. Nesrin Okay obtained her B.A. in mathematics from Bogazici University, Turkey, in 1987, and her M.Phil. and Ph.D. in economics from the City University of New York in 1992 and 1993, respectively. She is currently a Professor in the Department of Management and Financial Engineering at Bogazici University. Her current research interests include financial engineering, energy economics, econometric analysis of energy and emission series, volatility and risk modeling as well as the restructuring of the ESCO market in Turkey. Professor Okay gives short courses on the current status of the Turkish economy to various commercial banks as well as to financial and insurance companies.

Mr. Cian O'RIORDAN (Ireland)
PowerTherm Solutions
coriordan@powertherm.ie

Mr. Cian O'Riordan is a chartered engineer with an M.B.A. from University College Dublin. He is Managing Director of PowerTherm Solutions, a provider of energy and maintenance management services. He has 15 years of experience in the energy sector, including nine years in energy management. He has conducted well over 100 technical energy surveys of buildings and industrial facilities, and is particularly proficient in identifying cost-effective opportunities for energy savings through improved control of building services and lighting. He also works as a project manager assisting clients in implementing energy projects and management programs. He has provided advice to the Sustainable Energy Authority of Ireland on ESCOs and energy performance contracts since 2010.

Mr. Javier ORTEGA (Mexico)
Energy Efficiency Consultant
javier-os@hotmail.com

Mr. Javier Ortega Solis received his electrical engineering degree from the Mechanical and Electrical Engineering School (ESIME) of the National Polytechnic Institute (IPN).

He is an Energy Efficiency Consultant. He has developed energy efficiency studies for the German Agency for Technical Cooperation (GTZ) and has been contracted by the United States Agency for International Development (USAID) to develop projects for the National Housing Commission (CONAVI) and Nacional Financiera (NAFIN). He has also participated as a speaker on energy-saving topics in forums promoted by the Federal Electricity Commission (CFE) as well as in forums fostering the use of energy performance contracting.

Mr. Ortega acted for more than 11 years as Market Development and Pilot Programs Manager for the Trust Fund for Electric Energy Saving (FIDE) and was in charge of the design, development and implementation of programs to promote electricity savings in different sectors.

Mr. Boris PETKOV (Bulgaria)
Residential Energy Efficiency Credit Facility
boris.petkov@reecl.org

Mr. Boris Petkov currently serves as Project Manager for the provision of administration, marketing, technical as well as measurement and verification services in the framework of the Residential Energy Efficiency Credit Facility (REECL), a European Bank for Reconstruction and Development (EBRD) facility. He is an energy specialist with an energy engineering and business administration background who has contributed to the international energy efficiency dialogue with a view to formulating energy efficiency policies and measures in 51 country members of the Energy Charter. Mr. Petkov rolled out sustainable energy credit facilities, designed sustainable energy funds, established successful ESCOs and worked on a wide range of energy performance contracting advisory and energy management assignments. From 2005 to 2010, he served as the Bechtel/Nexant team's Project Manager of the Bechtel/Nexant team, the Ukrainian Energy Service Company (UkrEsco) Management Board Co-Executive as well as Finance and Business Development Advisor for a five-year assignment with the EBRD and the EC to establish and operate UkrEsco—the first energy efficiency financing vehicle for energy system upgrades and rehabilitation in the former Soviet Union.

Mr. Ettore PIANTONI (Italy)
Generale Energia S.p.A
epiantoni@genergia.it

Mr. Ettore Piantoni has more than 20 years of experience in the energy industry from various assignments involving project management, marketing, business development and general management. He holds a master's degree in chemical engineering from Politecnico di Milano, Italy. He is currently General Manager of Generale Energia S.p.A. (Genergia), an Italian ESCO, and he is chairman of the European (CEN-CENELEC) and Italian standardization committee (UNI/CTI- CEI).

Mr. Florin POP (Romania)
Energoeco
florin.pop@energoeco.com

Mr. Florin Pop started his career in 2001 as an energy efficiency specialist for EnergoBit Group. He has a deep understanding of energy efficiency projects. In 2008, Florin became General Manager of EnergoBit Group, managing the group's strategy, together with the board of directors and the management team. Additionally, since 2011, he has been the Administrator of EnergoBit ESCO, a company specializing in consultancy for energy efficiency projects and acting as one of the few ESCOs in Romania. This company recently contracted a loan from the EBRD in order to develop energy efficiency projects through the use of energy performance contracting, focusing mainly on the Romanian public sector.

He has been accredited as Certified Energy Manager by the Association of Energy Engineers and is a certified energy auditor from the Romanian Agency for Energy Conservation.

Mr. Frédéric ROSENSTEIN (France)
French Environment and Energy Management Agency
Frederic.Rosenstein@ademe.fr

Mr. Frédéric Rosenstein holds a master's degree in electrical engineering. After a first experience in industry, he joined the French Environment and Energy Management Agency (ADEME) in 2001 where he was responsible for smart grids and demand-side management (DSM). He then developed the activities of ADEME in the field of energy services. He is now in charge of energy performance contracting and smart buildings. He has participated in numerous international collaborations (International Energy Agency) and projects funded by the European Commission.

Mr. Alan RYAN (Ireland)
Sustainable Energy Authority of Ireland
Alan.Ryan@seai.ie

Mr. Alan Ryan holds a primary degree in building services engineering, a master's degree in environmental protection as well as an M.B.A. He is the Program Manager for the Public Sector Program of the Sustainable Energy Authority of Ireland (SEAI). The program provides a number of supports to public bodies in pursuit of the 33 percent energy efficiency improvement by the 2020 target. A key pillar of the program is 'Procurement, Funding and Financing,' assisting public bodies in identifying and implementing new ways of resourcing energy-saving projects. In his seven years at the SEAI, Alan has held program management roles supporting SMEs and large industries as well as promoting energy management best practices. His work at SEAI includes the development of energy performance contracting in Ireland.

Mr. Friedrich SEEFELDT (Germany)
Prognos AG
Friedrich.Seefeldt@prognos.com

Mr. Friedrich Seefeldt studied chemical engineering and energy technology at the Technical University of Berlin (Eng.). He joined the Berlin Energy Agency in 1996 where he developed and implemented energy performance contracting schemes under the Berlin "Energy Savings Partnership (ESP)." The experiences and model contract of these projects were used to develop energy performance contracting schemes in other European countries (Austria, Slovenia, Italy).

Since 2005, Mr. Seefeldt has been working for Prognos, a Swiss think-tank with more than 50 years' tradition of policy consultancy and applied economy research. He is responsible for the unit for energy efficiency, renewable energies and climate protection. He is also a member of the European Council for an Energy Efficient Economy (ECEEE).

Mr. SEELEY (Indonesia and South Korea)
William J. Clinton Foundation
cseeley@clintonfoundation.org

Mr. Seeley has been with the William J. Clinton Foundation since 1997. He is based in Bangkok and manages the Foundation's energy efficiency activities in buildings across Asia covering cities such as Tokyo, Seoul, Hong Kong, Ho Chi Minh City, Bangkok, Jakarta and Singapore. His work includes significant focus on the use of ESCOs and EPC as a delivery model for retrofitting existing buildings.

Prior to joining the Foundation, Christopher worked for a leading ESCO in Melbourne, Australia. He worked on both public and private sector projects to establish partnerships and he developed new projects and strategies to address the reduction of GHG emissions from the built environment. He is currently undertaking his Ph.D. in energy economics.

Mr. Sami SILTAINSUU (Finland, Norway)
Schneider Electric
sami.siltainsuu@schneider-electric.com

Mr. Sami Siltainsuu holds an MS in energy engineering from the Helsinki University of Technology. He has been working for the last ten years for Schneider Electric and is tasked with providing coaching and sales expertise including creating sales tools, sharing best practices and assisting in business planning. His experience also includes the use of energy performance contracting for project implementation in Finland, in neighboring countries as well as at the international level.

Mr. SILVA (Portugal)
ADENE
luis.silva@adene.pt

Mr. Luis Silva is the Director of ADENE. ADENE is the Portuguese energy agency and a non-profit organization with administrative and financial autonomy established in September 2000 by the Portuguese Ministry of Economy. ADENE focuses its activity energy efficiency. It is responsible for the development of the EPC concept in Portugal. Mr. Silva's contribution was made possible thanks to the collaboration of various ADENE experts directly involved in the development of EPC in the country.

Mr. Ivo SLAVOTÍNEK (Czech Republic)
ENESA a.s.
ivo.slavotinek@enesa.cz

Mr. Ivo Slavotínek graduated from the Czech Technical University in economics and energy management. He started his career for SEVEn (Energy Efficiency Center) in 1990 promoting energy efficiency business development in the Czech Republic.

After his involvement in SEVEn, he co-founded the first ESCO in the Czech Republic and negotiated the first EPC contracts in the public and private sectors.

Mr. Slavotínek has been responsible for the development and operation of energy holding company MVV Energie CZ since 1999. In 2006, he co-founded a new private ESCO, ENESA. Under his management, ENESA has become the EPC market leader in the country.

Mr. SOCEA (Romania)
EnergoBIt ESCO
tudor.socea@energobit.com

Mr. Socea Tudor holds an M.B.A. from MIP—Politecnico di Milano. He started his career as a Marketing and Sales Assistant with EnergoBit Prod. His professional background includes the positions of Commercial Manager for EnergoBit Prod and Branch Manager with EnergoBit Constanta. Since August 2011, Tudor has been holding the position of General Manager of EnergoBit ESCO, one of the few ESCOs in Romania.

Mr. Vasily STEPANENKO (Ukraine)
Ecological Systems
tn@esco.co.ua

Mr. Vasily Stepanenko has been CEO of "Ecological Systems" since 1991, one of the first ESCO operations in Ukraine. Since 2002, he has been the editor in chief of the electronic journal "ESCO," making him one of the best sources of information on energy performance contracting in Ukraine.

Mr. Terry STEWART (New Zealand)
Energy Efficiency and Conservation Authority
Terry.stewart@eeca.govt.nz

Mr. Terry Stewart is currently working with New Zealand's Energy Efficiency and Conservation Authority (EECA), managing commercial sector energy efficiency programs. He established a commercial electricity efficiency program for the Electricity Commission (EC) in 2007. The program, set up in consultation with the energy services industry, has been successful in getting uptake of electricity efficiency in commercial buildings and has been recognized as a best practice model in this area. The EC programs were transferred to EECA in late 2010. Since then, Terry has helped develop an EECA strategy for energy efficiency in the New Zealand business sector and has managed the development of new integrated programs. He has been following actively the use of energy performance contracting in New Zealand through his work at EECA.

Mr. Robert TAYLOR (China)
Energy Pathways, LLC
bobtaylor1@me.com

Mr. Bob Taylor is an independent advisor on energy development, focusing especially on energy efficiency in China. He has been closely involved in China since 1981. In his 27-year career at the World Bank, he worked in many countries in the fields of energy supply and demand analyses, rural energy development, electric power development, improved methods of coal use, urban heating, energy efficiency, renewable energy and GHG emissions abatement policies. He developed the World Bank's energy efficiency program and project portfolio in China between 1993 and 2006, where he was instrumental in supporting the development of energy performance contracting in the country. He worked as the Energy Sector Leader for East Asia and the Pacific during the 2006-2009 period. Since leaving the World Bank, Mr. Taylor has worked primarily on energy efficiency programs and projects in China in close collaboration with Chinese government agencies, experts and companies. He is President of his own consulting firm, Energy Pathways, LLC.

Dr. TOBOREK-MAZUR (Poland)
Cracow University of Economics
toborekj@uek.krakow.pl

Dr. Toborek-Mazur is a research worker at the University of Economics in Cracow where she obtained her Ph.D. in economic sciences (2000). She has been active in advising for ESCOs in Poland over the last few years.

Mr. Mike UNDERHILL (New Zealand)
Energy Efficiency and Conservation Authority
Mike.Underhill@eeca.govt.nz

Mr. Mike Underhill has a bachelor's degree in engineering and a master's degree in economics. He has completed the Advanced Management Program at Harvard and is a Fellow of the Institution of Professional Engineers New Zealand. He joined the Energy Efficiency and Conservation Authority (EECA) team as Chief Executive in May 2007. He has extensive management experience in the gas and electricity sector in New Zealand and overseas. Mr. Underhill has been following actively the use of energy performance contracting in New Zealand through his work at EECA.

Mrs. Therése UTSI (Sweden)
Schneider Electric
therese.utsi@schneider-electric.com

Mrs. Therése Utsi has an M.Sc. in economics (School of Business, Economics and Law, University of Gothenburg, Sweden). She was recruited in 2004 as an International Management Trainee at TAC AB, Sweden, and started out working in Dallas, Texas, USA, where she became familiar with the American ESCO market. Therése kicked off her career at the Stockholm office (Sweden) of Schneider Electric (former TAC AB) in 2005 as a Business Developer. She began by developing ESCO projects together with facility owners in the public sector. This gave her extensive experience with the public procurement process, addressing political forums under the framework of a gradually maturing ESCO market.

Mr. Lieven VANSTRAELEN (Belgium)
EnergInvest
lvanstraelen@energinvest.be

Mr. Lieven Vanstraelen has spent the last ten years of his career creating, developing and accompanying businesses in the area of energy efficiency, renewable energy, energy project financing and energy management. From December 2006 to March 2011, he was Managing Director of Fedesco, the first public ESCO in Europe and third-party investor for the 1,650 Belgian federal public buildings. At Fedesco, he was the driving force behind the creation of the Belgian "Knowledge Center for Energy Services and Third-Party Financing." Since April 2011, Mr. Vanstraelen has been co-CEO and shareholder of EnergInvest (Belgium) sprl, a management consultancy company in the EPC/ESCO sector. He is co-founder and President of BELESCO, the Belgian ESCO Association and co-founder of eden (the French energy innovation financing platform) and EMAB (the Energy Managers Association of Belgium). He is Senior Consultant at Fedesco's Knowledge Center as well as expert in the DSMIV Task XVI expert group on Competitive Energy Services of the International Energy Agency (IEA) and the EU-funded "European Energy Services Initiative" (EESI).

Mr. Arthit VECHAKIJ (Thailand)
Excellent Energy International Company Limited
arthit@eei.co.th

Mr. Arthit Vechakij has 12 years of experience in the ESCO business of Thailand. He drove EEI to be selected under the World Bank/ EGAT ESCO Pilot Project (2000). Moreover, Mr. Vechakij has successfully made EEI a well-known ESCO which developed several small/ medium-scale projects (USD 0.5-10 million investment). The latter included renewable cogeneration, energy efficiency and productivity improvement projects (investments by local/foreign financial institutions, renewable/energy efficiency private/public funds and venture capital companies).

Mrs. Alexandra WALDMANN (Slovenia)
Network:GREEN
networkgreen@gmx.eu

Mrs. Alexandra Waldmann has an M.Sc. in geography from the Catholic University of Eichstaett (Germany) and studied at the University of Nottingham (UK).

In 12 years of work experience before becoming co-founder and owner of two companies, she worked at the Berlin Energy Agency, promoting the use of energy performance contracting in Germany and abroad. Alexandra acquired extensive knowledge and experience in the fields of sustainable development, environmental management, climate protection at local level as well as in the use of energy performance contracting. Since her involvement with the Berlin Energy Agency, Alexandra is mostly dedicated to the field of energy performance contracting, developing and implementing renowned projects such as EUROCONTRACT.

Mr. James WAKABA (Kenya)
GVEP International
james.wakaba@gmail.com

Mr. James Wakaba is an engineer with 20 years' industry and management experience. He worked for five years in energy efficiency as a consultant conducting energy audits, providing energy policy consultancy, delivering training and participating in project management. He helped set up a center for energy efficiency and energy conservation at the Kenya Association of Manufacturers, being part of an initiative to create the first ESCO operation in Kenya. He also contributed to developing a curriculum for energy manager training at master's degree level at the University of Nairobi. He is currently working to increase energy access to rural and peri-urban poor in various African countries.

Dr. Ji XIAN (Japan)
Jyukankyo Research Institute Inc.
genki@jyuri.co.jp

Dr. Ji Xian, Ph.D., is a researcher at the Jyukankyo Research Institute Inc. She has a doctoral degree in environmental engineering (awarded by Kitakyushu University). Her research areas include energy consumption surveys in housing and commercial buildings, energy efficiency polices and strategies as well as the use of energy performance contracting in Japan.

Mr. Dominic YIN (Hong Kong)
HAESCO
dyeppa@gmail.com

Mr. Dominic Yin is Founder and current Chairman of the Hong Kong Association of Energy Service Companies (HAESCO). Mr. Yin is also CEO of the Greater China Sustainable Council as well as Chairman and CEO of Hong-Kong Eco-energy Limited (EESCO P2E2 Hong Kong Ltd.).

Mr. ZAKHAROV (Russia)
European Bank for Reconstruction and Development
alexeipzakharov@gmail.com

Mr. Zakharov is a Program Consultant with the Department of Energy Efficiency and Climate Change of the European Bank for Reconstruction and Development. He has more than eight years of policy-related advisory work, including for the three past years in the energy efficiency sector as an advisor to private sector companies and the International Finance Corporation (Word Bank Group) with a particular focus on public sector energy performance contracting.

Dr. Jean-Marc ZGRAGGEN (Switzerland)
Services industriels de Genève (SIG)
jean-marc.zgraggen@sig-ge.ch

Dr. Jean-Marc Zgraggen holds a degree in physics from the University of Geneva, a master of advanced studies in energy from the Ecole Polytechnique Fédérale de Lausanne (EPFL) and a Ph.D. in interdisciplinary sciences from the University of Geneva.

Since 2009, he has been heading the energy efficiency unit at Services industriels de Genève (SIG), the Geneva energy utility. His team provides clients with solutions to improve their energy efficiency. He is strongly involved in developing new energy services based on the ESCO model for SIG.

Appendix B

Esco Associations

Below is a list of ESCO associations active around the world. It is to be noted that the following list is not exhaustive.

Australia

Australasian Energy Performance Contracting Association (AEPCA)
Level 4, 40 Albert Road
South Melbourne 3025 Australia
Tel.: 03 8807 4650
http://www.climatechange.gov.au/government/submissions/~/media/submissions/
building-energy/aepca.ashx

Austria (and Germany)

Dachverband der Osterreichischen Contractoren (Umbrella Association of Austrian ESCOs)

Belgium

Belgian ESCO Association (BELESCO)
107 rue Joseph Coosemansstraat
Brussel 1030 Bruxelles
Tel.: +32 (02)737 91 19
Fax: +32 (02) 735 30 97
www.belesco.be

Brazil

Brazilian ESCO Association (ABESCO)
Avenida Paulista, 1313 - cjto 908
Brazil
Tel.: (11) 3549-4525
www.abesco.com.br/

Canada

Energy Services Association of Canada
34 King Street, Suite 600,
Toronto (Ontario) M5C 2X8
Tel.: (416) 644-1788
www.energyservicesassociation.ca

Chile

ANESCO CHILE
1363 Avenue Of. 1404 - Providencia,
Santiago Chile
Tel.: +562 433 4406
Fax: +562 264 0213 1404
www.anescochile.cl

China

China Energy Management Company Association (EMCA)
5th Floor, SINOCHEM Tower, A2, Fuxingmenwai Dajie,
Beijing,China (100045)
Tel.: 8610-63600181 ,63600015 ,63600457 ,63601432
Fax: 8610-63600459
http://www.emca.cn/bg/en/

Colombia

Consejo Colombiano de Eficiencia Energetica; Colombian Council for Energy Efficiency (CCEE)
Carrera 16 No. 86 A 53 Ofc. 404
Colombia
Tel.: +13108650283
http://www.cceecolombia.org/

Czech Republic

Czech Association of Energy Service Companies (APES)
APES, U Voborníků 852/10, 190 00
Praha 9, Czech Republic
Tel.: +420 286892687
Fax: +420 286892683

Europe

European Association of Energy Service Companies
Boulevard A. Reyers 80
1030 Brussels, Belgium
Tel.: +32 2706 8201
http://www.eu-esco.org/

EFIEES, European Federation of Intelligent Energy Efficiency Services
Rue Philippe le Bon 15
1000 Brussels, Belgium

Tel.: +32 (0)2 230 65 50
Fax: +32 (0)2 230 73 79
http://www.efiees.eu/en/qui_sommes_nous_site857.html

Finland (and Norway)

MOTIVA (EE org, not ESCO org)
http://www.motiva.fi/en/

France

the Federation of Energy-Environment Services, FEDENE (FG3E); Association of Energy Efficiency
rue de la Nursery 75008 Paris
Tel.: 01 44 70 63 90
Fax: 01 44 70 63 99
www.fedene.fr

Germany

ESCO Forum (former Bundesverband Privatwirtschaftlicher Energie-Contracting-Unternehmen e.V.) and Contracting Forum (German Electrical and Electronic Manufacturers' Association or Zentralverband Elektrotechnik- und Elektronikindustrie e.V. - ZVEI) Lyoner Straße 9
60528 Frankfurt am Main
Charlottenstr. 35/36
10117 Berlin
Tel.: (0) 30 306960-26
http://www.zvei.org/fachverbaende/energietechnik/esco_forum/kontakt/
http://www.energiecontracting.de/03_pressecenter/03_presse_kontakt.php?sektor=
Kontakt

Hong Kong

Hong Kong Association of Energy Service Companies (HAESCO)
Room 1801, Wing On Central Building,
26 Des Voeux Road Central,
Central, Hong Kong
Tel.: +(852) 2961-4863
Fax: +(852) 3007-1957
www.haesco.hk

India

Indian Council for Promotion of Energy Efficiency Business (ICPEEB)

http://www.icpeeb.org:8080/

Italy

Associazione Imprese di Facility Management ed Energia; Association of Facility Management and Energy Services Companies (AGESI)

Via Grigna 9, I-20155 Milan, Italy

Tel.: (+39) 023 925 200

Fax: (+39) 0239269016

www.agesi.it

Associanziaone Nazionale Societi Servizi Energetici (ASSOESCo)

HQ: Viale Bruno Buozzi, 19 / A -00197 Rome; Head Office: Loc Ribrocca, snc -

15057 Tortona (AL), Italy

Tel.: + 390131810346

Fax: +390131810270

www. assoesco.it

Federazione delle Associazioni Nazionali delle Industrie meccaniche ed affini (ANIMA-ITALCOGEN)

via Scarsellini 13 - 20161 Milan, Italy

Tel.: 80067530156

http://www.italcogen.it

La Federazione Nazionale Imprese Elettrotecniche ed Elettroniche (ANIE Federation)

Tel.: +39 02 3264.213

Fax: +39023264395

http://www.anie.it

Associazione nazionale delle Imprese Elettriche (ASSOELETTRICA)

via Benozzo Gozzoli, 24 -

00142, Roma, Italy

Tel.: +39 06 8537281

Fax: +39 06 85356431

http://www.assoelettrica.it

FEDERUTILITY
Piazza Cola di Rienzo, 80A 00192
Roma, Italy
Tel.: +39 06 945282.10 to 20
Fax: +39 06 94528200
http://www.federutility.it/

Associazione Italiana per la Promozione della Cogenerazione; Italian Association for the Promotion of Cogeneration (COGENA)
Via Isonzo, 34 Roma 00198 ,
Rome 00198, Italy
Tel.: +39 06 20369638
Fax: +39 06 20369376
http://www.federutility.it/

Federazione Nazionale delle Esco (FederESCO)
Via Po, 2 - Localita' Pian dei Mori, Italy
Tel.: +39 0577 392248
Fax: +39 0577 394285
www.federesco.org

Japan

Japanese Association of Energy Service companies (JAESCO)
Kioi-cho Fukuda Bldg. 3F, 3-29, Kio-cho, Chiyoda-ku, 102-0094
Tokyo, Japan
Tel.: 81-3-3234-2228
Fax: 81-3-3234-2226
http://www.jaesco.or.jp/english/

Macau

Macau Energy Saving Association (MESA)
1023 Avenida de Amizade, 1 andar AC Edf.
Nam Fong, Macau
Tel.: 853285976300
Fax: 85328974553
http://www.esamacau.org/

Malaysia

Malaysian Association of ESCO (MAESA)

Portugal

Associação Portuguesa de Empresas de Serviços de Energia (APESE)
Rua da Constituição, no. 2105, 2nd floor, fracção BU, 4250-170 PORTO, Portugal

Singapore

Energy Sustainability Unit
Block SDE 1, Level 3, Room # 03-05, Department of Building, School of Design and Environment, National University of Singapore, 4 Architecture Drive Singapore 117566
Tel.: +65 6516.3443
Fax: +65 6773.3837
http://esu.com.sg/
http://www.esu.com.sg/faqs.html#g1

South Africa

South African Association of Energy Services Companies (SAAEs)
PO Box 40684 GARSFONTEIN EAST
Tshwane 0060 South Africa
PO Box 40684
GARSFONTEIN EAST
Tshwane
0060
South Africa
Tel.: +27 (0)708 3835
http://www.esco.org.za/

South Korea

KAESCO—Korean Association of ESCO Companies
702, Sarnilplaza B/D 837-26 Yeoksam-dong
Gangnam-ku, Seoul 135-768. Korea
Tel.: 82 2 2052 5780
Fax: 82 2 2052 5779
http://www.esco.or.kr

Spain

Asociación de Empresas de Mantenimiento Integral (AMI)
Guzmán el Bueno, 21 - 4º dcha. 28015 Madrid
Tel.: 91 277 52 38
http://www.amiasociacion.es/

Asc. Empresas de Eficiencia Energética (A3e)
Doctor Arce, 14. Madrid 28002
Tel.: 917 610 250
www.asociacion3e.org

Asociación de Empresas de Servicios Energéticos (ANESE)
C/ Velázquez 53, 2ºI. 28001 Madrid
Tel.: 91 1310615
www.anese.es

Taiwan

6F.-3, No.48, Baoqiao Rd., Xindian Dist., New Taipei City 23145
Taiwan 24891
Tel.: 886 2 86650826
Fax: 886 2 86650825
http://www.taesco.org.tw/

Thailand

Thailand ESCO Association
Queen Sirikit National Convention Center, Zone D, 3rd Floor, 60 New Ratchadapisek Road, Klongtoey, Bangkok 10110
Tel.: 0 2345 1250-55
Fax: 0 2345 1258 to 9
http://www.taesco.org.tw/

United Kingdom

Energy Services and Technology Association (ESTA)
252A High Road, BENFLEET, Essex SS7 5LA
Tel.: 01268 569010
Fax: 01268 569737
http://www.esta.org.uk/

USA

National Association of Energy Service Companies (NAESCO)
1615 M Street, NW, Suite 800, Washington, DC 20036
Tel.: 202 822-0950
Fax: 202 822-0955
http://www.naesco.org

Appendix C

Glossary of Acronyms

ACRONYMS

AEE	Association of Energy Engineers
ADB	Asian Development Bank
AFD	French Development Agency
ANME	National Agency for Energy Conservation
BAU	business-as-usual
BOO	build, own, operate
BOOT	build, own, operate, transfer
CDM	Clean Development Mechanism
CEM	Certified Energy Manager
CHP	combined heat and power (cogeneration)
CIDA	Canadian International Development Agency
CMVP	Certified Measurement and Verification Professional
DBFMO	design, build, finance, maintain and operate
DBO	design, build, operate
DSM	demand-side management
EBRD	European Bank for Reconstruction and Development
EC	European Commission
EE	energy efficiency
EPC	energy performance contracting
ES	energy service
ESC	energy supply contracting
ESCO	energy service company
ESP	energy service provider
EU	European Union
EVO	Efficiency Valuation Organization
FBI	Federal Buildings Initiative
FDA	French Development Agency
FM	facilities management
GEF	Global Environment Facility
GHG	greenhouse gas
HVAC	heating, ventilation and air conditioning
IADB	Inter-American Development Bank
IFC	International Finance Corporation

IPMVP	International Performance Measurement and Verification Protocol
IRR	internal rate of return
JV	joint venture
LED	light-emitting diode
LFI	local financial institution
M&E	monitoring and evaluation
M&EPC	maintenance and energy performance contract
M&V	measurement and verification
MoU	memorandum of understanding
MUSH	municipalities, universities, schools and hospitals
NEEAP	national energy efficiency action plan
NGO	non-governmental organization
O&M	operation and maintenance
OECD	Organization for Economic Co-operation and Development
PDCA	plan-do-check-act
PJ	petajoules
PPA	power purchase agreement
PPP	public-private partnership
PV	photovoltaic
RE	renewable energy
RES	renewable energy source
RFP	request for proposals
RFQ	request for qualifications
SEC	Securities and Exchange Commission
SWH	solar water heating
SME	small and medium enterprise
SPV	special purpose vehicle
T&D	transmission and distribution
toe	ton of oil equivalent
TPF	third-party financing
TWH	terawatt hours
UN	United Nations
UNDP	United Nations Development Programme
UNFCCC	United Nations Framework Convention on Climate Change
USAID	United States Agency for International Development
VAT	value added tax
VSD	variable-speed drive
WRI	World Resources Institute

References by Chapter

CHAPTER 2

[1] Private conversations with PacifiCorp personnel

CHAPTER 4

DAWSON, Roger. *Secrets of Power Negotiating*. 1995. Career Press, Hawthorne, New Jersey

CHAPTER 5

Hansen, Shirley J. and James W. Brown, *Investment Grade Energy Audits: Making Smart Energy Choices*. 2004. The Fairmont Press, Lilburn, GA Making the Business Case

CHAPTER 6

Third World Center for Water Management

Energy Efficiency Absorbs Water, REW Energy Efficiency Markets Blog http://www.realenergywriters.com/ee-blog 5/12/2011

HANSEN, Shirley. *Sustainability Management Handbook*, 2011. The Fairmont Press, Lilburn Georgia.

http://finance.yahoo.com/news/Dow-Water-Process-Solutions-bw-201925564.html?x=0 6/16/2011

"Toiler Water Used for Drinking Getting over the "Yuck Factor,"

http://www.sustainablefacility.com/Articles/Industry_Watch/BNP_GUID_9-52006_A_10 4/21/2011

CHAPTER 9

Australia

[1] www.530collinsstreet.com.au

[2] www.nabers.com.au

[3] www.gbca.org.au

[4] www.cleanenergyfuture.gov.au

[5] http://www.ret.gov.au/energy/efficiency/eeo

[6] www.lowcarbonaustralia.com.au

[7] www.environment.nsw.gov.au/grants

[8] www.cbd.gov.au

[9] www.dtf.vic.gov.au

[10] www.climatechange.qld.gov.au
[11] www.melbourne.vic.gov.au/1200buildings

Austria

[1] www.IEADSM.org => Task XVI
www.contracting-portal.at
www.big.at/umwelt-soziales/contracting (Bundes-Contracting
www.esv.or.at, www.Grazer-EA.at
www.deca.at. the Austrian ESCO association

Brazil

[1] The estimate for 1996 is from Poole & Geller (1997), while that for 2009 is from Gonçalves & Associados (2010). Unfortunately, there are no surveys in between which provide a basis for estimates of total project volume. The survey performed for 2009 covered 32 out of a universe of 70 ESCOs. The high estimate simply extrapolates the values obtained in the survey. The low estimate assumes that the proportion of medium and large ESCOs in the survey is larger than it is in total, which was taken to be: small = 60%, medium = 20%, large = 20%.

[2] Based on estimates in Gonçalves & Associados (2010), the average payback period would be less than 2.5 months, although the estimate of savings achieved may be rather high (which would shorten the payback period).

[3] Large consumers have mostly shifted to the "free market," which became significant from 2003 and now accounts for 17 percent of total consumption. Many of them contracted supplies with eight-year contracts when there was excess capacity and prices were low. Now that these consumers are renewing their free market contracts, the prices are much higher.

[4] The results of this survey broadly confirm an earlier survey performed by ABESCO for the World Bank/UNEP in 2005 (ABESCO, 2005), which also highlighted the importance of the guarantees required to obtain credit.

[5] The BNDES is the ultimate source for almost all medium-term commercial bank debt financing in Brazil.

ABESCO *Análise dos Resultados da Pesquisa das Empresas de Serviços de Eficiência Energética no Brasil;* Report to the World Bank & UNEP under the project "Developing Financial Intermediation Mechanisms for Energy

Efficiency Projects in Brazil, China and India," February, 2005

Gonçalves & Associados: *Estudo do Perfil e Tendências do Mercado Nacional de ESCOs—Síntese dos Resultados*; Report to GIZ/Programa de Energia Brasil-Alemanha, August, 2010.

Poole, A.D. & Geller, H.; *The Emerging ESCO Industry in Brazil*, Instituto Nacional de Eficiência Energética and the American Council for an Energy Efficient Economy, Rio de Janeiro & Washington DC, 1997

Canada

[1] http://www.tbs-sct.gc.ca/pol/doc-eng.aspx?id=14494&section=text#appO Appendix O—Federal Building Initiative—Energy Management Published March 6, 1998

Chile

[1] PROFO (which stands for Proyectos Asociativos de Fomento or Associative Projects for Development) is a CORFO instrument that stimulates the association of companies with a common business idea so they can share information about markets and obtain access to competitive resources.

[2] Although the company Johnson Controls International (JCI) has annual net sales close to USD 30 million, as an ESCO, the company reports sales of USD 1.5 million, which classifies it within the ESCO market as a medium-sized company.

[3] Fundación Chile, September 2007, "Development of Proposals for Sector Instruments and Models for Clean Energy and Energy Efficiency" (Elaboración de propuestas de instrumentos y modelos sectoriales para energía limpia y eficiencia energética), prepared by Econoler International.

[4] PPEE, 2010, "Improving Financial Instruments to Support Chilean Energy Efficiency" (Mejorando los Instrumentos Financieros de Apoyo a la Eficiencia Energética Chilena), prepared by consulting firm Dalberg.

China

[1] The NDRC/WB/GEF China Energy Conservation Project Management Office published a definitive collection of 357 case studies of energy performance contracting projects implemented by the three pilot ESCOs in Energy Conservation Project Case Studies of Chinese ESCOs (in Chinese, China Economic Publishing House, 2006).

[2] See World Bank, China Energy Conservation Project Implementation Completion and Results Report (World Bank, December 2007).
[3] See World Bank, Second Energy Conservation Project Implementation Completion and Results Report (World Bank, December 2010).
[4] www.emca.cn

Colombia

UPME, 2002, Evaluación del potencial y estructura del mercado de servicios de uso racional y eficiente de energía.

CEPAL, 2009, Situación y perspectivas de la eficiencia energética en América Latina y el Caribe

Denmark

Aftale af 20. November 2009 mellem klima- og energiministerien og net- og distributionsselskaberne indenfor el, naturgas, fjernvarme og olie repræsenteret ved Dansk Energi, Dansk Fjernvarme, Foreningen Danske Kraftvarmeværker, HNG/Naturgas Midt-Nord, DONG Energy, Naturgas Fyn samt Energi- og Olieforum om selskabernes fremtidige energispareindsats. Location:

danskenergi.dk/~/media/Energieffektivitet/Underskrevet_aftale_2010.pdf.ashx

Bertoldi, P., Boza-Kiss, B., Rezessy, S. (2007), Latest Development of Energy Service Companies across Europe—A European ESCO Update. European Commission—Institute for Environment and Sustainability, Italy.

The Danish Energy Saving Trust, 2011: Første ESCO-projekt i den almene boligsektor. Location: www.savingtrust.dk/

Energibranchen (2008), Sælg resultater—ikke udstyr. Energy Performance Contracting—modeller for finansiering af energibesparelser. Dansk Industri, Copenhagen

IDA (Ingeniørforeningen i Danmark) (2010) Survey on energy savings in the municipalities with ESCO. November 2010

Jensen, J.O.; Oesten, P.; Balslev Nielsen, S. (2010) ESCO as Innovative Facilities Management in Danish Municipalities. Paper presented at the 9th EuroFM Research Symposium, EFMC2010, Madrid, Spain.

France

[1] Fédération des services énergie environnement (energy and environment service federation), http://www.fedene.fr.
[2] Les services d'efficacité énergétique, Club S2E, Guide à l'attention des clients privés et publics, www.clubs2e.org, June 2007.

[3] Mesure et vérification de l'efficacité énergétique, Club S2E, février 2009.

Germany

[1] Gesetz über die energetische Modernisierung von vermietetem Wohnraum"

Mietrechtsänderungsgesetz—MietRÄndG: Gesetz über die energetische Modernisierung von vermietetem Wohnraum" [Referentenentwurf BMJ vom 25.10.2011]

Prognos 2010:_Rolle und Bedeutung von Energieeffizienz und Energiedienstleistungen in KMU, final report for KfW Bankengruppe.

India

[1] BEE website (2011)

[2] http://www.powermin.nic.in/acts_notification/energy_conservation_act/index.htm

[3] WRI news release

[4] Rajiv Garg, "Implementation of Energy Conservation Act and BEE Action Plan," Available at http://www.energymanagertraining.com/Presentations2008/3L_2008Sep3_PulpandPaper/list.htm (accessed on October 22, 2008).

[5] WRI report POWERING UP-The Investment Potential of Energy Service Companies in India.

[6] EESL/WRI

[7] WRI NEWS

[8] http://www.energymanagertraining.com/NAPCC/main.htm

[9] BEE

Japan

Vine. E, Murakoshi. C and Nakagami. H: The evolution of the US energy service company (ESCO) industry: from ESCO to Super ESCO, Energy Vol.24,1999

Murakoshi. C, Nakagami. H and Sumizawa. T: Exploring the feasibility of ESCO business in Japan—demonstration by experimental study, In proceedings of the ACEEE 2000 Summer Study on Energy Efficiency in Buildings, 2000.8

Murakoshi C and Nakagami. H: Recent activities of Japanese ESCO industry. In proceedings of Improving Energy Efficiency in Commercial Building Conference 2004 (IEECB 2004), 2004.4

Murakoshi. C, Nakagami. H and Masuda. T :Detailed Analysis of the ESCO Market in Japan: Based on JAESCO Survey., in proceedings of the ACEEE 2004 Summer Study on Energy Efficiency in Buildings, 2004.8

Murakoshi. C and Nakagami. H: Current state of ESCO activities in Asia: ESCO industry development programs and future tasks in Asian countries. In proceedings of the ECEEE 2009 Summer Study, 2009.6

Morocco

[1] Average over the 2003-2008 period.

[2] www.adsmaroc.com

[3] Gestion de l'Energie dans les entreprises Marocaines

[4] www.siem.ca

[5] Published in Arabic in the Bulletin Officiel of October 24, 2011

Netherlands

www.agentschapnl.nl/esco

www.rotterdam.nl/groenegebouwen

New Zealand

BRANZ. (2010). Building Energy End Use Study Year 3. Wellington: Building Research Association New Zealand.

Energy for Industry. (2010). Case Study: Cogeneration Plant. Retrieved Sept 29, 2011, from Energy for Industry: http://energyforindustry.co.nz/experience/cogeneration-plant/

KEMA. (2007). New Zealand Electric Energy-Efficiency Potential Study Volume 1. Wellington, New Zealand: Electricity Commission.

Ministry Economic Development. (2011, July 13). New Zealand Energy Data File 2011. Retrieved September 15, 2011, from Ministry Economic Development: http://www.med.govt.nz/templates/MultipageDocumentTOC____46119.aspx

Ministry Economic Development. (2011, Aug). New Zealand Energy Strategy 2011-2021 and the New Zealand Energy Efficiency and Conservation Strategy 2011-2016. Retrieved Sept 15, 2011, from Ministry Economic Development: http://www.med.govt.nz/upload/77402/NZ%20Energy%20Strategy%20LR.pdf

Ministry Economic Development. (2011, June 11). Table 1.1b: Electricity Consumer Prices (Real 2010). Retrieved September 13, 2011, from Ministry Economic Development: http://www.med.govt.nz/templates/MultipageDocumentTOC____21628.aspx

Ministry for Environment. (2009, July). New Zealand's 2020 Emissions Target INFO422. Retrieved September 16, 2011, from Ministry for Environment: http://www.mfe.govt.nz/publications/climate/nz-2020-emissions-target/html/index.html

New Zealand Department of Statistics. (2011, September 13). Population Clock. Retrieved September 13, 2011, from Statistics New Zealand: http://www.stats.govt.nz/tools_and_services/tools/population_clock.aspx

NZ Dept. of Statistics. (2011). New Zealand in Profile 2011. Retrieved Sept 15, 2011, from New Zealand Department of Statistics 2011: http://www.stats.govt.nz/browse_for_stats/snapshots-of-nz/nz-in-profile-2011/population-demography.aspx

Philippines

[1] http://www.doe.gov.ph/neecp/escoslist.htm

[2] http://www.doe.gov.ph/EE/Accredited%20Companies.htm

[3] Lawrence Fernandez, Senior Manager and Head of Utility Economics, MERALCO.

[4] Republic of the Philippines
Leverage International, 1997, "Philippine New Commercial Building Market Characterization," Final Report, Manila; Busch, John and Deringer, Joseph, 1998, "Experience Implementing Energy Standards for Commercial Buildings and its Lessons for the Philippines," November, Lawrence Berkeley Laboratory Report No. 42146.

Portugal

[1] Energy Service Companies Market in Europe—Status Report 2010, by Angelica Marino, Paolo Bertoldi, Silvia Rezessy.

Russia

[1] Energy Efficient Russia Portal, http://energosber.info/news/detail.php?ID=65225, accessed Oct. 14, 2011.

[2] Article 72 (3) of the Budget Code.

[3] Municipal unitary enterprises (MUPs) that overwhelmingly operate municipal infrastructures in Russia (district heat generation assets, utility supply networks or street lighting) are exempt from this law although these enterprises are in most cases almost fully funded from the public budgets. Despite this fact, municipal authorities may mandate that a MUP procure goods and services under Law 94

regulations, and many do so.
[4] See Annex 6 to the Contract, available at http://www.economy.gov.ru/wps/wcm/connect/f6600c0045d341a0b89cfc293491a18d/contract.doc?MOD=AJPERES&CACHEID=f6600c0045d341a0b89cfc293491a18d.

Slovenia

[1] The municipality of Kranj following the model applied in the City of Berlin (Germany).
[2] http://www.aure.gov.si/eknjiznica/156-AN_1.pdf, p. 110.
[3] ZRMK, personal conversation.
[4] http://www.odyssee-indicators.org/publications/PDF/slovenia_nr.pdf

South Africa

www.eskomidm.co.za
www.esco.org.za
www.eskom.co.za/c/article/238/energy-efficiency

Spain

[1] Available at http://www.idae.es.
[2] Mainly, the Spanish Association of Enterprises of Complex Maintenance of Buildings, Infrastructures and Industries (http://www.amiasociacion.es), and, among others, the Association of Energy Savings Companies (http://www.anese.es) and the Association of Enterprises of Energy Efficiency (http://www.asociacion3e.org).
[3] ESCO definition in European Directive 32/2006: ESCO—a natural or legal person that delivers energy services and/or other energy efficiency improvement measures in a user's facility or premises, and accepts some degree of financial risk in so doing. The payment for the services delivered is based (either wholly or in part) on the achievement of energy efficiency improvements and on the meeting of the other agreed performance criteria.
[4] www.amiasociacion.es
[5] www.anese.es
[6] www.asociacion3e.org
[7] www.evo-world.org
[8] Three Iberostar Hotels in the Canary Islands, granted by the Government of the Canary Islands.

[9] Extracted from the Institute for Energy Diversification and Saving.

Switzerland

[1] http://www.swisscontracting.ch
[2] http://www.evo-world.org
[3] http://www.energho.ch
[4] "Statistique globale suisse de l'énergie 2010," Swiss Federal Office of Energy (SFOE), 2010, http://www.bfe.admin.ch.
[5] "Directive of the European Parliament on Energy End-Use Efficiency and Energy Services," http://europa.eu/legislation_summaries/energy/energy_efficiency/l27057_en.htm.
[6] "Action Plan 2008 for Energy Efficiency," Swiss Federal Office of Energy, http://www.bfe.admin.ch.
[7] "Energy Strategy 2050," Swiss Federal Office of Energy, http://www.bfe.admin.ch.
[8] "CO_2 Act," Swiss Federal Office of Energy, http://www.bfe.admin.ch.
[9] "Cost-Covering Remuneration for Feed-in to the Electricity Grid," Swiss Federal Office of Energy, http://www.bfe.admin.ch.
[10] http://en.wikipedia.org/wiki/2000-watt_society
[11] http://www.sia.ch/, see for instance SIA 380/1 (heating) and SIA 380/4 (electricity).
[12] http://www.bfe.admin.ch
[13] http://www.aenec.ch
[14] http://www.susi-partners.ch
[15] http://www.leprogrammebatiments.ch
[16] http://www.bfe.admin.ch/prokilowatt
[17] http://www.klimastiftung.ch/climate_foundation.html

Turkey

Aydin, L. and M. Acar, 2011. Economic impact of oil price shocks on the Turkish economy in the coming decades: A dynamic CGE analysis, Energy Policy 39: 1722-1731.

Erdogdu, E., 2010. Turkish support to Kyoto Protocol: A reality or just an illusion, Renewable and Sustainable Energy Reviews 14: 1111-1117.

Gund, L., H. Guven, and A.O. Memik, Book section on Turkey, pp.133-137, in Hansen, S.J., P. Langlois, and P. Bertoldi, ESCOs Around the World—Lessons Learned in 49 Countries, The Fairmont Press, 2009.

Marino, A., P. Bertoldi, S. Rezessy, and B. Boza-Kiss, 2010. Energy Service

Companies Market in Europe—Status Report 2010, European Commission Joint Research Center, Institute for Energy, EUR 24516 EN, 69-70.
Okay, E., N. Okay, A.E.S. Konukman, and U. Akman, 2008. Views on Turkey's impending ESCO market: Is it promising? Energy Policy 36: 1821-1825.
Okay, N., and U. Akman, 2010. Analysis of ESCO activities using country indicators Renewable and Sustainable Energy Reviews 14: 2760-2771.
Ozturk, M., C. Bezir, and N. Ozek, 2008. Energy market structure of Turkey, Energy Sources B 3: 384-395.
Ozyurt, O., 2010. Energy issues and renewables for sustainable development in Turkey, Renewable and Sustainable Energy Reviews 14: 2976-2985.
Talha Yalta, A., Analyzing energy consumption and GDP nexus using maximum entropy bootstrap: The case of Turkey Energy Economics 33: 453-460.
Toklu, E., M.S. Guney, M. Isik, O. Comakli, and K. Kaygusuz, 2010. Energy production, consumption, policies and recent developments in Turkey, Renewable and Sustainable Energy Reviews 14: 1172-1186.
Vine, E., An international survey of the energy service company (ESCO) industry 2005. Energy Policy 33: 691-704.

General References

ABESCO (coordinator Poole, A.D. &. Amaral, M.C.). Prepared for the World Bank/UNEP/UNF program *New Financial Intermediation Mechanisms for Energy Efficiency Projects in Brazil, China and India,* São Paulo, February 12, 2005.
ACEE *Summer Study, Successfully Advancing Energy Conservation Efforts in Mexico,* Rennè, Cohen, & Pérez, May 2006.
Administration of Seversk.(2006). Program of the regional development for administration unit of Seversk 2006-2009. Annex II. Project *Providing heat to municipalities*. The analysis of projects under investments of international financial organizations (in Russian).
Agence de l'environnement et de la Maîtrise de l'énergie (ADEME). (2006). Current situation of the Energy Efficiency Services market in France. *Country Overview*. EUROCONTRACT project.
Aidonis, A. & G. Markoginnakis. (2006). *Development of Pilot Solar*

Thermal Energy Service Companies (ST-ESCOs) with High Replication Potential. ST-ESCOs Market Analysis: Hellas. (Project Report of no. EIE/04/059/S07.38622).

ANEEL/SPE: *Manual para Elaboração do Programa de Eficiência Energetica.* (2008); prepared by the Superintendency for Research & Development and Energy Efficiency; approved by Resolution # 300 of February 12, 2008.

Angieliotti, R. (2010). *Asia ESCO conference 2010-Energy Performance Contracting* (EPC) Sharing the French ESCOs Experience, January 2010, from http://asiaesco.org/pdf/presentation/4-2.pdf.

Associazione Imprese di Facility Management ed Energia (AGESI). n.d. from www.agesi.it (partially in italian)

Austrian Energy Agency (E.V.A.). (2005). *Country Overview.* EUROCONTACT project.

Balance Energetico Nacional. (2006). *Dirrección Nacional de Energía y Tecnología Nuclear- Ministerio de Industria, Energía y Minería* from www.eficienciaenergertica,gub.uy.

Bates, S. (2010). *The Performance Contracting Advantage- Using Energy Savings to Fund Energy Infrastructure Improvements in Schools, Universities and Municipalities*, January 2010, White Paper. From http://www2.schneider- electric.com/documents/ buildings/ the_performance _contracting_advantage.pdf

BerliNews May 17, 2005. *European Energy Service Award 2005*, from http://www.berlinews.de/archiv-2004/3446.shtml [consulted August 5, 2006]

Bertoldi, P., Hinnells, M. & Rezessy, S. (2006a). Liberating the power of energy services and ESCOs in a liberalised energy market. In: *Proceeding of the International Energy Efficient Domestic Appliances and Lighting Conference (EEDAL`06),London, June 21-23, 2006.* Eds: Bertoldi, P., Kiss, B., Atanasiu, B. Ispra, Italy: European Commission, DG Joint Research Center.

Bertoldi, P., Rezessy, S., & Vine, E. (2006b) Energy service companies in European countries: Current status and a strategy to foster their development. *Energy Policy* 34: 1818-1832

Better Integration of Sustainable Energy (BISE). (2005). Reports by Countries: *Development of Municipal Energy Efficiency Networking Activities*. From http://www.bise-europe.org/IMG/pdf/National_reports_Bise.pdf [consulted August 28, 2006].

Bharvirkar, R., Goldman, C., Gilligan, D., Singer, T.E., Birr, D., Donahue,

P. et al. (2008). *Performance Contracting and Energy Efficiency in the State Governement Market*, November 2008. From http://eetd.lbl.gov/ea/emp/reports/lbnl-1202e.pdf.
Biasiotta, B.A. (2009) Johnson Controls inc. Using Performance Contracting to Implement ARRA Projects-2009. From http://www.johnsoncontrols.com/publish/etc/medialib/jci/be/white_papers.Par.85476.File.dat/arrawhitepaper.pdf
Bleyl-Androschin, J.W. & Schinnerl, D. (2008) *Comprehensive Refurbishment of Buildings through Energy Performance Contracting—A Guide for Building Owners and ESCOs*. From http://www.ieadsm.org/Files/Exco%20File%20Library/Key%20Publications/IEAdsm-TaskXVI_Bleyl,%20Schinnerl_Comprehensive%20Refurbishment%20of%20Buildings%20through%20EPC_081118_vers2.pdf
Bonfils, S., Langlois, P. & Polissois, G. (1997). *Energy Efficiency Projects and their Financing Mechanisms*, IEPF, Collection Cahier Prisme, Econoler International.
Center for Renewable Energy Sources (CRES). (2005a). EPC in Greece: *Current Situation. Country Overview.* EUROCONTRACT project.
Center for Renewable Energy Sources (CRES). (2005b). *ST-ESCOs newsletter*. Issue 4. From http://www.stescos.org/index.htm.
Ceresi, G. (2005). Role of ESCO in the industrial marketing in Italy: Siram experience. Presentation at *ESCO Europe Conference 2005*. October 4-5, 2005. Vienna.
Chabchoub, J. (2005). Country Summaries (Part 2) *The Environment for Energy Performance Contracting in Central Europe*. Monthly Balkan Energy Solutions Team (BEST) e-mail bulletin in power systems, renewable energy sources, electricity market and ecology
Chen, K., Zifeng, X. (2010). Energy Performance Contracting in China, October 2010. From http://www.kingandwood.com/files/20101122/File/China-2010-10-10-KC%20XZF.pdf.
Christyakova, O.N., Morin, A. & Pasoyan, A. (2006). *Removing Barriers to Residential Energy Efficiency in Southeast Europe and the Commonwealth of Independent States*. Kiev, Ukraine: Alliance to Save Energy.
Comisión Nacional de Energía, CNE. (2004) *Estimating Potential Energy Savings by Improving the Energy Efficiency of Different Consumption Sectors in Chile.*
Comprehensive resources energy board-of-inquiry energy-savings

committee. (2001). *Energy-saving committee report -a vout the state of the future measure against energy saving,* 2001.6.

CTEE (*Comitê Técnico para Eficientização do Uso de Energia*).(2001). *Plano Energia Brasil—Eficiância Energetica,* October, 2001

Cudahy, R.D. & Dreesen T.K. (1996). *A Review of the Energy Service Company (ESCO) Industry in the United States,* prepared for the Industry and Energy Department, The World Bank, Washington D.C., March 1996.

De Groote, W. (2006). ESCOs for households: a New Phenomena in Europe? In: *Proceeding of the International Energy Efficient Domestic Appliances and 93Lighting Conference (EEDAL`06),* London, June 21-23, 2006. Eds: Bertoldi, P., Kiss, B., Atanasiu, B. Ispra, Italy: European Commission, DG Joint Research Center."

Department of Energy. (2006). *Philippine Energy Plan*: 2006 Update. Manila, Philippines

Department of the Army USA (2008) *Department of the Army Policy Guidance for Implementation of an Energy Savings Performance Contract,* November 2008. From http://army-energy.hqda.pentagon.mil/docs/ESPC_policy_hdbk_v3_1108.pdf

Dreesen, T. EPS Capital. (March 2008).

Energy Conservation Center Japan. (2004). *The investigation report for the ESCO industrial spread in local authorities,* 2004.2.

Energy Conservation Center Japan (2005). *Investigation projects about superior ESCO project commendation system examination,* 2005.10.

IEEFP Mexico, EVO, ESP Capital & Pérez, M. (May 2006) *Energy Efficiency Financing Assessment Report.*

Energy Market and ESCO Market Assessment- Econergy. (2002).

EPA Clean Energy- Environment Technical Forum. (2008). *Energy Performance Contracting in State Facilities,* April 2008, from http://www.epa.gov/statelocalclimate/documents/pdf/background041008.pdf

ESMAP from http://www.esmap.org/activities/index asp?Sort=title&s=80

EU-Russia Energy Dialogue Technology Centre. (2006). Summary of the Seminar on ESCOs and Gas Flaring in *the Framework of the EU-Russia Energy Dialogue Moscow, Russia, October 26, 2006.*

European Bank of Reconstruction and Development (EBRD). (1998). EBRD and EU encourage energy saving in *Ukrainian small and medium-sized enterprises through loan to country's first Energy Service*

Company (ESCO). EBRD Press Release May 24, 2006. From http://www.ebrd.com/new/pressrel/1998/24may9.htm [consulted on 10 December 2006].

European Building Automation Controls Association. (2011). Energy Performance Contracting in the European Union, 2011. From http://www.euesco.org/fileadmin/euesco_daten/pdfs/euESCO_response_concerning_EPC.pdf

European Commission, DG Joint Research Center (EC DG JRC). (2005). *European Energy Service Companies Status Report 2005*. Authors: Bertoldi, P. & Rezessy, S. Ispra, Italy: EC DG JRC.

Fanjek, J. & Šteko, B. (2005). Energy efficiency project in Croatia. Presentation at *ESCO Europe Conference 2005*. October 4-5, 2005, Vienna.

Forsberg, A., Lopes, C., & Öfverholm, E. Öfverholm forthcoming. (2007). How to kick start a market for EPC—Lessons learned from a mix of measures in Sweden. In: *Proceedings of the European Council for Energy Efficient Economy 2007 Summer Study*. Stockholm: European Council for an Energy-Efficient Economy.

Fundación Chile. (2007). *Methodology to Identify and Assess Energy Services Companies (ESCO)*, final report within the framework of project BID-FOMIN, produced by Econoler International.

Fundación Chile. (2007a). *Study to Identify Market Potential and Approach*, final report within the framework of project BID-FOMIN, produced by Gamma Ingenieros S.A., Santiago.

Fundación Chile. (2007b). *Proposal of instruments and Sector Models for Clean Energy and Energy Efficiency*, status report, Project Inter-American Development Bank-FORMIN, Study conducted by Econoler International.

Garcia, A.G.P. (2008). *Leilão de eficiência energética no Brasil;* Doctoral thesis for the Federal University of Rio de Janeiro/COPPE, January, 2008

Geissler, M. (2005). EUROCONTRACT—Guaranteed Energy Performance. Standardised Energy Services for Europe's buildings. Presentation at *ESCO Europe Conference 2005*. October 4-5, 2005, Vienna.

Geissler, M., Waldmann, A & Goldmann, R. (2006). Market development for energy services in the European Union. In: *2006 ACEEE Summer Study on Energy Efficiency in Buildings—"Less is More: En Route to Zero Energy Buildings."*

Governement Property Group and the Energy Efficiency Council. (2011). Guidance Paper: *Integrated Energy Efficiency Retrofits and Energy Performance Contracting*, 2011. From http://www.eec.org.au/UserFiles/File/docs/Guidance%20Paper%20on%20integrated%20energy%20efficiency%20retrofits%20and%20EPC.pdf

Grim, M. (2006). The Austrian programme for private service buildings: ecofacility. In: *Proceedings of International Conference on Improving Energy Efficiency in Commercial Buildings (IEECB'06)*, Frankfurt (Germany), April 26-27, 2006. Eds. Bertoldi, P.& Atanasiu, B. Ispra, Italy: European Commission, DG Joint Research Center.

Hansen, S.J. (2005). *Making the Business Case for Energy Efficiency, International Energy Efficiency in Commercial Buildings Conference. 2004*, Frankfurt, Germany. Paper also presented as keynote to annual conference of Japan Association of Energy Service Companies, 2005.

Hansen, Shirley J. *ESCOs Around the World: Lessons Learned in 49 Countries*. 2010. The Fairmont Press, Lilburn, GA.

Hansen, Shirley J. *Performance Contracting: Expanding Horizons 2nd Edition*. 2006. The Fairmont Press, Lilburn, GA.

Hansen, Shirley J. and James W. Brown. *Sustainability Management Handbook*. 2011. The Fairmont Press, Lilburn, GA.

Herrera, Alice B., Ph.D. (2005) *Energy Efficiency Opportunities and Investment Requirements in the Philippines*. A Report Prepared for the Asian Development Bank.

Hinnells, M. (2006). Aiming at a 60% reduction in CO2: implications for residential lights and appliances and micro-generation. In: *Proceeding of the International Energy Efficient Domestic Appliances and Lighting Conference (EEDAL'06), London, June 21-23, 2006*. Eds: Bertoldi, P., Kiss, B., Atanasiu, B. Ispra, Italy: European Commission, DG Joint Research Center.

Hyponnen, S. (2006). Boosting efficiency with ESCO service. Presentation *at the European Conference on Developing the Energy Efficiency Market (DEEM)*. September 21-22, 2006, Budapest.

Institute for Building Efficiency. (2010). *Energy Performance Contracting in the European Union: Creating Common "Model" Definitions, Processes and Contracts*. September 2010, from http://www.institutebe.com/getmedia/143061e4-d526-42ce-9cdd-70191bead2f3/Issue-Brief—Energy-Performance-Contracting-in-the-EU—Part-2.aspx

Instituto para la Diversificación y Ahorro de la Energía (IDAE), n.d.

From www.idae.es(in Spanish) [consulted 16 July 2006].

Instituto para la Diversificación y Ahorro de la Energía (IDAE). (2005). *Energy Policies of IEA Countries: Belgium 2005 Review*. Paris: OECD/IEA.

International Energy Efficiency Financing Protocol IEEFP-Mexico-Bank Training Manual, EPS Capital, EVO Mónica Pérez, January, 2008.

Interview with Stephane le Gentil, spokesman of the European Association of Energy Services Companies. *Energy Performance Contracting for Public Buildings-The key to cost-effective energy efficiency in Europe*, May 2009. From http://www.epc-conference.org/files/presse/en/+Interview_le_Gentil_ESPC%20090602_.pdf

Irrek, W., Attali, S., Benke, G., Borg, N., Figorski, A., Filipowicz, et al. (2005). *PICO Light project*, SAVE Contract No. 4.1031/Z/02-038/2002—Final Report. Döppersberg, Germany: Wuppertal Institut.

Irrek, W, Thomas, S. & Benke, G. (2006). Internal performance commitments enabling a continuous flow of energy efficiency measures. In: *Proceedings of International Conference on Improving Energy Efficiency in Commercial Buildings(IEECB'06)*, Frankfurt (Germany), April 26-27, 2006. Eds. Bertoldi, P. and Atanasiu, B. Ispra, Italy: European Commission, DG Joint Research Center.

JAESCO survey. (2007).

Jyukankyo research institute: *Basic investigation of guideline policy in Eco-Energy City Osaka*, report, Osaka, 2001.3.

Jyukankyo research institute: Heisei 15 fiscal year. *The PFI practical use ESCO project introduction to governmental facilities, Ministry of Economy, Trade and Industry*, 2004.3.

Ketting, J. (2006). Energy Efficiency in Russia: A Chance to Excel or a Hard Lesson to Learn? *Russia Investment Review* 4: 94-95.

Langlois, P. & Robertson, Y. (2003). *Demand-Side Management from a Sustainable Development Perspective—Experiences from Québec (Canada) and India*, Econoler International Teri, IREDA, 2003.

Lima, L.E.A.; Ayres, C.M., Poole, A.D., Hackerott, C.F. & Campos, M. (2005). *Analysis of the Viability and Design of a Guaranteed Facility for Energy Efficiency Projects*; prepared for the World Bank/UNEP/UNF program "New Financial Intermediation Mechanisms for Energy Efficiency Projects in Brazil, China and India," August, 2005

Marçal, M.E. (2005). *Considerations for Structuring a Trade Receivables Fund ("FIDC") to Finance Energy Efficiency Projects in Brazil*; prepared for the World Bank/UNEP/UNF Program "New Financial Intermediation Mechanisms for Energy Efficiency Projects in Brazil, China and India, November 2005.

Marçal, M.E. & Magalhães, P.C. (2005). *Opportunities and Challenges in the Development of Financial Intermediation Mechanisms for Energy Efficiency Projects in Brazil*; prepared for the World Bank/UNEP/UNF program "New Financial Intermediation Mechanisms for Energy Efficiency Projects in Brazil, China and India," April, 2005,

Market study for UTE-USCO-Econoler International. (2008).

Marquez, Raymond A. (2005). *Comparative Analysis of ASEAN ESCOs*. A report prepared for the UNDP-GEF-DOE Philippine Efficient Lighting Market transformation (PELMAT) Project

Marquez, Raymond A. (2006). *Output Report in Assistance Provided for DBP Model ESCO Transaction Project and BDO Demo Project*. A Report prepared for the UNDP_GEF_DOE Philippine Efficient Lightning Market Transformation (PELMAT) Project.

Marquez, Raymond A. (2007). *ESCO Framework of Cooperation*.

Measurement and Verification Protocol Committee: *Study for* the *Measurement and Verification Protocol of the energy-saving effect, Energy Conservation Center Japan*, 2001.3.

Measurement and Verification Protocol Committee: *The Measurement and Verification Protocol Guideline of the energy-saving effect, Energy Conservation Center Japan*, 2002.3.

Millin, C., Bullier, A. (2011). *Energy Retrofitting of Social Housing through Energy Performance Contracts-A feedback from the FRESH project*: France, Italy, United Kingdom and Bulgaria, January 2011. From http://eaci-projects.eu/iee/page/Page.jsp?op=project_detail&prid=1869&side=downloadablefiles.

Ministry of Environment: *The statement-of-principles description data about promotion of the contact which considered curtailment of emission of greenhouse gas in the nation, independent administrative agencies, etc.*, 2007.12.

MOTIVA, Oy. (2005). *Country Overview*: Finland. EUROCONTRACT project.

MOTIVA, Oy. n.d. from www.motiva.fi(information on ESCOs is in Finish).[consulted January 30, 2007].

Murajda, T. (2005). Energy efficiency contract in district heating domain—Elementary schools in Petrzalka by C-TERM spol. s.r.o. In: *Proceedings of the Energy Efficiency Potential in Buildings, Barriers and Ways to Finance Projects in New Member States and Candidate Countries*. Tallin, Estonia July 2005.Eds: Paolo Bertoldi and Bogdan Atanasiu. Ispra, Italy: European Commission, DG Joint Research Center.

Murakoshi, C., Nakagami, H. & Sumizawa, T. (2000). The investigation research on evaluation of a ESCO experimental project, in *proceeding of the 16th conference on Energy, Economy and Environment, Japan Society of Energy and Resources*, 2000.1.

Murakoshi, C., Nakagami, H. & Sumizawa, T. (2000). Exploring the feasibility of ESCO business in Japan- demonstration by experimental study, in *Proceeding of the ACEEE 2000 Summer Study on Energy Efficiency in Buildings*, 2000.8.

Murakoshi, C., Watanabe, T., Akashi, Y & Nakagami, H. (2007). *Study on the development circumstances and the characteristics of ESCO business in Japan*, J. of Architecture and Urban Design, Kyusyu University No.12, P91-101, 2007.7.

Murakoshi, C., Watanabe, T., Akashi, Y & Nakagami, Y. (2008). Study on the characteristics of ESCO business in Japan, Architectural Institute of Japan, *Journal of Environmental Engineering* No.622, 2008.2.

MURE-Odyssee. (2006a). *Energy Efficiency Profile: Luxembourg*. Also available on-line: www.mure2.com.

MURE-Odyssee. (2006b). *Energy Efficiency Profile: Spain*. Also available on-line: www.mure2.com.

New construction subcommittee: *International Performance Measurement & Verification Protocol* Volume 3, Efficiency Valuation Organization (EVO), 2006.1.

Nexant. (2004). *Contratos de Desempenho para Serviços de Eficiência Energética no Setor Público do Brasil*: Questões Jurídicas e Possíveis Solucões; report to the Brazilain Ministry of Mines and Energy with support from USAID, January, 2004

North Carolina Guide to Energy Performance Contracting for K-12 Schools. Local Governments &Community Colleges, October 2008. Prepared by Waste Reduction Partners, Land-of-Sky Regional Council with grant funding by the USA EPA. From http://wastereductionpartners.org/phocadownload/userupload/Resources/NC%20Guide%20to%20Energy%20Performance%20Contracting%20with%20bookmarks%20on.pdf

NREL. (2011). The Energy Performance Contract: What's In it ?, August

2011. From http://apps1.eere.energy.gov/tribalenergy/pdfs/tribal_business/epc_0811.pdf

Optima Energia.(March 2008).

Ortega, O. (2011). Les contrats de performance énergétiques, mars 2011. From http://www.developpement-durable.gouv.fr/IMG/pdf/Rapport_definitif_contrats_performance_ energetique-2.pdf.

PAD—Project Appraisal Document—World Bank—Uruguay Energy Efficiency Project.

Peterson, S. (2009). *Evaluating how Energy Performance Contracting impacts the business case for investing in building energy efficiency improvement*, London 2009. From http://www.eu-esco.org/fileadmin/euesco_daten/pdfs/TowardsNET-ZERO-euESCO.pdf

Poole, A.D. & Meyer, A.S. (2006). *Brazil Country Report*; prepared in English and Portuguese for the project "Developing Financial Intermediation Mechanisms for Energy Efficiency Projects in Brazil, China and India," World Bank, August 2006,

Poole, A.D. &. Poole, J.B.N. (2003). *Summary of Results of the Survey of Brazilian Energy Efficiency Service Providers*; prepared for the CIDA program "Greenhouse Gas Emissions Reduction in Brazilian Industry" (GERBI), Rio de Janeiro, December, 2003,

Programa País de Eficiencia Energetica (2006-2007)

Programa País de Eficiencia Energetica (2005-2006) Energy Efficiency Directory of Chile.

Programa País de Eficiencia Energetica Strategic Plan (2007-2015) if Programa País Eficiencia Energética de Chile.

Pujol, T. (2004). The Barcelona solar thermal ordinance. In: *Proceedings of Annual Conference of Energie-Cités: Working in Synergy with the Private Sector?* Martigny, Switzerland, April 22-23, 2004.

Racolta, S. (2005). The UNDP/GEF Energy Efficiency Financing Team in Romania. In: *Proceedings of the Energy Efficiency Potential in Buildings, Barriers and Ways to Finance Projects in New Member States and Candidate Countries*. Tallin, Estonia July 2005.Eds: Paolo Bertoldi and Bogdan Atanasiu. Ispra, Italy: European Commission, DG Joint Research Center.

Rezessy, S., Dimitrov, K., Urge-Vorsatz, D., & Baruch, S. (2006). "Municipalities and energy efficiency in countries in transition. Review of factors that determine municipal involvement in the markets for energy services and energy efficient equipment, or how to augment the role of municipalities as market players.

Energy Policy 34(2): 223-237."

Riley, B. (2010). *Alabama Energy Performance Contracting Guide*, rev. 10, 2010. From http://www.adeca.alabama.gov/C7/Performance%20Contracting/Document%20Library/AEPC%20Manual%20revised%2010-2010.pdf.

Rodicsm G. (2005). ESCOs in the Hungarian Energy Market. In: *Proceedings of the Energy Efficiency Potential in Buildings, Barriers and Ways to Finance Projects in New Member States and Candidate Countries*. Tallin, Estonia July 2005. Eds: Bertoldi, P. & Atanasiu. B. Ispra, Italy: European Commission, DG Joint Research Center.

Russian Energy Efficiency Demonstration Zones (Rusdem). n.d. from http://www.rusdem.com /Pages /index.htm [consulted 17 November 2006].

Saffet, B. (2007). A New Era in Energy Efficiency in Turkey. *Energy Review* 9: 2-4.from http://www.turkishweekly.net/energyreview/TurkishWeekly-EnergyReview9.pdf [consulted March 5, 2007].

Sehovic, H. (2005b). BiH Experience in Energy Efficiency Energy Efficiency Financing. Presentation at *the Energy Efficiency Investment for Climate Change Mitigation*.

Sellier, D., Falque-Masset, M.L., Baumgatner, G., Jacquot, R. (2011). *Le Contrat de Performance Énergétique, un outil efficace pour l'Île-de-France*, Mars 2011. From http://www.fedarene.org/documents/publications/Others/arene/cpe-idf.pdf.

Seven7. (2008). *Energy Performance Contracting-Method Description and Application*, April 2008. From http://www.ecnetwork.info /SEC-Tools/SEC_Tools_-_Guideline _on_ EPC_ methodology[1].pdf

Shonder, J., Morofsky, E., Schmidt, F., Morck, O., Himanen, M. (2010). *Best Practice Guidelines for Using Energy Performance Contracts To Improve Government Buildings*, Annex 46, May 2010. From http://www.ecbcs.org/docs/Annex_46_ESPC_Best_Practices.pdf.

Sorrel, S. (2005). *The Contribution of Energy Services Contracting to a Low Carbon Economy*. Tyndall Centre Working Paper, Environment & Energy Programme SPRU (Science & Technology Policy Research), Freeman Centre.

ST-ESCO project. (2006a). *ST-ESCOs Market Analysis: Austria*. Project Document. Project no. EIE/04/059/S07.38622.

ST-ESCO project. (2006b). *ST-ESCOs Market Analysis: Spain*. Project Document. Project no. EIE/04/059/S07.38622.

Taylor, R.P., Govindarajalu, C. Levin, J., Meyers, A.S., & Ward, W.A.

(2008). *Financing Energy Efficiency: Lessons from Brazil, China, India and Beyond*; ESMAP/World Bank, 2008.
Travaux publics et Services gouvernementaux Canada. (2010). *Energy Performance Contracting- Energy Efficiency Retrofit Measures for Governement Buildings Workshop*, 14 August 2010. From http://www.pertan.com/ORNL_govenergy/Morofsky.pdf.
USAID. (2005). Credit Guarantees Promoting Private Investment in Development. *Year Review 2005*. Washington: USAID.
UTE-USCO-M. González. (2008).
Vegel, M. (2006). Eurocontract. European Platform for the Promotion of Energy Performance Contracting. Presentation at *the ESCO Europe 2006 International Conference*, Prague, September 26-27, 2006.
Ver, A.A. (2007). ESCO Association of the Philippines. Presented to *the 2nd Asia ESCO Symposium*, Tokyo, Japan. February 1-2, 2007
Vine, E. (2005). An international survey of the energy service company (ESCO) industry. *Energy Policy* 33: 691-704.
Wilkins, A. (2011). Lessons from North America: Presentation to *the Sustainable Energy Authority of Ireland Financing Retrofit*: Public Sector: New Instruments for Delivering Energy Efficiency, May 27, 2011, from http://www.seai.ie/News_Events/Previous_SEAI_events/Anne%20Wilkins.pdf
Ministerio de Industria, Energia y Minería Dirección Nacional de Energia. Proyecto de Efficiencia Energetica, Uruguay. Proyecto BIRF MULT 53298, from http://www.efficientlighting.net/doc/20070108(2).pdf.
Zachariev, D. (2005). ESCO in Bulgaria: Projects, market, barriers. In: *Proceedings of the Energy Efficiency Potential in Buildings, Barriers and Ways to Finance Projectsin New Member States and Candidate Countries*. Tallin, Estonia July 2005.Eds: Bertoldi, P. & Atanasiu, B., Ispra, Italy: European Commission, DG Joint Research Center.
Žídek, O. (2005). Energy Performance Contracting in the Czech Republic—history, present and future development. Presentation *at ESCO Europe Conference 2005*. October 4-5, 2005, Vienna.
Zobler, N., Philips, M. & Diamond, M. (2009) *Energy Performance Contracting Financing Options-EPC Toolkit for Higher Education*, April 2009. From http://www2.presidentsclimatecommitment.org/ documents/ ccitoolkit/ Energy_Performance _Contracting_Financing_Options.pdf

Index